The Competitiveness of Tropical Agriculture

The Competitiveness of Tropical Agriculture

A Guide to Competitive Potential with Case Studies

Roger D. Norton

with

Amy Angel
Ricardo Argüello
Álvaro Balcázar
Héctor Martínez
Henry Samacá
Anne Turner

ELSEVIER

AMSTERDAM • BOSTON • HEIDELBERG • LONDON
NEW YORK • OXFORD • PARIS • SAN DIEGO
SAN FRANCISCO • SINGAPORE • SYDNEY • TOKYO
Academic Press is an imprint of Elsevier

Academic Press is an imprint of Elsevier
125 London Wall, London EC2Y 5AS, United Kingdom
525 B Street, Suite 1800, San Diego, CA 92101-4495, United States
50 Hampshire Street, 5th Floor, Cambridge, MA 02139, United States
The Boulevard, Langford Lane, Kidlington, Oxford OX5 1GB, United Kingdom

Notices

Knowledge and best practice in this field are constantly changing. As new research and experience
broaden our understanding, changes in research methods, professional practices, or medical
treatment may become necessary.

Practitioners and researchers must always rely on their own experience and knowledge in
evaluating and using any information, methods, compounds, or experiments described herein. In
using such information or methods they should be mindful of their own safety and the safety of
others, including parties for whom they have a professional responsibility.

To the fullest extent of the law, neither the Publisher nor the authors, contributors, or editors,
assume any liability for any injury and/or damage to persons or property as a matter of products
liability, negligence or otherwise, or from any use or operation of any methods, products,
instructions, or ideas contained in the material herein.

Library of Congress Cataloging-in-Publication Data
A catalog record for this book is available from the Library of Congress

British Library Cataloguing-in-Publication Data
A catalogue record for this book is available from the British Library

ISBN: 978-0-12-805312-6

For information on all Academic Press publications
visit our website at https://www.elsevier.com/

Working together
to grow libraries in
developing countries

www.elsevier.com • www.bookaid.org

Publisher: Nikki Levy
Acquisition Editor: Nancy Maragioglio
Editorial Project Manager: Billie Jean Fernandez
Production Project Manager: Lisa Jones
Designer: Victoria Pearson

Typeset by TNQ Books and Journals

Contents

Part III
Case Studies in the Competitiveness of Tropical Agriculture

Part IV
Concluding Remarks

Preface

This book has been written with the aim of bringing together diverse strands of analysis of agricultural competitiveness and their applications in the tropics, as well as providing a framework for assessing quality issues, in the hope of providing a reference that is useful to researchers, students, practitioners in the field, and policymakers. It is a topic that is fundamental to economic development concerns and, as the examples in the book show, can lead to worthwhile policy dialogs.

Many persons have contributed to the development of this book. It is a deeply collaborative effort. Extensive contributions were made by the colleagues recognized on the title page. The case studies are adapted from papers researched and written with them. The basic material for Chapter 7 was written with Álvaro Balcázar, Chapter 8 with Anne Turner, Chapter 9 with Amy Angel, and Chapter 10 with Ricardo Argüello, Héctor Martínez, and Henry Samacá. Carlos Federico Espinal provided suggestions and important inputs for the studies that evolved into Chapters 7 and 10. Jaime Forero Álvarez and Tomás León Sicard also provided material for the original study underlying Chapter 7.

Charles Richter made useful comments on drafts of the original study for Chapter 10, and Sandra Acero Walteros and Elkin Pardo provided excellent assistance in managing large files of data for competitiveness calculations in that study.

The material for Chapter 8 benefitted from suggestions by Christophe Ravry and Loraine Ronchi of the World Bank and by Martin Webber, Carlton Jones, and Grant Cavanaugh of J.E. Austin Associates, Inc., and from contributions by Viateur Bicali and Sulaiman Kyambadde. Antonio Cabrales, the then President of the Salvadoran Foundation for Economic and Social Development (FUSADES), conceived and sponsored the work that here is adapted as Chapter 9. The United States Agency for International Development (USAID), the UN Food and Agriculture Organization, and the World Bank sponsored the initial work on the study that was adapted into Chapter 7, and the USAID sponsored the study that gave rise to Chapter 10. Thanks to Keith Willett of the Department of Economics at Oklahoma State University, whose invitation to this author to give a seminar provided an opportunity to start developing these ideas, and to Nancy Maragioglio, Billie Jean Fernandez and Lisa M. Jones of Elsevier for superb support for the concept of this publication and its development. I am grateful to Jenni Simonsen of the Texas A&M University library system for outstanding bibliographic assistance.

A very special thanks goes to my family for their continuous encouragement and support for the work on this book: to my wife Fabiola and to my children Hillary, Heather Rachel, and Jose Crisanto.

Part I

Concepts, Issues, and Policy

Chapter 1

The Context and Scope
of the Book

1.1 THE CONTEXT OF AGRICULTURAL COMPETITIVENESS

Agriculture feeds the world, and it is at the same time a major source of livelihood in tropical countries. It is the largest single employer in low-income countries, and it is often the only means of sustenance for the poorest segments of the population. These roles will continue to be important, and world demands on agriculture will only increase in the coming decades. In a widely cited projection, the United Nations Food and Agriculture Organization estimated that global food production will have to increase by 70% between 2005/2007 and 2050 (U.N. Food and Agriculture Organization, 2009, pp. 12–13), and by twofold in developing countries to feed growing populations.[1] Even if this target is achieved some countries will remain food insecure (Alexandratos & Bruinsma, 2012). In spite of the pressures for large increases in the volume of agricultural production, the demands for quality in foods also continue to increase and the competitiveness of agricultural markets is ever more pronounced, especially for exports. Hence national policies oriented toward increasing food production need to take into account quality requirements and to be cognizant of which products are likely to be the most competitive internationally. Project decision makers at the local level similarly have to be aware of these considerations. From the perspective of producers, in the long run, the competitiveness of their outputs will determine whether their marketable production can be sustained.

On the policy side, the expansion of international trade opportunities throughout the world offers both challenges and new opportunities for developing countries. Undoubtedly this process has led to shifts in production patterns within these economies and will lead to more shifts, especially in sectors like agriculture that have significant participation in trade. To assist the flows of resources to the products with the greatest economic potential, and provide support for the transitions out of others, it is important to have assessments of where a country's strongest competitiveness lies in the long run.

Apart from being a supplier of food and source of income for rural populations, the agricultural sector is also a strong driver of growth for the entire

The Competitiveness of Tropical Agriculture. http://dx.doi.org/10.1016/B978-0-12-805312-6.00001-5

economy, which is another reason for supporting the realization of its competitive potentials. Over the years, economists have developed different conceptions of the contributions of agriculture to economic growth, but it is now understood that agriculture is the sector with greatest multiplier effects on the rest of the economy. This was first demonstrated by the early work of Block and Peter Timmer (1994) for Kenya. Timmer later summarized multiple findings on agriculture's contributions to growth in the following words (2009: 39–40):

> *Virtually all of the studies have concluded that the "agriculture multiplier" is significantly greater than 1, especially in relatively closed, "nontradable" economies of the sort found in rural Africa, where it is often between 2 and 3. But even in the more open economies of Asia, where rice is more tradable than most African staple foods and local prices more easily reflect border prices, the agriculture multiplier is close to 2 in the early stages of agricultural modernization when productivity gains are fastest.*

Furthermore, it has been found that agricultural expansion contributes more than other sectors to reduction of both urban and rural poverty, as noted in the work of Mellor (2000), Timmer (1997), and Ravallion and Datt (1996). De Janvry and Sadoulet (2010) developed new evidence on this issue and summarized their findings as follows:

> *Results show that rural poverty reduction has been associated with growth in yields and in agricultural labor productivity, but that this relation varies sharply across regional contexts. GDP growth originating in agriculture induces income growth among the 40% poorest, which is on the order of three times larger than growth originating in the rest of the economy. The power of agriculture comes not only from its direct poverty reduction effect but also from its potentially strong growth linkage effects on the rest of the economy.*

On the basis of a broad survey of the literature on this topic, Grewal, Grunfeld, and Sheehan (2012) stated that:

> *Most studies have concluded that growth in the agriculture sector is highly beneficial for poverty reduction, although some recent studies have also found that the importance of agriculture diminishes as economies grow and become more diversified (p. 14).*

They add that the poverty-reducing effect of agriculture varies by country and is greater the more intensive a country's agricultural production is in the use of unskilled labor (2012: 31).

Although agriculture has been providing these benefits to entire economies its continued growth has come to depend increasingly on competing well in the international marketplace. In the first 15 years of this century the sector's total exports have grown at a very rapid pace in all developing regions of the world, much more rapidly than domestic food consumption in these regions. Between 2000 and 2013, according to FAOSTAT data, in Africa they grew from US$13.5 billion to US$49.7 billion (21% annual growth rate); in South America, from US$34.0 billion to US$165.7 billion (30% rate); in Central America, from US$12.6 billion to US$37.4 billion (15% rate); and in Asia, from US$64.4 billion to US$281.5 billion (26% rate). This remarkable trend has been occurring in a period when sanitary and phytosanitary standards for food imports have been progressively tightened, which has made market access more demanding.

Being competitive in all senses of the term is now more important than ever for agriculture in the tropics including for smallholders, and agricultural competitiveness in turn is important for the development of the economy. Smallholders cultivate many of the higher value export products because they are intensive in the use of labor, and therefore if their production is competitive in world markets, expanded opportunities for trade may help them rise out of poverty. Cocoa is one of many examples of these kinds of crops (Shriver, 2015): "Cocoa sets itself apart as a signature smallholder crop; 90% of all cacao worldwide is produced by small farmers (1–5 hectare averages)." Vanilla is another classic example; its vines require hand pollination, and therefore it is highly labor intensive. Cape gooseberry or *uchuva* (*Physalis peruviana* L.), cultivated in Colombia, East Africa, and other places and exported to Europe, is yet another example. Based on their work at the Colombian National Agricultural Research Agency, Zapata, Zapata, Saldarriaga, Londoño, and Díaz (2002) report that the areas with the most important cultivation of cape gooseberry are smallholder areas in which cropping activities are basically done with family labor.

Peru is an example of a country that has experienced an extraordinary agricultural takeoff driven by high-value export products. In 19 years, its exports of fruit and vegetables grew 20-fold, from US$60 million in 1990 to US$1.2 billion in 2009 (Meade, Baldwin, & Calvin, 2010). Asparagus, grapes, paprika, mango, artichoke, avocado, bananas, citrus, and onion are the crops leading this expansion, all high-value crops.

1.2 THE SCOPE OF THIS BOOK

Development partners and investors as well as policymakers often have to choose the crops for which they will attempt to develop markets or provide agricultural research and advisory services and other kinds of stimulus, and accordingly they need a sense of which crops may be the most competitive in the long run. Crops that are perceived to have competitive potential on the basis of evidence at the farm level may face hurdles further along the value chain in having that potential realized. In addition, some of the material in the case studies discusses how sector policies affect competitiveness and how they can be reoriented in the light of the areas of competitive strength, to provide incentives for the more competitive products, which are the ones with the best growth prospects.[2] Competitiveness underpins the sector's growth. So not surprisingly a rich and multifaceted policy dialog can develop out of competitiveness analysis.

In this context the purposes of this book are (1) to provide a guide to the main methodologies used to date for calculating competitiveness, as well as to the literature on these topics; (2) to present a two-track procedure that assesses different dimensions of competitiveness including a framework for assessing product quality and identifying issues to be resolved regarding quality; (3) to illustrate the procedures with quantitative and qualitative evaluations of competitiveness in Colombia, Rwanda, and El Salvador, including approaches for compiling and presenting results in a way that best helps project decision makers; (4) to illuminate value chain issues both at the product and at the sector levels, the latter including rarely found insights into export marketing concerns from the viewpoint of marketing agents in Colombia; (5) to illustrate the development of policy dialogs on the basis of competitiveness analysis; and (6) to summarize lessons learned about agricultural competitiveness in the tropics, its assessment, uses of the assessments, and its roles in the development context.

A few of the policy themes addressed are the following. The first Colombian case study includes calculations of the economy-wide effects on employment and incomes of modest shifts toward a more competitive national cropping pattern. It also presents discussions of how prevailing policies for research and other programs affect competitiveness and recommendations for a sector strategy to enhance their contributions to competitiveness. The study for Rwanda develops recommendations for overcoming obstacles to competitiveness all along the value chains and the extent to which each crop's production should be encouraged in newly developed watershed areas. The study for El Salvador includes calculations of the net resource transfers between sectors, away from agriculture, brought about by declining real agricultural prices and competitiveness, which in turn had resulted from strengthening of real exchange rate, and it offers suggestions for strategies tailored to that country that can strengthen its agricultural competitiveness.

The kinds of analyses reported in this book help identify competitive potential at the local level (Chapter 10), which helps define priorities for development projects, and problem areas by crop value chain that need improvement for becoming competitive. This kind of diagnosis of issues and their importance can be one of the most useful contributions of competitiveness analysis. Sensitivity analyses carried out in El Salvador and Colombia illustrated the impact on competitiveness of several variables, including yields and prices, rural wage rates, the cost of land and capital, and exchange rates. As these examples show, carrying out sensitivity analyses can be as useful as calculations of competitiveness under base year conditions, if not more so.

Analyses in the book cover 58 tropical crops and livestock products and temperate zone crops grown in the tropics, plus a large number of additional fresh and processed products evaluated through a method based on customs classifications of international trade data. The high-value crops included range from tea and coffee to bananas, avocadoes, macadamia, cashew, tree tomatoes, pineapple, dragon fruit, blackberries, mango, cacao, passion fruit, lulo, rubber, and others. Although the conditions surrounding each product differ substantially in different regions of the world, many of the issues identified in these analyses occur in other countries as well. It is hoped that the case studies can serve as points of departure or checklists for product analyses in other countries, in addition to illustrating methodological and policy issues.

ENDNOTES

1. Although the categories "tropics" and "developing countries" do not completely overlap (with numerous exceptions like industrialized Singapore and the nontropical but developing Central Asian countries), for convenience they are used interchangeably in this book because the main focus is on countries that are both tropical and developing.
2. An example of policy recommendations to support fuller realization of a product's inherent competitiveness is found for the case of rice in Pakistan in Ilyas, Mukhtar, and Javed (2009).

Chapter 2

Introduction

2.1 COMPETITIVENESS CONCEPTS

Competitiveness is subject to different interpretations, and a producer's competitiveness changes over time. Firms and product lines that once seemed to demonstrate a high degree of competitiveness in a given country sometimes retrench or collapse rather quickly because of changes at the micro or macrolevel. These changes can include new market conditions, failure of producers to keep abreast of market trends, or inadequate attention to costs and productivity issues and also to technological and marketing advances by competitors. Macrolevel forces can include changes in policy measures such as tariffs, export taxes, and price controls. Above all they may include movements in the exchange rate, whether induced by policy or by booms in foreign exchange earnings in sectors such as mineral extraction.

A product may be competitive on the domestic market but not on the international market. This most often occurs for products that are low in value in relation to their weight, which translates into high unit transportation costs that shield domestic producers from their external competitors and effectively make the product nontradable (e.g., alfalfa). It can also occur for products that have a very short shelf life and hence are limited to the domestic market, such as lulo (also called naranjilla, *Solanum quitoense*) in Colombia and neighboring countries. In addition, lower quality portions of harvests often will not have an international market and thus become nontradable, as has been observed for tomatoes, clementines, and other citrus in Morocco (Azzouzi, Laytimi, & Abidar, 2007, p. 177). Being competitive only on the domestic market also can characterize producers that are favored by government interventions such as support prices or high import tariffs. For the purposes of this book, the term competitiveness usually refers to international competitiveness: exportable or potentially so and being able to compete successfully against imports or to export without incurring losses.

Although some studies try to rate entire countries by competitiveness, the concept has greater clarity and precision when applied to producers, or groups of producers in the same industry, facing similar conditions in input and product markets. Factors at the countrywide level can influence the competitiveness of a wide range of products, but competitiveness itself refers to the

The Competitiveness of Tropical Agriculture. http://dx.doi.org/10.1016/B978-0-12-805312-6.00002-7

ability to produce an item or service and sell it at a profit. Profitability is one basic indicator of competitiveness, a necessary condition, but it is not sufficient.

Competitiveness is determined by two fundamental conditions: (1) the capacity to produce at costs sufficiently low relative to the product's price (efficiency in production) and (2) the ability to consistently satisfy market requirements. For this last condition, producers have to respect quality standards and be reliable suppliers, especially for high-value products. These two dimensions of competitiveness may be called the *cost dimension* and the *quality dimension*. They correspond to Porter's distinction (1990, p. 39) between the cost focus and the product differentiation focus as routes to competitiveness.

In agriculture, product differentiation may be achieved through one or more of the following strategies: (1) high quality; (2) identification of special production locations or attaining denomination of origin; (3) products tailored to particular consumer demands or requirements of processors (e.g., varieties of sorghum required for brewing beer); (4) timing of market entry and specialized logistics, such as the ability of Colombian cut flower producers to time flower blooming and ship within hours to major markets for holidays; and (5) production of a range of related products, as in the case of a dairy industry that solidifies its market position by producing yogurts, cheeses, and special milk drinks in addition to milk itself. (The diverse dimensions of quality itself are reviewed in Section 2.4.7.)

For entering many markets producers also have to meet minimum quantity requirements. This is a challenge for smallholders and obliges them to form producer associations or cooperatives or seek other ways to bulk up the products offered to the market. They may enter into contract farming or collaborate in marketing with nuclear farms or make other associative arrangements. Hence for producers who can be competitive by the cost-price criterion, the three additional requirements for competitiveness are satisfying *requirements for quality, quantity, and reliability.* In the words of Sarker and Ratnasena (2014, p. 521), "Competitiveness can be viewed as the ability to produce and sell products in a competitive environment that meets consumer demand in terms of price, quality, and quantity and at the same time, ensures sustained profits for the farms." Participating in well-organized value chains is one way to meet these requirements.

Porter stresses factors in the economic environment of a firm or industry that help determine profitability and hence competitiveness, including the bargaining power of suppliers and buyers (1990, p. 35). These forces are attributes of a product's value chain, and value chain analysis is at the heart of the second of two tracks used in this study to evaluate competitiveness. The information and financial flows characteristic of tightly integrated value chains, and their organizational structure, can help ensure product quality as well as helping smallholders meet the quantity and reliability requirements of high-value markets.

Some products are relatively homogeneous, and therefore the cost-price dimension is more important than the quality dimension. Most grains, root crops, oilseeds, and industrial products such as cotton and sugar are examples, although basic quality requirements always must be satisfied. These products are often referred to as *commodities*. Other products for which the quality dimension is paramount may be called *value products*. Examples include coffee, tea, cocoa, spices, cheeses, and many fruits and vegetables. In reality the difference between commodities and value products is not sharply drawn. They are end points of a continuum because quality standards are present to a greater or lesser degree in all products. For example, corn produced by smallholders in Central America often varies substantially in grain size, percentage of broken grains, presence of impurities, humidity, and other quality characteristics, so much so that sometimes it cannot compete with corn imported in large, uniform lots for processing by concentrates plants and flour mills. At the same time it may be competitive in consumer preferences in local markets for fresh corn. In this regard corn can be both a commodity and, to a degree, a value product.

Competitiveness is a dynamic concept. The basic requirements for sustaining it over time are increasing productivity and continuously improving quality to meet evolving market specifications. Productivity in turn has both physical and economic sides. In crop agriculture, the physical side means yields per unit of land or per unit of labor. Improvements in the economic side refer to reductions cost per unit of output, which are brought about in part by yield increases. However, improvements in economic productivity of a farm operation also can come about through increases in the unit value of its outputs, which means productivity is also linked to the quality dimension. Value increases result from improvements in quality in the broadest sense including, for example, a higher level of food safety; use of varieties better aligned with consumer preferences; better product selection, handling, and packaging; more timely delivery to consuming locations; and a higher degree of processing. Consumer preferences can extend to production methods that are environmentally friendly as well, especially for crops like coffee, and these methods in turn help ensure the sustainability of competitiveness.

As international markets become more sophisticated and demanding, sustaining competitiveness increasingly refers to the quality dimension. Maintaining and increasing quality in turn requires investment in human capital and developing connectivity: enhancing the capacity to improve products and to gain continuous access to information on technologies and markets. It requires an ability to adapt to changing conditions, and that in turn requires constant learning on the part of producers and those involved in all segments of the value chain. For producers, it means developing a dialog with buyers and processors that goes well beyond the issue of price and includes information on technologies of production and postharvest management, production finance, and coordination with other suppliers as well. In an illustration of continuous changes in crop varieties to meet shifting market tastes, managers of a flower farm near Medellín, Colombia, told the author that every six months they change to new flower varieties with seeds purchased from specialized flower breeders.

For small- and medium-scale producers, the challenges of competitiveness may require the development of social capital—group efforts—as well as individual capacities. Cooperation is needed because much learning about new technologies and access to inputs and finance takes places through groups, and also because only through groups can small producers offer the quantities demanded by market representatives. This is one route to satisfying the three keys to competitiveness, applicable to farms of all sizes, namely, quantity, quality, and reliability of supply. The higher value lines of production are the most demanding in these respects, but they also are the ones that offer the best avenues out of poverty for smallholders.

2.2 COMPETITIVENESS VERSUS COMPARATIVE ADVANTAGE

In the short-run competitiveness, may be achieved with the aid of subsidies and other government policies, but these policies may not be sustainable in the long run. Therefore sustainable or long-run competitiveness is a concept that necessarily abstracts from government interventions. It represents *inherent or underlying competitiveness* of an enterprise or line of production. One way of expressing this concept is *profitability calculated at economic prices*, that is, calculated with input and output prices adjusted to eliminate the effect of government policies that may be transitory.

In the context of international trade, inherent competitiveness is *comparative advantage*. It is comparative rather than absolute because it refers to those lines of production in which a country is *relatively more efficient*. As David Ricardo pointed out two centuries ago (1817), even if country A can produce all goods more efficiently than country B, both will benefit if B exports its relatively more efficient goods to A. These are the goods in which B has a comparative advantage, and by importing them country A can free up its resources to specialize in its own most efficient lines of production. A corollary is that (in the absence of government interventions that change prices) the goods exported from each country will be profitable over the long run. When a country's production patterns are aligned with its comparative advantage, it is maximizing efficiency, that is, it is following its most efficient growth path.

In Ricardo's analysis, the concept of opportunity cost was basic to comparative advantage. It is the concept of what other goods could have been produced with the same resources. "A country is said to have a comparative advantage in the production of a good if it has a lower opportunity cost of producing the good (in terms of foregone production of other goods and services) than do other countries" (World Bank, 2009).

In contemporary analysis of comparative advantage, instead of using two countries that are trading partners, as Ricardo did, international markets are taken to represent other countries in general. Also, to abstract from policy interventions on product markets value added is measured at international prices: output valued at the corresponding international prices less purchased inputs valued at their international prices. Thus *comparative advantage* is a measure of the country's inherent ability to compete on world markets, and *competitiveness* may depend in part on the policy context.

The limitation of guiding decisions by the criterion of competitiveness is that it may not be sustainable because it is generated by policy intervention, and that policy may be taking resources away from other goods and hence it may be lowering an economy's overall efficiency. As expressed in the cited World Bank study, "Competitiveness does not explicitly consider opportunity costs to the country of transfers and subsidies that affect the direct costs of production. In the short run, a country can be competitive in a particular activity if that activity is subsidized with resources drawn from elsewhere in the economy or from donors, but unless those transfers lead to long-run declines in the opportunity cost of carrying out the activity, that competitiveness will not be economically sustainable" (2009, p. 9).

However, the *cost dimension* measures reported in the case studies in this book purposely refer to competitiveness rather than comparative advantage. In attempting to evaluate the opportunity costs, it can be difficult to disentangle the multiple and reinforcing (or mutually contradictory) effects of policies on product, input and factor prices, and how permanent they may be. The approach of this book is to correct prices for obviously transitory policy interventions but retain those that will be present for the long run. It also includes evaluation of competitiveness under alternative assumptions about policies and hence about long-run prices. This question is discussed more extensively in Section 4.2.

The *quality dimension* that receives considerable emphasis here is equally relevant to competitiveness and comparative advantage. A challenge for value chains on the quality side is to find ways to overcome obstacles to the realization of an inherent comparative advantage: ways to remove the technical and management barriers to a product's becoming fully competitive on international markets. As mentioned, the two-track methodology presented here aims to help policy makers and development experts identify those barriers and find solutions to them.

Passion fruit in Colombia exemplifies a product that has a comparative advantage but for which efforts are needed to overcome the barriers to its realization. The crop's competitiveness was evaluated by the cost-price measure in 16 producing areas of Colombia, and it was found to be competitive in

6 of those areas (Chapter 10). The relative lack of distortions in its factor and input markets suggests that passion fruit produced in those areas has a comparative advantage. However, analysis of its value chain revealed that the barriers to its export—to the realization of its comparative advantage—include inadequate control of diseases and insect infestations (with producers typically reacting only when the problem has become quite serious), lack of washing the product after harvest, inappropriate packing, and poorly organized value chains that result in long waits before the product is collected and hence lead to product spoilage and loss.

In summary, the work reported here may be regarded as efforts to measure long-run competitiveness, or approximations to comparative advantage, and to provide guidance on how to realize it. Frequently, agricultural development policies tend to provide incentives for those products that have little or no comparative advantage. It is important to bear in mind that promoting realization of the agricultural sector's comparative advantage can have substantial benefits in terms of incomes and employment for all sectors together. Calculations of those potential benefits are presented as one of the results of the Colombian case studies.

2.3 NATIONAL FACTORS INFLUENCING AGRICULTURAL COMPETITIVENESS

Measures of international competitiveness and comparative advantage have been constructed from various conceptual foundations. Since the time of David Ricardo a basic pillar has been the idea that relative factor endowments of countries and their intensity in production are the main determinants of comparative advantage and hence are likely to strongly influence competitiveness in international trade. Heckscher and Ohlin (1991) codified it in their well-known theorem that attempted to explain international trade patterns in terms of relative resource endowments—of capital and labor. It states that a country exports those goods that use most intensively its abundant factors and imports those goods that use most intensively its scarce factors, and, because of the law of diminishing returns to a factor, over time factor prices will tend to equalize in the trading partners. Vanek (1959) added land endowments to the list of factors that explain international trade patterns. In an application to agriculture of the Heckscher−Ohlin−Vanek framework, Sanderson and Ahmadi-Esfahani (2011) adjusted national factor endowments of trading countries by their productivity and then applied climate change scenarios to productivities to assess the effects of climate change on comparative advantage in grains and livestock.

For agriculture, the most important natural factor endowments are land, labor, climate regimes, accessible supplies of water, and geographical proximity to major international markets. Seasonality can also provide a natural advantage, as in the case of Chile that exports fruits to Northern Hemisphere countries in seasons when they cannot produce them. However, measurement of these factors is not always straightforward for a number of reasons, among them their heterogeneity. Land, for example, comes in varying qualities and endowments per farm, and it also varies by climatic zone within a country and how well endowed it is with complementary infrastructure such as irrigation, electricity, and access to all-weather roads. In Colombia, the existence of substantial amounts of arable land at different altitudes and climate zones is often cited as a national advantage because it enables the country to produce a wide variety of crops and often harvest them throughout the year. It also possesses soils and altitudes appropriate for cultivating coffee, the world's largest traded product after petroleum.

To the natural factor endowments we may add built factors—infrastructure—and other factors created over time by human actions: agricultural research capacity, availability of credit for farmers, policy environments conducive to investment, average levels of education in rural areas, market structures for agricultural inputs, business experience in the agricultural sector, and other human factors. Policy environments can include regimes that affect the management of resource endowments, such as land tenure and the incentives it provides to production, and water management regimes, as well as many others. These environments also can have negative effects on competitiveness. A disadvantage of a historical nature that Colombia shares with many other Latin American countries is the tradition of large landholdings operated at low levels of intensity and the resulting social schisms and distrust in rural areas. In addition to the inequitable land distribution bequeathed to present generations, it has inhibited the development of active land rental markets, which potentially could constitute a major channel of access to land that could be put to productive use by the rural poor. The history of landlessness, land invasions, and land reforms has created an uncertain situation in which landowners are hesitant to rent out their land for fear of losing it. Overcoming the negative aspects of this legacy is a major challenge for agricultural and rural policy.

Infrastructure is often quite deficient in developing countries with marked cost consequences for farmers and enterprises wishing to export. Dethier and Moore (2012) report that "a majority of studies conclude that infrastructure generally has a significant impact on growth, particularly in developing countries." Also, Webb (2013) finds a very strong positive effect on rural incomes in the Peruvian Andes from investments in connectivity. He points out that in that region per capita incomes increased by 1.4% per year from 1900 to 1994 and then by 7.2% from 1994 to 2011 (2013, p. 214). He comments: "The explanation of the rural takeoff can be summarized, then, in the following way: the central driver would have been the sudden transformation and improvement of the communications platform initiated during the 1990s, which included roads and other infrastructure for freight and people, together with the arrival and rapid multiplication of telecommunications, both telephones and internet, reinforced by the rapid advance of rural electrification, literacy and access to the national identity card [that allowed access to many services]" (Webb, 2013, p. 220).

Brazil has been a prime example of the constraints imposed on competitiveness and economic growth by inadequate infrastructure. Amann, Baer, Trebat, and Villa (2014, p. 3) report that:

> *it is painfully evident that Brazil's economic transition from middle-income status is still very much a work in progress. Nowhere is this more apparent than in the field of infrastructure, the problems relating to which have drawn unfavorable attention from domestic and foreign investors, policymakers and commentators. The infrastructural challenges facing Brazil…[extend] across transportation, energy, telecommunications, sanitation and housing. Indeed, as this paper will argue, infrastructural problems lie at the heart of Brazil's growth constraint and go some way to explaining why that growth has failed to keep pace with that generated by other large emerging market economies, notably China.*

Through careful regression analysis, they conclude that a 1% increase in state-level spending on transport infrastructure would be associated with a 6.6% increase in per capita incomes (2014, p. 13).

Over time, agricultural research is another key to improving productivity, and it generally has shown very high returns. Estimates of the internal rate of return to agricultural research by region of the world, in many studies, have been consistently high. Hurley, Pardey, and Rao (2014) compiled 2681 evaluations of the returns to agricultural research and development (R&D) covering the period 1958—2011 and calculated the following median internal rate of return (IRR) by region: Sub-Saharan Africa, 35%; Latin America and the Caribbean, 41%; and Asia and the Pacific, 53%. Mean estimates were higher owing to the long right tail of the distribution of returns—the rightward skew of the distribution. Similarly high IRRs can be seen in most of the individual country studies, for example, a median of 71.8% for 28 studies in India (Pal & Byerlee, 2006).

However, the Hurley—Pardey—Rao study and another one by the same authors (Hurley et al., 2014) point out deficiencies in using the IRR as an appropriate measure of the returns to R&D investment, in particular the existence of multiple numerical solutions, the fact that the IRR calculation assumes that beneficiaries of the investments can reinvest the benefits at the same high rate of return, and its discounting of investment costs also at the same high rate of return. These concerns lead the authors to propose the use of the "modified internal rate of return," which uses different, and much lower, discount rates on investment costs and for returns on reinvestments. The consequence is a drastic reduction of calculated rates of return, but they still remain more than high enough to justify additional investment in agricultural R&D. For a sample of 270 estimates of returns to US agricultural R&D, the mean rate of return falls from 39.0% to a range of 9.8—14.3%, for a plausible range of both discount rates. The median estimate falls from 67.9% to a range of 13.6−17.8%. Nevertheless, in the studies that have also reported benefit−cost ratios (632 studies reviewed by Hurley et al. (2014)) the mean of those ratios was calculated at 22.9 and the median at 10.5. In another country example, estimates of the modified internal rate of return to public agricultural investment in Uruguay came in at 24% and the marginal benefit−cost ratio at 48.2 (Bervejillo, Alston, & Tumber, 2011). Overall, the value of agricultural R&D seems unquestionable for dynamic competitiveness.

Evidence from the Agricultural Science and Technology Indicators (ASTI) of the International Food Policy Research Institute shows that the productivity of agricultural research is beginning to be recognized in tropical countries. After a long period of decline, investments in agricultural research in Sub-Saharan Africa appears to be rising, particularly those from the private sector: "Following a decade of stagnation during the 1990s, public agricultural R&D spending in SSA increased by more than one-third in real terms during 2000−2011, rising from $1.2 billion to $1.7 billion in 2005 constant pur-chasing power parity (PPP) dollars." Also, "private agricultural R&D in Africa is likely to grow faster than public sector R&D" (Pray & Nagarajan, 2013, pp. 3, 2).

A similar recovery of spending on agricultural research has been docu-mented by the ASTI project for Latin America and the Caribbean: "2004 marked an important turning point in terms of R&D spending, with overall levels rebounding rapidly. By 2013, the region as a whole spent $5.1 billion on agricultural R&D, in 2011 PPP prices. . . representing a 75 percent increase over levels recorded in the early 1980s" (Stads, Beintema, Perez, Flaherty, & Falconi, 2016, p. 7).

For East Asia and the Pacific, the project reports that "during 1996−2008, agricultural R&D spending in Asia−Pacific increased by 50%, from $8.2 billion to $12.3 billion in 2005 PPP prices. The main driving countries of this region-wide growth were China and India. China's agricultural R&D spending rose from 1.6 to 4.0 billion PPP dollars (in 2005 prices) over this period, largely as a result of government reforms that promoted innovation in agricultural science and technology and which opened new funding opportu-nities. India's level of investment also increased substantially during this time due to increased government commitment to agricultural R&D...In line with growing financial resources for public agricultural R&D, research spending by private firms has also increased in many of the low- and middle-income countries of Asia−Pacific. As with public spending, China and India lead in private agricultural research investment...Agricultural R&D spending in Cambodia and Vietnam quadrupled between 1996 and 2008, and Bangladesh and Malaysia also reported significant increases" (Flaherty, Stads, & Srini-vasacharyulu, 2013, pp. 2, 4).

In addition to these specific kinds of national factors, the institutional and policy framework as a whole can support the long-run competitiveness of products, provided that policies do not favor one sector at the cost of under-mining another. This "'enabling environment' comprises policies, institutions and support services that form the general setting under which enterprises are created and operate" (da Silva & de Souza Filho, 2007). It determines the ease of creating enterprises and whether potentially competitive products are encouraged or discouraged by policy instruments.

2.4 ISSUES IN EVALUATING COMPETITIVENESS

2.4.1 Reliance on Costs and Prices

Competitiveness has different facets, and its evaluations come in a variety of shapes and forms, and accordingly various kinds of issues can arise. A com-mon issue is a tendency to base assessments of competitiveness only on cost-price measures. Procedures for such assessments are a major part of this book, and the concern for comparing production costs and market prices is under-standable, especially in light of fluctuating or downward moving international prices trends for some of the higher value products. For example, between 1992 and 2001 the real price index for fruit exports declined by about 7%, and for cut flowers and medicinal plants it declined by about 28% (Hallem et al., 2004). The expansion of exports of these kinds of products is reflected in part in price competition, and therefore to stay in the market, producers and ex-porters have to monitor price trends and control costs. Nevertheless, to base competitiveness assessments only on the cost-price dimension ignores the many hurdles that block the road to market, as illustrated by the above-mentioned case of Colombian passion fruit and others brought out in the case studies.

Tree tomato or tomatillo (*Cyphomandra betacea*) in Rwanda is another of the cases described here. The crop appears to have a competitive cost structure. There is demand for it in fresh form in Europe, and trial export shipments have been made. However, as Turner and Norton (2009) point out those shipments were found to have high levels of pesticide residues, so the best option is to export in organic form. For organic tree tomato to succeed, the government would have to delineate zones for organic production. The crop does respond well to organic pesticides. The Rwandan Government has emphasized that success in exporting tree tomatoes will depend on installing a cold chain, developing professional relationships with buyers, becoming a reliable sup-plier, and complying with export certification requirements as well as obtaining organic certification, implementing a traceability system, and giving attention to the quality of packaging.

Another bottleneck is the practice for postharvest handling of tree tomato. After being harvested, the fruit should be washed, disinfected, and waxed, but most producers are not accustomed to doing that. For transportation, the fruit should be placed in containers that are rigid with separations inside, and plastic boxes are the international standard for that purpose. These containers also should be washed and disinfected after every use. The challenges of complying with quality requirements are numerous.

In conclusion, cost-price measures reveal only part of competitiveness and the other part depends on how the multiple actors all along the value chain handle their roles in the chain. It involves many issues of different kinds.

2.4.2 Use of Producer Subsidy Equivalents as Indicators of Competitiveness

Calculations of the effect of net producer subsidies (net of taxes) on producer prices, or producer subsidy equivalents, show the net policy stimulus to the production of the goods in question (see Tsakok, 1990, p. 96ff). They are often used in international trade negotiations as indications of the degree of policy support to producers in various countries. Also, sometimes it is assumed that products with a high level of producer subsidies can compete in the international marketplace only because of that support, i.e., that they are inherently uncompetitive—that they have a comparative disadvantage. This may often be the case because uncompetitive producers tend to lobby hardest for government support, but it is not necessarily true. Economic protection, whether measured by coefficients of economic protection or by producer subsidy equivalents, is a concept distinct from comparative advantage and competitiveness. For this reason the methodologies reviewed in this book provide direct measures of comparative advantage and competitiveness. The temptation to use protection levels as indicators of competitiveness or lack thereof is illustrated in the assessment of the international competitiveness of vegetables in India (Vanitha, Kumari, & Singh, 2014) that uses the nominal protection coefficient (NPC) as one of its measures, under the assumption that an NPC value less than 1 indicates competitiveness.

2.4.3 Spatial Variations of Competitiveness Within a Country

Obtaining the data for calculations of cost-price competitiveness can be difficult and time-consuming, and carrying out interviews along value chains equally so. Detailed estimates of production costs are sometimes available only for one locality. For these reasons, often competitiveness is assessed at only one spatial point within a country. This has been the case for the first Colombian study reported here and for the one for El Salvador. However, the competitiveness of a given product can vary substantially across the regions of a country, owing to differences in climate, altitude, soils, transportation costs, access to irrigation, storage infrastructure, and other variables. These differences show up sharply in the second of the Colombian studies in this book. Thus whenever possible it is useful for competitiveness evaluations to be carried out for multiple locations in a country. Investors take locational considerations into account in their evaluations, and policy makers need to be equally aware of them in evaluating the prospects for different product lines.

Investments in transportation and storage infrastructure can make a difference in these costs and therefore can have effects on competitiveness by region in a country, so the evaluations can be based on scenarios that take into account plans for significant projects of those types. In the case of Rwanda, an example is the effect on avocado growers' future prospects of plans for an East African regional railway.

2.4.4 Competitiveness, Comparative Advantage and Shadow Pricing

Since comparative advantage assessments try to abstract from policy interventions that may be temporary, efforts are usually made to adjust factor and product prices to levels that would obtain in the absence of the interventions—shadow prices. In the case of resources used in production, shadow prices are intended to reflect the opportunity costs of those resources. There is extensive literature on the calculation of shadow prices and their relation to macroeconomic policy regimes.[1] For example, the initial debate over the shadow price of foreign exchange for project evaluation concerned the options of using existing exchange rate policy or a shadow exchange rate that would be obtained under an "optimal" macroeconomic policy (Balassa, 1974). Various practices have emerged for calculating the shadow exchange rate. For example, the Asian Development Bank calculates it by adjusting the prevailing exchange rate for the effect of net trade taxes and subsidies (Lagman-Martin, 2004).

The topic of shadow pricing methods is beyond the scope of this book, but the approach taken here is to use factor prices consistent with what can be termed *long-run competitiveness*. Corrections are made for policy distortions that are of doubtful sustainability, but it is recognized that some market distortions may be rather permanent, such as the effects of the Dutch disease syndrome on exchange rates (Vollrath & Vo, 1988) or the presence of some monopolies or monopsonies in small markets.

In the calculations of long-run competitiveness for El Salvador, prices of imported goods were corrected to remove tariffs that would not have been sustainable under World Trade Organization (WTO) accords and regional trade agreements. The price of irrigation water was raised to reflect the true cost of maintenance and operation of irrigation systems, since eventually the tariffs would have to reach that level in order to make the systems sustainable. Regarding the cost of credit, some nongovernmental organizations (NGOs) were charging low interest rates that would not be sustainable in financial markets, and some commercial farmers were able to access external loans (in a dollarized economy) at low interest rates not available to the vast majority of producers. At the other extreme, microfinance organizations tended to charge much higher interest rates to cover risks. An attempt was made to cut through this complex web of interest rates and posit a long-run real return to capital, on average for the sector, to be used as the cost of credit in the calculations.

A central issue in costing farm production is the wage rate to be used for the cost of labor, which is the main input for production by low-income farmers. Early theorizing on agricultural development sometimes assumed a zero cost of labor since clearly there was excess or underemployed labor in rural areas. However, in no circumstances will the supply of labor be positive at zero wage, so that option can be discarded. For hired field labor, the prevailing wage rate has been used, which is lower than the official minimum wage in most tropical countries. A more difficult issue concerns the valuation of family labor. It can be argued that in the short run, once the farm's planting decision has been made, the opportunity cost of family labor is not equal to the wage offered for casual labor in nearby towns because family labor is immobile—some family members cannot leave the farm until after the harvest. A decision to leave the farm and migrate to a city would be made after the harvest. Yet family members will value leisure and the time needed to perform household tasks, so the opportunity cost of their time on the farm must be positive, even though it will be below the market wage rate. Experiments with a mathematical programming model that simulated market equilibria in Mexican agriculture (Bassoco & Norton, 1983) suggested that family labor's opportunity cost was about 50% of the prevailing rural wage. Jooste and van Schalkwyk (2001) used a figure of 60.9% of the market wage as the true opportunity cost for unskilled rural labor in South Africa. Varying approaches to this issue are illustrated in the case studies in this volume. For the Salvadoran calculations reported here, it was assumed that on average 25% of farm labor will be from the farm's own family, and half of the rural wage rate was used for costing family labor, which meant that the cost of average unit of labor was 87.5% of the rural wage rate.

Regarding the opportunity cost of agricultural land, the usual procedure is to use the annual land rental cost or, in the absence of a robust rental market, the annualized purchase price for agricultural land. Obviously this valuation will vary according to the land's fertility and proximity to markets, whether irrigation facilities are present or not, whether the land is cropland or pasture land, and other factors. In his work on comparative advantage coefficients for crops in Bangladesh, Shahabuddin has chosen to use the prevailing rural wage rate for costing both family and nonfamily labor, and regarding the land market he points out that the opportunity cost of land may vary by cropping season and may be a function of the crop rotations that are practiced:

> ...in order to meaningfully interpret these estimates as an indicator of comparative advantage, it is necessary to know the nature and scope of

competition or complementarity in the choice of crops. Although most non-rice crops compete for land in the dry boro season, there is not always a one-to-one substitution between the two crops. In some cropping patterns, the substitution of one dry-season crop for another may also entail changes in the choice of crops in other seasons, because of overlapping crop-growing seasons and agroclimatic factors. In such a case, the appropriate...comparisons would be among the year-round cropping patterns rather than among individual seasonal crops (2000, p. 40).

This experience in Bangladesh constitutes an alert that costs of production may have to be allocated over multiple crops, in both cropping systems and rotations.

More than trying to specify exact values for shadow prices, it can be valuable to perform sensitivity analyses on them, to see how responsive the results are to their variations. Some of the sensitivity analyses can correspond to policy scenarios. In the case of Colombia, at the time of the studies the Central Bank estimated that the exchange rate was overvalued because of the "Dutch disease" effect of large amounts of cocaine exports, so sensitivity analyses were performed to understand the effect of exchange rate variations on crop competitiveness. In the second Colombian study, sensitivity analysis was carried out on the wage rate, the cost of transporting products to the nearest market, the cost of capital, the discount rate (needed for the calculations for perennial crops), and product prices. Variations in product prices could be interpreted as representing changes in the exchange rate, tariff policies, or market factors.

One of the things revealed by this sensitivity analysis was that, for a number of crops, the competitiveness of a crop depended on paying the prevailing rural wage, which was significantly below the official minimum wage for rural areas. Paying the official wage would have made many crops uncompetitive. However, the prevailing wage of course more accurately reflected the labor supply—demand situation in rural areas, and it is more likely than the officially mandated wage rate to represent the long-run cost of labor.

2.4.5 Annual Crops vs. Perennial Crops and Livestock

The literature on the cost-price criterion for competitiveness is based on annual production cycles, whether in crops or for manufactured goods. The formula needs modification for the case of perennial crops and livestock, for which the returns may come years after the investment. That modification, which brings in questions of appropriate time discount rates, is developed and discussed in Section 4.2.5. One of the empirical challenges can be projecting accurately the yield curve over time for perennial crops, usually rising and then eventually falling, and also for animal-based products. For tree crops, yields often demonstrate the following pattern over time: "(1) an early period of no yield, normally occurring in year one through year 3, (2) a period of increasing yield at an increasing rate, (3) a period of increasing yield at a decreasing rate, and (4) a period of decreasing yields" (Mahrizal, Nalley, Dixon, & Popp, 2014, p. 2). In the case of cocoa and coffee trees, after 25 years the yields are normally considered to be sufficiently low that replanting should be considered or, in the case of coffee, tree rejuvenation through selective pruning and/or soil amendments. A difficulty smallholder farmers face is the reluctance to uproot old trees or insufficient resources for replanting. At the other end of the spectrum, passion fruit vines are usually replaced after 3−5 years, although in some commercial operations they keep yielding, with adequate crop management, for up to 10 years. For competitiveness analysis, having accurate information on yields and costs during the first 5−7 years is much more important than for 15−25 years out from planting dates because of the discounting factor.

2.4.6 Data on Farm Production Costs and Transportation Margins

Obtaining reliable and sufficiently representative data on farm production costs and yields, as well as transportation margins, is one of the major challenges faced in competitiveness analysis. The costs need to be disaggregated by type of input and, for perennial crops, by year. Surveys developed for collecting farm budget data are expensive and time-consuming so usually existing, less systematic sources are used. Agricultural development banks often are a principal source of information on these data since they need to know the details of the production costs of their clients. The research units of Central Banks sometimes compile this kind of information, and extension services also do it in some cases. Sometimes structured interviews can be set up with experienced local extension agents to compile the required data, although of course there always is a margin of error in such information, which depends heavily on recall and personal experiences.

Production cost profiles are often developed for agricultural projects of international development partners. However, often those profiles tend to show projected or desired costs, rather than actual practice, and therefore it is generally preferable to avoid using that source of information. When data are available on actual practices, it can be useful to differentiate cost structures by type of technology utilized, e.g., not mechanized, semimechanized, and fully mechanized. The more mechanized or input-intensive technologies may not be the most competitive, depending on the cost of labor and other inputs and the corresponding yields.

Attempts are being made to fill this gap in internationally comparable farm budget data. Hemme, Uddin, and Ndambi (2014) developed estimates of the cost of production of milk for 46 countries using a standard costing methodology for two farms per country. However, since their presentation does not disaggregate costs by the type of input, these data would not be usable in that form in analyses of competitiveness that include sensitivity analysis on policy variables and other influences on factor and input prices. Also, this kind of comparative cost analysis across countries needs to be done for high-value export crops, for which cost data are scarcer than for agricultural commodities. Overall, much remains to be done on the data side, including developing widely applicable methods for allocating overhead costs among crops in mixed cropping systems, and developing estimates of the variations in costs over large numbers of farms that are producing the same output. An implication for the work of agricultural extension services is the need to advise farmers in cost accounting and its value for their farm planning, which will make them more successful as farmers in addition to generating better data for future analysis.

For transportation costs, usually it is necessary to interview local traders and other participants in value chains. A useful shortcut to estimating total costs of transportation, handling, and storage at stages in a value chain can be to compare farm gate prices with rural wholesale prices and then urban wholesale prices. However, those prices will reflect existing inefficiencies in the corresponding stages of the value chains, and obtaining price information can be a challenge in itself, especially for niche products. In the end, it often proves necessary to carry out field interviews with stakeholders. In these cases, it is important that the interviewers themselves have substantial familiarity with the production and marketing conditions in the locales they are working in to collect data.

2.4.7 Determinants of Quality

Much of the material in this book emphasizes that quality is at the heart of competitiveness, but it can be a multifaceted and elusive concept. As illustrated by the case of tree tomato in Rwanda, it can embrace even the packing materials used to get the product to market. In the final analysis, the market—the consumer—determines what quality is and whether a given product meets quality standards. And these standards evolve over time, sometimes rapidly. Nevertheless, there are basic attributes of quality that cut across products and markets. They include the following:

1. *Food safety and sanitary and phytosanitary requirements for export.* These include freedom from or minimal levels of residues of pesticides and other harmful contaminants and dangerous bacteria. These requirements have become stricter in recent years. For example, the new Food Safety Modernization Act in the United States significantly raises the bar for the import of food products into the United States. It includes "a new requirement that all facilities that manufacture, process, pack or hold food products have documented preventive control programs in place. The preventive control requirements will go above and beyond a typical Hazard Analysis and Critical Control Point plan, and will include requirements

related to recall plans, sanitation, employee hygiene training, environmental monitoring, allergen control program, current good manufacturing practices, supplier approval and verification activities, corrective actions... and records" (Leavitt Partners & Eurofins, 2012). The food import regulations for Europe and other countries also include requirements on traceability, labeling, and packaging.[2] Traceability has become a key component of food safety certification systems. A thorough review of food identity preservation and traceability, as well as national food safety requirements in several countries, may be found in Bennet (2009). Traceability is examined in detail in the work by Dabbene, Gay, and Tortia (2013), who suggest that its implementation requires rethinking the organization of a supply chain. In addition to official food safety standards, the private sector in developed countries has defined standards that increasingly govern imports from developing countries, standards such as GlobalG.A.P (www.globalgap.org), where the G.A.P. stands for Good Agricultural Practices, and Safe Quality Food (SQF), www.sqfi.com. The latter consists of standards developed by the Safe Quality Food Institute, an arm of the food retailers institute Food Marketing Institute.

2. *Getting the genetics right.* Genetic quality and varieties preferred by consumers and marketing entities is another fundamental attribute of food quality. Taste, size, color, smell, texture, ease of use, and other characteristics can determine market preferences for varieties. In this sense, the search for product quality often begins with plant breeders. In a well-known example, the predominant place of Hass avocadoes in the market shows how diverse factors may influence market preferences. Other avocado varieties have excellent taste and creamy texture, such as the Macarthur avocadoes, but Hass tends to be preferred by buyers because its skin is tough enough to withstand the stresses of shipping, it has a relatively long shelf life, its growing season is all year around, and it has good taste attributes even if not the very best.

In sorghum, a brewery in Uganda persuaded more growers to adopt the *Epuripur* variety, well suited for brewing beer, by offering guaranteed prices through purchase agreements. In this way, the number of farmers cultivating that variety rose from 350 to 8000 in 4 years (COMPETE Project, 2009). More generally, "it is important that sorghum breeders recognize that food quality is critically important and is an essential part of grain yield. This has proven true in Honduras where *Sureño*, an improved sorghum, has been adopted by farmers because it has good tortilla-making qualities and a sweet juicy stalk that improves its forage quality" (Rooney, 2010).

Cassava is another example of a crop for which genetic development is important for its marketability. In many parts of Africa, cassava plantings have been devastated by cassava mosaic disease, cassava brown streak virus, and Ugandan cassava brown streak virus, which generate dry rot in the roots and thus make the crop unusable (Legg et al., 2014). There is an urgent need for development of resistant varieties and their multiplication and dissemination to farmers.

3. *Appearance, taste, uniformity of product, and degree of ripeness.* These factors are very important for marketing. They are related to seed selection by farmers and to crop management and harvesting techniques. There are 19 recognized grades of cardamom in the international market, on the basis of color, size, shape, and uniformity, and the top grades can sell at prices more than three times the price of a low grade. Oddly shaped carrots may have acceptable taste but will not sell as well as straight ones. In southern Honduras, smallholder cantaloupe producers in contract farming arrangements for export learn to rotate cantaloupes in the field at regular intervals to ensure uniformity of the patterns on the fruit's outside. Kaniwa (*cañihua*), a relative of quinoa, is grown at high altitudes in Peru and Bolivia and has strong nutritional qualities. However, its entry into broader markets is hampered by the fact that *cañihua* plants in a field ripen at different rates, making it difficult to harvest them in volume. This problem occurs for many crops when farmers use seed retained from previous harvests, and that practice also can lead to harvests with products of different sizes and colors and greater vulnerability to crop diseases.

Harvesting practices themselves are relevant to the uniformity of the harvested batches. In coffee, smallholders sometimes are tempted to harvest immature beans along with mature ones to sell a larger quantity, and this lowers the overall quality of the lots they sell.[3] This phenomenon occurs with other crops as well, often because of the needs of smallholders for immediate cash. Harvesting strategies affect the shelf life of products. Timing the harvests correctly can avoid overripe or underripe products and the price discounts associated with them. Postharvest strategies are relevant here too. For many fruits and vegetables, it is important to take the field heat out of the products soon after harvest and dry them, to reduce the chances of spoilage before the produce reaches markets. For some products, insufficient postharvest drying can lead to premature sprouting, which severely undermines marketability. Taste is subjective and determined by market acceptance standards, and it is related to both genetics and crop management. There are some objective measures, such as degrees Brix[4] in sweet fruit and the coffee rating system based on cupping developed by the Specialty Coffee Association of America. More recently, World Coffee Research developed a sensory lexicon that identifies 109 attributes of coffee's taste; see http://worldcoffeeresearch.org/read-more/news/174-world-coffee-research-sensory-lexicon. Efforts are underway to develop more complete "flavor wheels" for cocoa as well.

4. *Resistance to diseases and infestations.* Plant diseases and infestations of course can reduce harvests, thereby lowering unit yields and raising unit costs, in addition to the costs imposed by strategies for defending the

plants. In addition, they can lead to an inferior product for consumers. An example is cardamom produced in Guatemala, the world's largest producer and exporter of that crop. Increasingly the cardamom harvests in that country are affected by infestations of the tiny insect thrips (*Sciothrips cardamomi*) that leave scars across the harvested seeds, and this problem sharply lowers the cardamom's price in its main markets in the Middle East, where buyers judge the product by appearance (see Milian, 2014, for further information.).

5. *Humidity and impurities.* Low humidity and the absence of impurities are important dimensions of product quality and are best controlled in the postharvest management immediately following the harvest. Humidity may not be a concern for the final consumer, but it is relevant for wholesalers who store and ship the product, and for whom high humidity can mean product loss through development of molds and spoilage. Both processing and final consumers are sensitive to the presence of bits of sand and other impurities that can be present in some lots of grains like rice. For these reasons, product drying and sorting and grading are important postharvest activities on farms or near farms for many products. In the case of vanilla, proper harvesting for humidity control is crucial for quality: "Processors offer higher prices for better quality beans, and one method to achieve quality gains is through harvesting at appropriate moisture content levels" (Komarek, 2010, p. 238).

6. *Handling, storage, transportation, and packaging.* Proper management in this area is crucial for maintaining product quality and for minimizing food loss. For the case of pineapple in Rwanda, Turner, and Norton (2009) found "farm-to-market transportation of pineapples is still carried out in rudimentary fashion in most cases and, together with the lack of a cold chain, this results in considerable losses of fruit and lowering of its average quality when it reaches markets. At least one person involved in the trade in a managerial capacity said pineapple supplies are insufficient because the production areas are far from Kigali and the fruit arrive in the capital in very poor condition. Because of the state of rural access roads, in some cases pineapple can be obtained in better condition from Uganda, and also from DR Congo, for parts of the country located closer to those borders. Effectively, transportation conditions have created a market for pineapple that is semi-segmented spatially. In addition, these circumstances create a relatively large amount of fruit suitable only for processing as opposed to consumption in fresh form, so in relation to demand there can be a surplus of pineapple juice and jam on the local market while fresh supplies are in balance with demand." More broadly, Gustavsson, Cederberg, Sonesson, Otterdijk, and Meybeck (2011, Annex 4) estimate that in sub-Saharan Africa 51% of production of fruits and vegetables and 38% of production of roots and tubers is lost or damaged beyond usefulness in the stages of postharvest handling and storage, processing and packaging, and distribution.

7. *Environmental attributes and treatment of labor.* In addition to environmental monitoring requirements in the new US Food Safety Modernization Act, many product markets require some kind of environmental seal of approval, e.g., of the Forest Stewardship Council, Rainforest Alliance, or their market acceptance and price may be enhanced by environment-friendly production practices, e.g., bird-friendly coffee. The Fair Trade seal guaranteeing humane treatment of employees enhances product value on international markets, and many private buyers are moving in the direction of imposing requirements in that regard. However, the importance of these attributes in the marketplace can sometimes be overshadowed by other dimensions of quality. A recent survey-based analysis of determinants of coffee prices by the Transparent Trade Coffee Project of Emory University's Goizueta Business School reported that TTC Insights (2015):

> *Factors mentioned on at least 50% of the websites describe a farm's location, processes, varietals and elevation, as well as the given name(s) of the farmer or his/her family. Another relatively common practice is to list prior awards or accolades; most commonly, a positive result in a Cup of Excellence competition. One can make the case that each of these attributes gives roasters an opportunity to positively differentiate a specialty coffee based on expected bean quality and a quality-driven farm identity. And:*

> *At the other end of the continuum are two factors that emphasize environmental contributions and farm workers. When communication drifts toward emphasizing these more social factors, there is evidence of lower average prices. It is not that these factors are unimportant. Instead, it seems that* **specialty coffee markets are currently places where market valuations – and therefore prices – are driven by markers of expected coffee quality rather than indications of positive social impact.**

8. *Special characteristics.* Some markets offer a premium for products with special characteristics, such as organic products and coffee produced by women's cooperatives, and for products with a denomination of origin. However, as with other improvements in quality or perceived quality, these certifications and those related to environmental attributes and labor concerns can be expensive from the viewpoint of smallholders. Often international NGOs underwrite the cost of the certifications during the lifetime of projects, but that raises questions about the sustainability of the certifications, which must be renewed at regular intervals. Melo and Hollander (2013) describe an experience with cocoa in Ecuador in which the cost of certification impinged heavily on producer profits and in spite of donor preferences had to be abandoned after 15 years.

Many tropical products are high in antioxidants and other healthy attributes, because of the intense natural competition for space, light, and nutrients in tropical ecosystems. These special characteristics are part of their market appeal, but the other characteristics mentioned have to be preserved also in order for these inherent benefits to be appreciated in the marketplace.

Trieinekens, van der Vorst, and Verdouw (2014, p. 506) provide an illustration of the complexity and detail of quality determinants, for the case of pork. Many or all of them may enter into consumers' assessment of product quality. They classify quality attributes of pork as intrinsic and extrinsic. The intrinsic attributes are:

- Sensory
  - Tenderness
  - Color
  - Marbling
- Health
  - Safety (zoonosis)
  - Food additives
  - Antibiotics
  - Residues
- Convenience
  - Packaging
  - Shelf life
  - Preparation characteristics

The extrinsic attributes are:

- Animal welfare
  - Farm production system
  - Transportation
  - Slaughter
- Ecological footprint
  - Farm manure and waste management
  - Transportation
  - Slaughter products (high and low value)
- Origin and authenticity
  - Production location
  - Community impact
  - Farm production system
  - Processing system

Other variables also enter into determination of product quality, but the aforementioned eight general characteristics, or most of them, are essential quality attributes for almost all agricultural products.[5] Lack of quality in these dimensions can block a product's entry into markets or reduce its price. Many of them enter into buyer's grading standards that are directly related to price. In the case of attributes (vii) through (ix), not having these characteristics may not always represent an obstacle to entry into markets but having them can improve a product's price.

ENDNOTES

1. For especially comprehensive reviews of shadow pricing methods, the interested reader may wish to consult Squire (1989) and Florio (2014).
2. A quantification of the trade effect of food safety requirements in importing countries found it to be more important than import tariffs in those countries, for Chinese exports (Chen, Yang, Yang, & Findlay, 2008).
3. In one of the more advanced producer cooperatives in Guatemala (ASOBAGRI) it was found that 89% of the coffee farmers reported that they separate the green coffee berries and the overripe ones before processing the berries that are at the appropriate level of ripening. Almost certainly a significantly lower proportion of producers do this in the smallholder sector as a whole (CRECER and Grameen Foundation, 2015).
4. One degree Brix signifies 1 g of sucrose in 100 grams of solution.
5. The challenges of producing quality crops, from agronomic strategies to postharvest management, are well illustrated and documented for coffee in the book by Oberthür, Läderach, Pohlan, and Cock (2012).

Part II

Methodologies for Evaluating Competitiveness

Chapter 3

International Trade and Prices as Measures of Competitiveness

Relative prices and international trade patterns tell a good deal about comparative advantage. From a static viewpoint, if a product's domestic price is significantly higher than the international price equivalent (after adjusting for transportation cost differentials and port handling costs), the suspicion arises that it may not have a comparative advantage and that only a high price on the domestic market allows it to remain profitable. Although that is not always the case, if the product enjoys a favorable price differential in the domestic market in this sense and is not exported, and significant quantities of it are imported, then it almost certainly does not have comparative advantage in production. Imports from its competition could be expected to increase.

On the other side, if a product's domestic price is close to the international price or below it, and there are no imports of it, then it is likely to have a comparative advantage. If it is exported under these conditions, then it is almost certain to enjoy a comparative advantage. Ball, Butault, San Juan, and Mora (2010) use relative agricultural prices between the U S and European countries, adjusted for purchasing power parity, to evaluate the competitiveness of agriculture in these trading partners. They also look at productivity over time as indicators of trends in competitiveness. In their framework, relative inflation rates and exchange rate movements play a major role in determining competitiveness.

For these reasons, the analysis of comparative advantage or long-run competitiveness sometimes begins with a review of relative prices and with the net trade balance of each product. There are many trade-based indicators of competitiveness, and this review concentrates on the ones that are most widely used, beginning with the simplest ones. For trade variables, the simplest indicator of possible comparative advantage is that a product is exported. Since trade data are almost always examined in the form of product groups, at least down to the six-digit level in trade classifications, a more accurate indicator of possible competitiveness is that net exports ($e_j - m_j$) are positive, where e_j signifies exports of the product group, and m_j, imports of the product group. However, these static observations of trade do not convey a lot of information.

The Competitiveness of Tropical Agriculture. http://dx.doi.org/10.1016/B978-0-12-805312-6.00003-9

It is more illuminating to look at their time trends, that is, whether net exports are increasing or decreasing over time.

The first Colombian study reported in this book uses multiple indicators of comparative advantage. Among the static measures the simplest one is an extension of the net export balance in the form of the relative trade balance (RTB) for a product or product group, which may be expressed as:

$$RTB_j = (e_j - m_j)/(e_j + m_j) \tag{3.1}$$

This indicator is calculated for groups of products and shows whether a set of related products is more export or import oriented as a whole.

The values of RTB_j fluctuate over time even when trends are apparent, so inferences were based on ranges of values. If $0.33 < RTB_j < 1.0$, then for the product group the country is considered to be a net exporter. If $-1.0 < RTB_j < -0.33$, then the country is considered to be a net importer of the product group. For cases in which $-0.33 \leq RTB_j \leq 0.33$, then either there is little international trade in the product group in this country's case or the trade goes both ways, imports and exports, in significant quantities. Trends in the RTB indicator were also examined for more clues about potential comparative advantage or long-run competitiveness. A disadvantage of the RTB measure is that it erases quite a bit of potentially useful information. Grubel and Lloyd (1975, as mentioned in Vollrath & Vo, 1988) pointed out that, for 10 major countries that are part of the Organization for Economic Co-operation and Development, at the 3-digit Standard International Trade Classification (SITC) level this net trade measure removed 63% of the countries' total trade.

Another weakness in the RTB measure is that a country may be a net exporter of many product lines and not all of them will necessarily have a *comparative advantage*. Alternatively, it can be a nascent exporter of several product lines, so that their RTB values fall in the indeterminate range, when in fact some of them do have a comparative advantage. Balassa (1965) was the first to suggest taking into account a country's export share in total international exports, both in the aggregate for all products and for the particular product under consideration, to ascertain what he called *revealed comparative advantage* (RCA). He defined this measure as:

$$RCA_j = (e_j/E_j)/(e_t/E) \tag{3.2}$$

where E_j signifies total world exports of product j, e_t stands for total exports of the country, and E stands for the world total of all kinds of exports. This concept is a measure of the degree of a country's specialization in a product on the international scene. Comparative advantage is indicated when RCA >1.0, i.e., when the country is more specialized than average in exports of product j.

To illustrate this concept with arbitrary numbers, if a country's total exports represent 10% of all world exports, but its coffee exports represent 30% of world coffee exports, then the RCA for coffee would be 3.0 (30%/10%), a relatively high value. But the RCA measure alone is not sufficient to say definitively whether comparative advantage is present. It does, however, indicate whether a product is competitive under the existing policy regime. The RCA measure can also be viewed in the dynamic context to see whether competitive products are increasing their competitiveness or losing it, and whether currently noncompetitive products are moving toward competitiveness over time. This has been done in the first Colombia case study presented later in this book.

RCA is widely used. Examples of applications of the RCA procedure include Serin and Civan (2008), to evaluate the competitiveness of olive oil, tomatoes, and fruit juice in Turkey, and Kumar and Rai (2007) to evaluate the competitiveness for tomatoes and tomato products in India. For Turkey, the former found strong competitiveness in olive oil and fruit juice but not in tomatoes. The latter found lack of Indian competitiveness in tomatoes and tomato products. A more recent application of the RCA formula for India (Vanitha, Kumari, & Singh, 2014) found that of seven vegetables, only onions are competitive. For Nepal, Salike, and Lu (2015) made RCA calculations for both agricultural and manufactured products and generally found agriculture to be the more competitive sector, especially for cardamom, ginger, lentils, medicinal herbs, tea, and juice (all high-value, smallholder products) but not for sugar. Kiet and Sumalde (2006) used RCA calculations, among others, to find that shrimp exports from Vietnam are highly competitive, but they indicate the need for attention to quality issues.

Karim and Ismail (2007) use RCA calculations for a number of agricultural products in Egypt, Sudan, and Kenya to assess the potential for greater trade within the region. If the countries are highly competitive in different products, then the potential for more trade is indicated, which turns out to be the case for many products. Mwasha and Kweka (2014) use RCA to show a competitive advantage for Tanzania in coffee, tea, mate, spices, and tobacco. In Brazil, Dorneles, Dalazoana, and Schlindwein (2013) have used RCA to confirm the comparative advantage of soybean products produced in Mato Grosso. Dynamic interpretations of the RCA measure are discussed in the case study in Chapter 7 of this volume.

From the viewpoint of competitiveness evaluations, a basic limitation of the RCA is that it cannot be used for ranking products in order of competitiveness. Amplifying the points of Hoen and Oosterhaven (2006), Sarker and Ratnasena (2014, p. 524) make the point in the following words:

> The [RCA] has a lower limiting value of zero but the upper limit is infinity or undefined. Therefore, as an index, it is asymmetric. The [RCA] can only indicate whether or not a country has comparative advantage in a commodity or sector. Its magnitude has neither the ordinal property nor the cardinal property and can generate misleading results particularly for countries with a small market share in the world export market. Due to these inadequacies, the same value of [RCA]

might signify different levels of comparative advantage for different countries or commodities and hence, cannot be used to compare comparative advantage across countries or commodities.

Dalum, Laursen, and Villumsen (1998) and Laursen (1998) proposed a variant of the RCA measure to correct for this lack of symmetry. It is termed symmetric revealed comparative advantage (RSCA) and is calculated as:

$$RSCA_j = (RCA_j - 1)/(RCA_j + 1) \qquad (3.3)$$

It ranges in value between -1.0 and $+1.0$, where any value above 0 indicates the presence of a comparative advantage. This measure facilitates conclusions about a product's degree of comparative advantage or comparative disadvantage. However, its restricted range of values does not mean it has a normal distribution, and it may suppress useful information contained in the extreme values of the RCA.

A practical advantage of the RCA or RSCA is that it does not require information on production costs for its calculation, only data on international trade performance (SITC data), albeit for both the country's exports and total world exports. It generally is possible to apply it to a large number of products or product groups. The reliance on trade data alone is especially useful for agroindustrial products, for which cost-price measures of competitiveness usually cannot be calculated owing to lack of public availability of information on cost structures of private firms.

Another criticism of the RCA and RSCA measures is that they do not take into account imports. Hence Vollrath and Vo (1988) proposed use of what they term the *revealed competitiveness index*, which incorporates imports but avoids the problems associated with using net trade. It does this by giving imports and exports separate roles in the formula. They also proposed the use of logarithms to resolve the problem of asymmetry of the RCA measure. Their revealed competitiveness index is defined as:

$$RC_{j,n} = ln\{ [(e_{j,n}/E_j)(e_{a,n}/E)] [(m_{j,n}/M_j)(m_{a,n}/M)] \} \qquad (3.4)$$

where the subscript j indicates the commodity, n the country, a the sum for all goods, and capital letters indicate world totals, for good j with the subscript and for all goods without the subscript. The RC value can be either positive or negative. They point out that this "measure of competiveness uses both export and import data because competitive advantage, like the concept of comparative advantage, is determined by both relative supply and relative demand" (1988:3). Their illustrative applications of the RC measure refer to entire countries, not industries or products, and one of the inferences drawn from the applications is that the Dutch disease syndrome undermines agricultural competitiveness, which has been pointed out on the basis of other analyses (e.g., Norton, 2004).

Another problem of the RCA indicator is that its mean is not 1.0 across countries for the same commodity, as would be expected since comparative advantage in one country should be offset by comparative disadvantage in another. Also, the mean is unstable with respect to the number of countries or commodities.

Yu, Cai, and Leung (2009) proposed a trade-based measure of competitiveness that overcomes many of the problems discussed, including an unstable and inappropriate mean[1] and an asymmetric distribution.[2] Their measure, the normalized RCA (NRCA), starts from a hypothetical situation of comparative advantage neutrality for product j in country n. It measures how much the product's actual trade performance deviates from what it would be with a neutral comparative advantage. Following the exposition of the NRCA by Sarker and Ratnasena, in a situation in which no country had a comparative advantage in any commodity (the comparative advantage neutral condition), the exports of good j of country n would be:

$$\widehat{e}_{j,n} = e_n E_j / E \qquad (3.5)$$

where e_n denotes total exports of country n. Then the difference between actual exports of j from country n and the comparative advantage neutral exports becomes:

$$\Delta e_{j,n} = e_{j,n} - \widehat{e}_{j,n} = e_{j,n} - e_n E_j / E \qquad (3.6)$$

Normalizing by world exports[3], the indicator called the normalized RCA is defined as:

$$NRCA_{j,n} = e_{j,n} / E - e_n E_j / E^2 \lessgtr 0 \qquad (3.7)$$

A positive value of $NRCA_{j,n}$ implies that commodity j has a comparative advantage in country n and a negative value, that it does not.

Sarker and Ratnasena calculated Eq. (3.7) for Canadian exports of wheat, beef, and pork for each year from 1961 to 2011 and found that wheat consistently has a comparative advantage in Canada and pork does not, and beef is on the borderline. In addition, they hypothesized that comparative advantage should be relatively stable over time and found that the coefficient of variation of the normalized RCA is much less than the coefficient of variation of the RCA. Hence they conclude that the former should be a more reliable indicator of comparative advantage. That may be a reasonable conclusion when the productive environment is largely governed by permanent factor endowments such as land, water, climate and geographical proximity to major export markets. However, when other elements of the enabling environment discussed in Section 2.3 above undergo evolution, then comparative advantage itself may indeed change. Using RCA calculations, Le (2010) found a marked shift in Vietnam's comparative advantage in as short a period as 5 years, away from agricultural products and toward labor-intensive manufacturing.[4]

Comparative advantage theory predicts specialization in product lines according to a country's resource endowments. Trade-based measures such as the RTB, the RCA, the RSCA, the RC, and the NRCA coefficients employ this relationship in the opposite causal direction, using *observed* degrees of specialization as indicators of underlying comparative advantage. Although the RCA coefficient has been used in one of the case studies reported here, one of its disadvantages is that it may not be applicable to all the crops with potential comparative advantage, because even at a detailed level of disaggregation trade data include residual categories, the ubiquitous not elsewhere classified category. This practice affects fruits and vegetables more than other product groups, to such a degree that the trade-based measures, including the NRCA, cannot be applied to many of the high-value products in that group.

Cost-based measures of long-run competitiveness are presented in Section 4.2. They have the advantage of product specificity, and of identification with particular growing locations within a country and particular technologies of production, but the farm budget data they are based on may be subject to greater errors than the trade data used in the RTB, RCA, RC, and NRCA measures. Hence both kinds of measures have their roles and limitations.

ENDNOTES

1. Laibuni, Waiyaki, Ndirangu, and Omiti (2012) used an international product specialization index, which has a mean value of 1.0, as well as RCA calculations, to assess the competitiveness of Kenya's cut flower sector vis-à-vis competitors for market share in different importing regions of the world.
2. Hoen and Oosterhaven (2006) also developed a trade-based comparative advantage measure, called the additive RCA, which corrects for these issues.
3. Other normalization procedures are explored in Bebek (2011).
4. The history of development of trade-based comparative advantage measures, along with a systematic evaluation of 10 variants of the RCA measure, but not including the NRCA, can be found in Vollrath (1991).

Chapter 4

Price−Quality Tradeoffs and Multitrack Evaluations of Competitiveness

The rationale for using multiple tracks in evaluations is to capture both the efficiency and quality aspects of competitiveness. The trade-based coefficients such as RCA and RC implicitly incorporate both aspects precisely because they indicate revealed or ex post comparative advantage or ex post long-run competitiveness, as pointed out by Yercan and Isikli (2007). A cost−price criterion of competitiveness for primary producers, a measure of the efficiency aspect, may be regarded as an ex ante indicator of comparative advantage or competitiveness, because issues downstream in the value chain may hinder its realization. Hence a trade-based measure of competitiveness does not require complementary analysis in another track, but a cost−price measure does need to be complemented with the quality criteria that are addressed in various stages of value chains.

Quality issues are pervasive in tropical agriculture, especially in the segments of fruit and vegetables. Turner and Norton (2009) reached the following conclusions for Rwandan agriculture after the intensive review of product value chains reported in the case study in Chapter 8 of this book:

> For the domestic market, a pervasive issue is lack of sufficient quality in fresh fruits and vegetables. A major buyer for supermarkets stated he would purchase more products locally, rather than importing them, if quality were higher. This and other evidence indicate the need for intensive, on-farm technical assistance… It has been estimated that only 10% of horticulture producers have received technical assistance, and even that has not always been of a continuing, hands-on variety… The technical assistance should cover harvest and post-harvest operations also. The post-harvest phase is equally important for ensuring product quality. An obvious example of the need for improvements in this area is the practice of tossing tomatoes together into rustic baskets that were designed to carry harvested potatoes. The result is an inferior product when it reaches the buyer's hands, no matter how good it was when picked from the plants.

The Competitiveness of Tropical Agriculture. http://dx.doi.org/10.1016/B978-0-12-805312-6.00004-0

39

Quality is even more important for export markets. In addition to meeting consumer expectations regarding taste, appearance, consistency, packaging, and other characteristics, obtaining certifications is increasingly important for penetrating export markets. These can include organic, Fair Trade and other types of certification at the farm level and HACCP and ISO, as well as organic and Fair Trade, certifications at the processing level. Rwanda has advantages for organic production because of the traditionally low use of agro-chemical inputs. However, this advantage has not always been well exploited. For example, a sample of tree tomato fruit shipped to the UK was rejected because of pesticide residues in excess of the limits.

In view of the potential role that organic products can play in Rwandan agricultural exports, the government may wish to [provide] support… by reaching a consensus with farmers and decreeing organic production zones, in which only organic production will be carried out. Clearly, a farmer cannot earn organic certification if the neighboring farm is using pesticides that could be carried across farm boundaries by the wind, so usually zones that embrace multiple farms have to go the organic route together. Perhaps more importantly, certifying all the farmers in a given geographical area vastly increases the ease of setting up an Internal Control System that is by far the most cost-effective way of certifying numerous small-scale farmers.

The quality issue is more important and more widespread in fruits, vegetables, tea and coffee than in, say, grains and legumes. Hence, quality has to be given priority in the development of this sub-sector. In overall terms, Rwanda's climate and small farm size (high ratio of labor to land) lends it a comparative advantage in fruits, vegetables, tea and coffee, which are high-value products, vis-à-vis many other kinds of crops, but to fully realize this comparative advantage it will be necessary to pay close attention to quality concerns.

Markets can vary in their relative emphases on quality and price for products coming from their suppliers. For extreme cases, Fig. 4.1 shows markets for which one or the other criterion prevails. The vertical axis is a measure of quality, and the horizontal axis represents prices from suppliers (their costs). On this latter axis prices decline with movement to the right. In a price-dominant market, only products offered with prices low enough to lie to the right of the vertical line $M^{(p)}$, and above the minimum acceptable quality q^*, will be of interest to the market. Conversely, in a quality-dominant market, only products above the horizontal line $M^{(q)}$, and to the right of the maximum acceptable price p^*, will be acceptable.

Many products have both a minimum acceptable quality level and a maximum acceptable price, e.g., specialty coffee. For those products the requirement for them to be marketable is that they lie in the area to the northeast of point A in Fig. 4.1, between the lines $M^{(p)}$ and $M^{(q)}$.

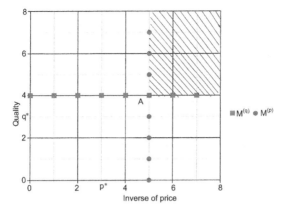

FIGURE 4.1   Price—quality space.

Quality, in light of its multiple defining characteristics mentioned earlier, may be an unavoidably ordinal concept for some products and therefore it may not always be possible to assign cardinal values to it along the vertical axis. Cardinality of consumer preferences is not required to establish the basic axioms of consumer demand theory, but cardinal quality measures are useful in establishing tradeoffs between quality and costs (quality and supplier prices). For grains and oilseeds, the US Department of Agriculture has established basic grades. For corn, they are based on product weight, percentage of heat-damaged and otherwise damaged kernels, and amount of foreign matter present. Special grades also exist. However, such well-defined standards generally do not exist for tropical fruit and vegetables and other tropical products. Genetic differences between varieties and degree of ripeness at harvest, for example, are not readily assigned numerical values.

Nevertheless, numerical valuations of quality may exist for some characteristics, such as the percentage of bruised or rotten fruit in a shipment. To frame the issue faced by farmers and value chains, it is assumed for the moment that quantitative indexes of quality, or of some key determinants of quality, exist. If price is taken to be one dimension of desirability from the consumption viewpoint, then a consumer's indifference curves involving quality may be visualized as in Fig. 4.2. Under Lancaster's innovative work (1966) that recasts consumer theory in terms of preferences for characteristics of goods rather than for goods themselves, these indifference curves are convex to the origin, as in classical consumer theory. For the sake of illustrating the issues, aggregate market indifference curves in the quality—price space are used as shown in Fig. 4.2. In their application of Lancaster's approach, Mark, Brown, and Pierson (1981) showed that such aggregate

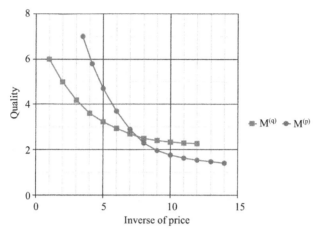

FIGURE 4.2   Market indifference curves.

indifference curves do exist, with a negative relationship between price and quality, for the case of the quality attributes of beers:

> *In sum the analysis yields an aggregate demand function representative of discriminating consumers… not content to take beer quality and price as a secondary part of the consumption package (1981:37).*

One of the curves in Fig. 4.2 $M^{(p)}$ illustrates a price-driven market, where larger amounts of quality may be sacrificed to obtain a lower price, and the other $M^{(q)}$ illustrates a quality-driven market, where the quality requirements will be reduced only for large decreases in price.[1] If in an extreme case in which all dimensions of quality were taken to be subsumed in price, so that the units of the vertical axis are monetary units (a good of $7 quality will be bought at $7), then the market indifference curves in Fig. 4.2 become downward-sloping straight lines.

The production possibilities frontier for a producer (or for an entire value chain) can be introduced into the analysis. In Fig. 4.3 producer 1, with production possibilities frontier $X^{(1)}$, is able to attain higher quality but producer 2, with frontier $X^{(2)}$, is able to produce at a lower cost (at a lower price to buyers). Correspondingly, the buyer–seller equilibrium E_1 will be high-quality, high-cost for producer 1, and lower quality, lower cost for producer 2, point E_2.

Fig. 4.4 shows the choices facing a producer who wants to enter a more competitive market, either because it is becoming necessary to maintain sales or because it will be helpful to increase sales volume. When compared with the producer's present price–quality situation, that new market (on a higher indifference curve) will be characterized by a higher quality for the

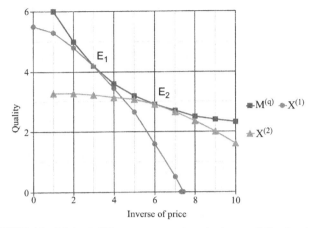

FIGURE 4.3 Market indifference curve and production possibility frontiers.

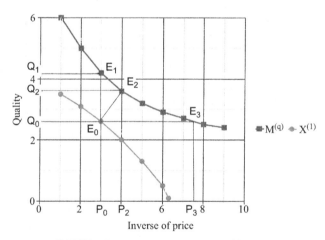

FIGURE 4.4 Paths to improved competitiveness.

same price, or a lower price for the same quality. For a producer located initially at point E_0, the choices are (1) moving to point E_1 through improvements in quality alone, at the same cost of production, (2) improving quality while simultaneously lowering costs by moving to point E_2, or (3) moving to point E_3 through reducing costs of production while retaining the same quality level.

The lesson of Fig. 4.4 is that trying to compete on the new market on the basis of price alone, going to point E_3, may require very large reductions in the cost of production, perhaps unattainable, and that trying to compete on the basis of quality improvements alone, with the same cost structure (point E_1), may require large changes in quality, also perhaps unattainable. Moving to F_2 would require smaller cost savings and smaller improvements in quality, and hence may be more feasible, although the producer would have to work on both aspects of the operation.

An analysis like that illustrated in Fig. 4.4 can be carried out separately for each relevant dimension of quality (one quality dimension at a time on the vertical axis), so that calculations may be made for each quality-improvement challenge faced by a producer, for example, regarding the costs associated with strategies for reducing the percentage of impurities in each batch of product, and the costs required for strategies for harvesting several times in the same field to ensure more uniform ripeness.

The next section presents the cost—price criterion of competitiveness used in the case studies, and the section after that discusses the quality criteria used in two of those studies. These two sets of criteria constitute the two tracks of a methodology that have been found useful in evaluating competitiveness. The first one is a quantitative criterion that depends on the collection of numerical data on farm budgets, costs of transportation and product transformation, and border prices (international prices at a country's ports). The second set of criteria is largely qualitative, and the methodology consists of identifying the quality issues in a value chain and the stages of the chain where they arise. Seeking remedies for the identified issues would then require cost calculations and tradeoff analyses, such as those illustrated in Fig. 4.4.

ENDNOTE

1. Market indifference curves may sometimes reflect processing characteristics of primary products, as in the case of sorghum mentioned earlier, as well as pure consumer preferences.

Chapter 5

Track 1 Methodology: Cost-Price Measures of Competitiveness

The most straightforward approach to measuring international competitiveness, often employed, is to compare the costs of production across countries for the same product, adjusted for transportation and marketing costs and translated into a common measure via exchange rates. This can be helpful, as in the case of the study of rice production costs in Southeast Asia by Bordey et al. (2015) that showed Philippine unit costs to be higher than those of Thailand and Vietnam. However, as the same study showed, part of the cost advantage for those countries was attributable to the free provision of irrigation water. For this reason, an assessment of long-run competitiveness (LRC) has to take into account subsidies that may be transitory, as well as other factors such as, in some countries, exchange rate distortions and tariffs and nontrade barriers. Such an assessment starts with a disaggregation of costs into imported and domestic costs.

5.1 IMPORTED VERSUS DOMESTIC COSTS

The basic data requirement for an assessment of LRC on cost-price grounds is information on production costs and international prices. Setting aside for the moment the possible effects on prices of policy interventions, the fundamental relation of competitiveness is that production costs be less that the product's price. Disaggregating costs between those that correspond to national factors of production and those that represent imported inputs, this relation may be written as:

$$f \cdot \sum_k a_{kj} v_k + \sum_i a_{ij} p_i \leq p_j \qquad (5.1)$$

where f = exchange rate (dollars per unit of domestic currency); a_{kj} = required use of national factor k for each unit of good j produced; v_k = prevailing price of factor k (in domestic currency); a_{ij} = required use of imported (or tradable) input i for each unit of good j produced; p_i = prevailing price of input i, expressed in dollars, including tariffs for its import, port handling costs, and transport costs and marketing margins up to the farm or other spatial point of

The Competitiveness of Tropical Agriculture. http://dx.doi.org/10.1016/B978-0-12-805312-6.00005-2

production; and $p_j =$ prevailing price of product j in dollars; in the case of import substitutes, p_j is the cif import price (import price including cost, insurance, and freight) of the equivalent product placed in the country's main port including port handling fees, whereas in the case of exports, it is the free on board (fob) export price in that port. (This is the price quoted by a seller that includes all charges up to placing the goods on board a ship in the departure point designated by the buyer.)

The costs of production include a normal rate of return to the capital invested in the productive undertaking. If this return were not included, the equation would not represent sustainable competitiveness. Equally, when a product has to pass through a marketing chain or be processed before competing with an import or being exported, the costs of those activities should be included in Eq. (5.1).

Eq. (5.1) is a static measure of competitiveness, and it applies to the existing scale of production. If the volume of production were to change, then for the equation to remain valid constant returns to scale would have to be assumed. This assumption is probably a realistic approximation for most smallholder agriculture. The basic distinction in this equation is between domestic (national) factors of production, whose costs (returns) are represented by the first term in the left-hand side of the equation, and importable (tradable) inputs to production, whose costs are represented by the second left-hand term. Accordingly, the unit value of the output j may be decomposed into its components:

$$\text{value added per unit of output} + \text{cost of intermediate inputs per unit of output} = p_j$$

When Eq. (5.1) is written as a strict inequality, the difference between its two sides corresponds to excess profits or earnings that inflate value added above and beyond the level that would be obtained in a competitive equilibrium. In this case, the producer of j would be considered highly competitive, until enough competition arose that Eq. (5.1) was driven to hold as an equality.

Although land is a domestic factor of production, and therefore its returns are included in value added, the distinction between domestic and tradable is not always entirely clear-cut. For example, invested capital may be partly owned by foreigners, and the repatriation of those earnings would not be part of domestic value added. Nevertheless, this distinction is a basic one and in most circumstances can be defined without ambiguity. For tradable inputs, even if they are produced domestically, the defining characteristic is that international prices play a role in the formation of their prices, and that domestic transaction costs are not so high as to make them nontradable. This distinction can also lead to complications because, at one extreme, subsistence wages are related to the cost of staple foods in the diet, and the prices of these can be influenced by international prices, but such indirect effects are ignored here in defining tradable and non-tradable inputs.

5.2 COMPARATIVE ADVANTAGE AND LONG-RUN COMPETITIVENESS

Eq. (5.1) may be rewritten as

$$f^* \cdot \sum_k a_{kj} v_k^* + \sum_i a_{ij} p_i^b < p_j^b \qquad (5.2)$$

where the symbols have the same meaning as before but the asterisk signifies shadow prices of resources and the superscript b signifies border prices (international prices in a form relevant to the country) for tradable goods, possibly adjusted for certain distortions as explained below. Eq. (5.1) is the condition for current competitiveness in international markets, whereas Eq. (5.2) is the better indicator of inherent or LRC. It can be a requirement that entrepreneurs look at when deciding on long-run investments. When the shadow prices reflect true opportunity costs of resources, and border prices are adjusted to eliminate effects of policy interventions, then Eq. (5.2) is a condition for comparative advantage in product j.

In this book, the term LRC is used more than the term comparative advantage because some kinds of policy interventions may remain for the long term and hence some of the corresponding factor and product prices may be the same in Eqs. (5.1) and (5.2). In other cases, the v_k^* used here differs from v_k, as in the case of the foreign exchange rate, the aforementioned implicit wage for family labor, the price of irrigation water, and border prices are adjusted to remove effects of tariffs, but generally caution is used in assuming that future opportunity costs will diverge from present factor prices.

Eq. (5.2) can be rearranged so it reads as follows:

$$DRC_j = f^* \cdot \sum_k a_{kj} v_k^* \bigg/ \left[p_j^b - \sum_i a_{ij} p_i^b \right] < 1.0 \qquad (5.3)$$

Following Bruno (1972), when the shadow prices reflect true opportunity costs of factors of production without policy interventions, Eq. (5.3) is the condition that must be satisfied for a product to have a comparative advantage. The left-hand side of the equation is called the domestic resource cost coefficient (DRC). DRC represents the cost in domestic resources of earning or saving a (net) dollar of foreign exchange, through exporting or substituting for imported products. When some more or less permanent policy interventions are allowed to affect the shadow prices, then this equation is a condition for LRC rather than comparative advantage, and the left-hand side may be termed the LRC. The distinction between prices of domestic factors and prices of tradable goods in Eqs. (5.1)–(5.3) facilitates operational identification of which factor or product prices should be modified in the analysis to remove the effects of short-run policies.

The lower the value of the DRC or LRC, the greater the comparative advantage or LRC. Thus the calculations of these coefficients can serve to develop a hierarchy of products according to their inherent competitiveness. Gorton, Davidova, and Ratinger (2000) applied both DRC calculations and revealed comparative advantage in examining the competitiveness of Bulgaria and the Czech Republic versus the European Union. The measures largely coincided except that cereals in these two countries were competitive according to their DRCs but not according to RCA measures because of trade restrictions, which indicates another limitation of the RCA in addition to the ones discussed earlier.

5.3 FURTHER CONSIDERATIONS REGARDING LONG-RUN PRICES

Some basic elements of public policy are here to stay. For example, if a country levies taxes on the farm inputs of diesel and electricity, it is highly unlikely that those taxes will disappear in the future. Since government budgets are permanent features of an economy, taxes of one kind or another have to be permanent features. And in the tropics, commodity-based taxes are much easier to collect than income-based taxes. In the words of the International Monetary Fund, "The base for an income tax is hard to calculate [in developing countries]" (Tanzi & Zee, 2001). And "Indirect taxes or taxes on consumption are a mainstay of the revenue base [in developing countries]" (Carnahan, 2015).

Thus shadow prices, or "economic prices" that eliminate all taxes that change relative commodity prices, cannot yield scenarios that would be realistic in the future, and competitiveness calculated on that basis can be misleading. In the analyses reported here, those kinds of taxes are retained in the projections of future prices, whereas more transitory effects of policy are eliminated. The resulting prices may be called "sustainable policy prices" for lack of a better term. Developing shadow prices of this nature is one of the features of the analyses in this book.

For both products and purchased farm inputs, the sustainable policy prices do not include import tariffs because they can be repealed in the future. By the same token, it can be useful to employ estimates of long-run international prices that do not reflect the distortions in international markets caused by subsidies in the more developed countries. These subsidies have tended to depress prices in low-income countries to a significant degree sometimes: the US Farm Bill "provides commodity price support and other payments to US farmers that tend to suppress world prices and distort market conditions" and "by subsidizing their own farmers at the current magnitude, European agricultural produce (primarily beef, poultry, and tomatoes) can be sold in Africa at prices so low that African producers cannot even compete in their own countries despite the advantage for low-cost production" (Asenso-Okyere, 2013).

Another key shadow price is that of capital, i.e., the interest rate for farm loans. Some considerations regarding this interest rate have been mentioned earlier, and conclusion from these studies has been that an appropriate long-run opportunity cost of capital for farmers is 8%. Sensitivity analyses have been conducted around that rate to see how competitiveness calculations respond to variations in it.

In summary, the observed prices that have been replaced with shadow prices at least in sensitivity analyses in these studies are the rural wage for family labor, the exchange rate, the fee for irrigation water, the cost of capital, some imported products and inputs (because of tariffs and international price distortions), and some export products (because of international price distortions). Significant export incentives have not been observed in the policies of these countries.

In addition to these procedures for handling prices, the fourth case study in this volume (Chapter 10 for Colombia) introduces other considerations into the cost accounting of production. Quantifications of risk are included, along with farm administration costs and a minimum required return on investment. The reasoning is if these costs cannot be covered, then production of that item will not be viable in the long run. By the same token, if these costs are included and the product's assessment still indicates competitiveness, then the users of the results can be more confident that the product is truly competitive even taking into account the usual possibilities of errors in the data.

5.4 COST ESTIMATES: THE TERRITORIAL DIMENSION AND FARM BUDGETS

Marketing and other value chain activities consist of shifting a product spatially, changing its physical and/or chemical form through processing, and providing appropriate packing. Competitiveness measures have to take into account this element, even though cost calculations may be made at the primary production level. Two procedures have been followed when, for a given product, a single Track 1 competitiveness indicator has been calculated for the country, one for export products (or potential exports), and one for import substitutes. For exports, the calculations take into account transport, marketing and processing costs between the farm and the port of departure for international shipments. In this way the sum of all costs of production and product management can be compared with the fob export price p_j^b. For each product, the most representative production zone has been selected as the geographical basis for estimating these postharvest costs. In the case of calculations for different zones in a country, the postharvest costs have been estimated in this sense separately for each zone.

For the case of import substitutes, the costing of domestically produced items is not carried up to a port but rather to a representative national wholesale market, which is taken to be that of the capital city. Likewise, for the imported counterpart, the costs of unloading from a ship, port management and transportation, plus marketing margins to the same national wholesale market, are added to the cif import price. That total cost is taken to represent p_j^b for import substitutes.

For the production cost estimates, the importance of finding reliable farm budget data that are disaggregated by type of input has been mentioned. In the Colombian studies, the initial list of products was passed through two data filters: first, were farm budget data available? And second, were those data sufficiently reliable? To make the latter decision, data sets were compared across different crops and different locations in the country. Some factors such as the (annualized) cost of land and the cost of labor and certain inputs should have been comparable for different crops and, in most cases, for different locations. Farm budget vectors that were outliers in this regard were rejected. Sometimes all the data for a given crop had to be rejected and the crop dropped from the list for which competitiveness estimates were to be developed. In addition, the personal field experience of Colombian (and Salvadoran) colleagues proved invaluable in assessing the reliability of farm budget data. Thus human judgment inevitably enters the assessment process from the very beginning.

For some crops alternative technologies of production were evaluated in the cases of El Salvador and Colombia because the degree of competitiveness may vary sharply over technologies. This consideration raises the interesting possibility that the Track 1 methodology may be used to evaluate whether technologies that are *recommended* by projects are indeed more profitable than the ones commonly used by producers and if they stand up to competitiveness tests. This is important because the estimates of returns made in evaluations of agricultural development projects, prior to their approval, frequently are based on specified production technologies as well as cropping patterns. Rarely if ever are these recommended technologies assessed for competitiveness, and such assessments could be a valuable addition to a priori project evaluation procedures.

5.5 THE MULTIYEAR CRITERION FOR COST COMPETITIVENESS

To be able to analyze the LRC of perennial crops and livestock, Eq. (5.3) has to be modified so that it incorporates discounted future flows of costs and benefits. The first step in making this modification is to convert Eq. (5.3) into total values rather than unit prices, by multiplying it by the output level x_j:

$$x_j \cdot \left(f^* \cdot \sum_k a_{kj} v_k^* + \sum_i a_{ij} p_j^b \right) < p_j^b \cdot x_j \qquad (5.4)$$

which may be written as:

$$z_j^n + z_j^m < y_j \qquad (5.5)$$

where, z_j^n is the value of national factors used in the production of j; z_j^m is the value of imported inputs used in the production of j; y_j is the value of the output of j; and all the variables z are valued at sustainable policy prices. For crop agriculture, usually output x_j is normalized as yield per hectare or per unit of some other land measure.

Eq. (5.5) is simply the condition of LRC expressed in terms of total values instead of unit costs and prices. Now, when there is a flow of products and inputs over time the condition in Eq. (5.5) has to be valid in present value terms. If the operator PV(h) represents the present value of h, then Eq. (5.5) may be transformed into

$$PV\left(z_j^n\right) + PV\left(z_j^m\right) < PV\left(y_j\right) \tag{5.6}$$

Making the annual discount rate r explicit, Eq. (5.6) may be written as:

$$\sum_t \left(z_{j,t}^n \big/ (1+r)^{t-1}\right) + \sum_t \left(z_{j,t}^m \big/ (1+r)^{t-1}\right) < \sum_t \left(y_{j,t} \big/ (1+r)^{t-1}\right) \tag{5.7}$$

where the symbol t represents the number of years after the base year. To arrive at the version of Eq. (5.7) applied in this book, we can return to the equation format in unit quantity requirements of inputs and prices of inputs and outputs, and make the same transformation, which gives the following result as the requirement for LRC of product j, where the time index t starts at 1:

$$\left[\sum_t \left(\left(f^* \cdot \sum_k a_{kj,t} x_{j,t} v_k^*\right) \big/ (1+r)^{t-1}\right)\right.$$
$$\left. + \sum_t \left(\left(\sum_i a_{ij,t} x_{j,t} p_i^b\right) \big/ (1+r)^{t-1}\right)\right] \big/ \sum_t \left(p_i^b x_{j,t} \big/ (1+r)^{t-1}\right) < 1.0$$
$$\tag{5.8}$$

Eq. (5.8) has been applied to the data to evaluate LRC of multiannual production systems, where the $x_{j,t}$ represents the year-by-year yields and the unit input requirements "a" also vary by year. For example, typically the investments in seedlings and irrigation systems are made in the first year or years, whereas agrochemicals are applied in all years in varying amounts.

In addition to these parameters that can vary over time, the value of Eq. (5.8) also depends on the discount rate r. In the long run and in the absence of significant distortions in the economy, it would be expected that the value of r would converge over time to the opportunity cost of capital, and that this latter concept would approximate the real rate of growth of the economy.[1] Therefore, in the empirical evaluations of comparative advantage and LRC, sensitivity analysis with respect to the value of r has been conducted with values around the long-run opportunity cost of capital.

The discount rate has to be distinguished from the short-run opportunity cost of capital for farmers. The imperfections in rural credit markets are well known, and although activities like land titling can reduce them, they are going to persist for a considerable amount of time. In the words of Hoff and Stiglitz (1995):

> *Rural credit markets do not seem to work like classical credit markets are supposed to work. Interest rates charged by moneylenders may exceed 75% per year... neither the traditional monopoly [by moneylenders] nor the perfect markets view can explain other features of rural credit markets which are at least as important and equally puzzling as high interest rates:*
>
> - *The formal and informal sectors coexist, despite the fact that formal interest rates are substantially below those charged in the informal sector.*
> - *Interest rates may not equilibrate supply and demand: there may be credit rationing...*

Hence the short-run opportunity cost of capital for rural areas is always higher than the long-run opportunity cost for the entire economy. The calculations in this volume use a base value of 8% in real terms for the short-run opportunity cost, and sensitivity analyses are conducted around the value. However, farmers pay their input costs in current prices, and that includes credit, so under an inflation rate of 2%, for example, the short-run cost of capital becomes 10%: $k = w_k + \delta$, where the first term on the right signifies the real short-run cost of capital in rural areas and δ signifies the expected rate of inflation over typical project lifetimes. That 10% is then applied to the cost of all purchased inputs (working capital), for it also corresponds to the farmer's opportunity cost of his or her savings invested in the farm. The value of k, the 10% in this example, is applied equally in future years as the cost of working capital, whether borrowed or not, and then it is discounted by the real rate of discount. If there were firm grounds for expecting a significant and permanent change in the rate of inflation in the future, then δ would be changed and that would modify the future value of k. In effect, the discounting with a *real* discount rate is applied to a time sequence of future years' snapshots of costs and benefits in *current* prices of each year.

5.6 LIMITATIONS AND INTERPRETATIONS OF THE LONG-RUN COMPETITIVENESS MEASURE

The LRC indicator has a number of limitations. First, it is a static measure anchored in the prices and production technology of the base year and it is well known that making improvements in production methods over time is one of the keys to sustaining competitiveness. Second, as per the previous discussion, taken alone, the LRC neither incorporates quality considerations nor the timeliness and reliability of supplies brought to the market. For these reasons an LRC less than 1.0 is not a sufficient condition for a product to be acceptable in markets.

An empirical limitation, discussed earlier, is that calculations of LRC depend on the farm budgets that are available and whether they are sufficiently representative and accurate. Even if they contain acceptable data in these senses, there always is a spectrum of producers, some more efficient and some less so. Therefore results indicating that a product has a long-run competitive advantage do not necessarily mean that is true for all its producers. Vice versa, if an LRC indicates lack of competitiveness, there still may be some producers who are competitive in that product.

Given these limitations, it is always advisable to carry out sensitivity analyses. First, it is important to know if the results are robust under possible errors in the data, or lack of representativeness. Second, it is valuable to know which parameters are most crucial in determining competitiveness, in the sense that modest variations in them could affect the results substantially. From another perspective, numerical analysis of this kind always brings with it a reminder to be assiduous in seeking out all types of costs that a producer may incur. Some of them may not be very obvious, such as the cost of watching over a crop in the field to avoid theft (a surprisingly common concern), the implicit cost of crop loss in storage or transit to the nearest sale point, and the cost of private technical assistance for some producers.

Although it may not be clear for a given crop or livestock activity which technology of production is the most representative, the LRC calculations lend themselves to the exploration of the economic implications of possible variations in the technologies. Also, regarding the uncertainty surrounding parameter values, sensitivity analysis can be conducted on those, to determine which parameters are most critical for competitiveness and therefore which ones need the most careful field documentation.

The LRC can serve as an instrument to facilitate dialogs with producers, showing them the degree of competitiveness associated with each production configuration and encouraging them to seek the technological variants or alternative crops and types of livestock that are most competitive. As agricultural specialists identify promising crops and livestock products in specified areas, the LRC methodology can be applied to determine their potential for becoming competitive economically. Viewed in this way, the application of the LRC methodology can be a process that continues over an extended period and involves many participants. The initial calculations then would be only the starting point of such a process.

Use of competitiveness analysis does not suggest that in the future the sector will become completely specialized in its most competitive products and completely go out of production of all the others. As noted, a favorable LRC value does not mean that all producers of that output are competitive, or would be competitive in the absence of policy interventions. Supply curves always slope upward, and the points along them usually represent hundreds or even thousands of producers, ranging from the most competitive to the least competitive for that product. In the case studies in this book, comparative advantage is calculated for representative technologies (costs of production) for each department, but it is likely that some producers of each crop and livestock product have more efficient cost structures and/or better yields than the ones used for these calculations—and by the same token others are less efficient than these results suggest.

However, experience has shown that the sector's product composition does tend to move over time in the direction of those products with a comparative advantage or LRC. To accelerate the sector's growth and rate of employment creation, economic policy may wish to provide greater encouragement to the more competitive products, for example, in the form of directing more agricultural research to them or supporting temporary marketing initiatives and explorations. The LRCs may provide approximate guidance to policy in this sense. The message for those producers higher up on supply curves, i.e., those with higher marginal costs of production, is that they should pay special attention to increasing their productivity (the yield—cost relationship) to ensure the continued viability of their production lines. In general, their struggle to remain profitable may be more difficult than it is for other producers, but under favorable production conditions and with improvements in their operations they may be successful.

ENDNOTE

1. See, for example, Solow (1970).

Chapter 6

Track 2 Methodology: Value Chains and Quality Criteria

6.1 THE NATURE OF VALUE CHAINS

Track 2 analysis provides indications of how competitive a product may be according to quality criteria and how close to, or far from, it is to fulfilling its competitive potential, however strong or weak that may be. It is based on a number of quality criteria and identifies issues that limit a product's potential. By the same token, it provides guidance on problems that need to be resolved for a product to emerge into a competitive space. Quality evaluations necessarily require mapping of a product's entire value chain since issues affecting competitiveness can arise at various points along the chain. As pointed out in Norton (2014, p. 490):

> *Satisfactory responses are needed for questions all along the chain, from seeds to markets: Are the varieties that farmers plant are the same that the market demands? Is there sufficient volume of disease-free, high-quality planting material? Do farmers cultivate and harvest according to best practices (including harvest timing and selection of the products ready for harvesting)? Do postharvest practices maintain quality (drying, threshing, storage, classification and selection, and initial processing in some cases)? Are transport links and facilities available, including cold chain facilities, that will prevent damage to the products when moved to markets or marketing agents? Are plant or animal hygiene standards met all along the chain? Do processing facilities meet international standards? And so forth. When one or more of these questions is answered negatively but the potential appears to be present, it has to be asked if policies and programs can overcome the bottlenecks at a reasonable cost.*

The demands of competitiveness have been a major catalyst for the development of agricultural value chains in the tropics. In the words of Reardon, Barrett, Berdegué, and Swinnen (2009, p. 1725), "there has been rapid agrifood industry restructuring in the 1980—2000s. Among companies in the restructured segments, there has been significant...shift from public to private standards (and) shift from spot market relations to vertical coordination of the supply chain using contracts and market inter-linkages...This modernization been adopted

The Competitiveness of Tropical Agriculture. http://dx.doi.org/10.1016/B978-0-12-805312-6.00006-4

to reduce costs and increase quality to strategically position companies in a sharply competitive context."

The linkages embodied in value chains in fact are much more than supply chains or logistical sequences. On the product side, they are characterized by a sequence of stages, each involving transformation of the product in space and time and perhaps in physical form. Value is added in each stage. Although products move downstream in a value chain, from producers through marketing intermediaries, transportation agents, and processors, to final markets, there are important flows in the opposite direction. Payments necessarily move upstream from final markets through stages to producers, but also knowledge, inputs, and finance move in that direction.[1] Knowledge about production technologies, market opportunities, and market requirements is transmitted from the markets and processors, often through marketing agents, to producers. "In the food sector, supermarkets in many Latin American and Asian countries have initiated total quality management programs for perishables like fresh fish, meat and vegetables" (Trienekens, 2011, p. 52). This sharing of knowledge is what makes a value chain viable, creates trust, and enables the cooperation that facilitates the solution of problems as they arise among the different components or participants in a value chain. Value chains exist for multiple reasons, perhaps mainly because the participants have imperfect information about other links in the chain and thus need partners, and they need to minimize uncertainty and transaction costs (Prowse, 2012, pp. 28–29). Another important reason for their existence pointed out by Prowse is to help ensure compliance with standards and to provide traceability.

Fig. 6.1 illustrates the principal kinds of flows in a value chain. The multiplicity of areas of interaction along a chain is vital. Value chains do not always survive, and those that are univariate—that concentrate only on price negotiations between buyer and seller—are less likely to thrive in the long run because they are centered on a relationship in which the participants are fundamentally opposed and which offers no opportunities for cooperation that will benefit all. Experience has shown that value chains are more durable not only when they are multivariate but also when (1) they concern value products rather than commodities, (2) primary producers have fewer alternative buyers that they may be tempted to jump to in the short run if the price is more attractive, and (3) the market is reasonably stable so that contracted prices do not deviate far from a spot price (Wiggins & Keats, 2013).

A more detailed look at the product flows in a generic value chain is provided in Fig. 6.2. Value chains display a great detail of diversity, and some of them include elements not shown in this diagram, but it shows the role of different kinds of inputs and postharvest processing steps, as well as the routes to different kinds of final uses or markets, from home consumption to supermarkets and export markets. A value chain begins with one primary product, or a set of closely related primary products, and may end with many

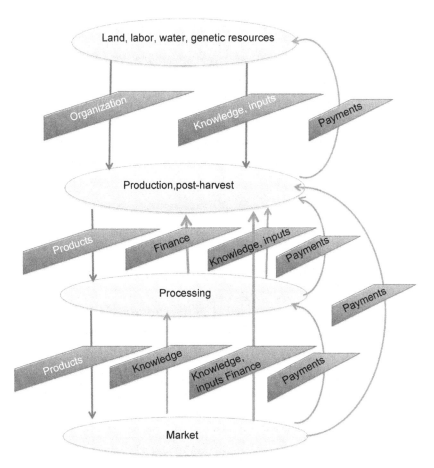

FIGURE 6.1   Value chain overview.

more products derived from processing. In the present context, a value chain is defined on the basis of the product it begins with.

The flow of product does not have to obey a linear sequence from one stage of a value chain to others, but rather some flows may skip stages, whereas others go through all of them. Some of the manifold ways in which the different domestic agents and markets can interact are well illustrated by the case of Cape gooseberry in Uganda, as shown in Fig. 6.3. That diagram also introduces into the picture the transformation of one primary product into multiple processed products, which is a frequent occurrence.

In other cases product flows may be concentrated on local, national, or export markets, but not all three. Trienekens (2011, p. 54) uses this distinction to attempt a threefold taxonomy of value chains, from supplying the local

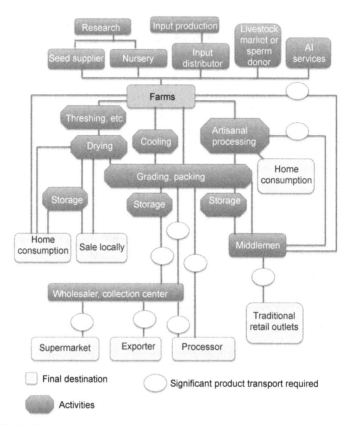

FIGURE 6.2   Illustrative agricultural production value chains: actors and activities.

market to the national market to export markets, in increasing order of so-
phistication and product quality:

> *An example of an A system is the production of cassava or sorghum by small
> local West African producers for local markets. Often these products enter into
> complex distribution networks for local markets in different places. The B-system
> can be characterized as the local middle to high-income chain. These producers
> aim at the emerging supermarket sector in many countries of the tropics. Most of
> the volume in these chains is delivered by small/medium size producers, orga-
> nized in cooperatives and/or linked in subcontracting arrangements. Micro
> producers deliver inputs on demand to balance demand and supply in this system
> (buffer function). Although the production volume produced by B-systems is
> smaller than that of A-systems, the value generated is larger. B-systems
> increasingly produce according to national and sometimes international retail
> quality and safety standards. An example of a B-system value chain is the pro-
> duction of vegetables in Kenya for modern South African retailers operating in*

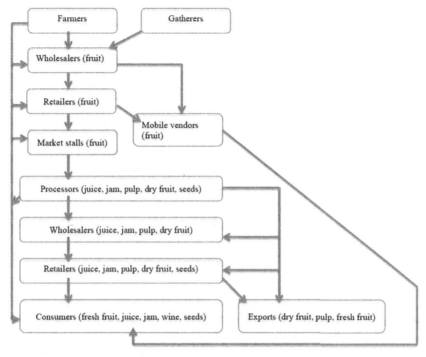

FIGURE 6.3   Cape gooseberry value chain in Uganda. *Adapted from Barirega (2014).*

Kenya (Reardon, Timmer, & Berdegue, 2004). Finally, the C-system can be characterized as the export chain. It is completely focused on export, although low quality or rejected products are sold at the national, in many cases retail, market. The trend is towards increasing economies of scale and foreign direct investments. Export chains tend to become more integrated and with fewer actors. Although volumes are small compared to local markets, the value added is relatively high. An example of a C-system value chain is the South African table grape chain that focuses on export (Trienekens & Willems, 2007), or the international flower value chains with production in Kenya and Ethiopia.

However, modern value chains are characterized, a fundamental factor differentiating them from traditional value chains is that prices and quantities to be delivered are negotiated in advance of the harvest. In Section 10.5 of this book the main issues affecting Colombia's value chains for high-value products is explored in more detail, including several aspects of the exporting process itself.

A major concern about agricultural value chains in the tropics is food waste in the various stages of value chains before the consumer level. In developed countries a larger share of the waste occurs in households and restaurants, but in poorer countries most of the waste occurs on farm and in postharvest management, processing, and distribution. From the viewpoint of competitiveness and profitability, food loss reduces the price received by farmers for their harvests. The same forces that cause food loss can degrade food products by the time they reach their markets, occasioning a further discount to price. Tables 6.1 and 6.2 show estimates of the percentage of harvests lost by region of the developing world and by stage of the value chain.

As the tables show, the production and postharvest stages are significant sources of food loss. The causes of losses in these stages can be very numerous. Threats to product integrity are varied and range from birds, rodents, and insects to molds and humidity to breakage, spillage, and contamination in the course of handling and processing. Specifically, for the case of corn as an example:

- Heat is a threat on farm and in harvesting and processing.
- Rain is a threat on farm (when it is irregular or torrential) and to the drying process.
- Humidity can affect crops in transport, storage, and the shelling and cleaning processes.
- Contamination can occur at any stage especially in cultivation practices and storage.
- Insects, birds, rodents, and mold can enter the crop during growing, field drying, harvesting, postharvest drying, and in storage.
- Bacteria and rancidity can enter during drying and transport and in storage.[2]

Looking at the conditions that all types of harvested food products experience in their journey along a value chain, Bourne (2014, p. 340) identifies the following five primary causes of food loss from the viewpoints of biological, chemical, and physical processes:

1. Biological and microbiological: Consumption or damage by insects, mites, rodents, birds, larger animals, and microbes such as fungi and bacteria.
2. Chemical and biochemical: Undesirable reactions between chemical compounds that are present in the food such as the Maillard reaction, fat oxidation, and enzyme-activated reactions as well as contamination with harmful substances such as pesticides or obnoxious substances such as lubricating oil.
3. Mechanical: Spillage, abrasion, bruising, excessive polishing, peeling or trimming, puncturing of containers, defective seals on cans, or other containers.
4. Physical: Excessive or insufficient heat or cold and improper atmosphere.
5. Physiological: Sprouting of grains and tubers, senescence in fruits and vegetables, and changes caused by respiration and transpiration.

Bourne then identifies the "secondary" causes that lead to the occurrence of the primary causes of loss and comments that they usually are the result of

TABLE 6.1 Estimates of Food Losses by Stages of Value Chains, Africa and West and Central Asia (% of Product)

Crop Group	Production	Postharvest Handling and Storage	Processing and Packaging	Distribution	Consumption
Sub-Saharan Africa					
Cereals	6	8	3.5	2	1
Roots and tubers	14	18	15	5	2
Oilseeds, pulses	12	8	8	2	1
Fruit, vegetables	10	9	25	17	5
Meat	15	0.7	5	7	2
Fish and seafood	5.7	6	9	15	2
Milk	6	11	0.1	10	0.1
North Africa and West and Central Asia					
Cereals	6	8	2–7	4	12
Roots and tubers	6	10	12	4	6
Oilseeds, pulses	15	6	8	2	2
Fruit, vegetables	17	10	20	15	12
Meat	6.6	0.2	5	5	8
Fish and seafood	6.6	5	9	10	4
Milk	3.5	6	2	8	2

U. N. Food and Agriculture Organization (2011).

TABLE 6.2 Estimated Food Losses by Value Chain Stages, South and Southeast Asia, Latin America (% of Product)

Crop Group	Production	Postharvest Handling and Storage	Processing and Packaging	Distribution	Consumption
South and Southeast Asia					
Cereals	6	7	3.5	2	3
Roots and tubers	6	19	10	11	3
Oilseeds, pulses	7	12	8	2	1
Fruit, vegetables	15	9	25	10	7
Meat	5.1	0.3	5	7	4
Fish and seafood	8.2	6	9	15	2
Milk	3.5	6	2	10	1
Latin America					
Cereals	6	4	2–7	4	10
Roots and tubers	14	14	12	3	4
Oilseeds, pulses	6	3	8	2	2
Fruit, vegetables	20	10	20	12	10
Meat	5.3	1.1	5	5	6
Fish and seafood	5.7	5	9	10	4
Milk	3.5	6	2	8	4

U. N. Food and Agriculture Organization (2011).

deficient human management of the production, processing, and marketing process. These causes are (Bourne, 2014:341):

1. inadequate drying equipment or poor drying season;
2. inadequate storage facilities to protect food from insects, rodents, birds, rain, and high humidity;
3. inadequate transportation to get the food to market before it spoils;
4. inadequate refrigerated or frozen storage (for perishables);
5. marketing systems that fail to connect potential buyers with producers; and
6. legislation: the presence or absence of legal standards can affect the eventual retention or rejection of a food for human use.

Bourne links the main specific causes of food loss to the category of food, as follows (Bourne, 2014:341):

Cereals: fungi, insects, vertebrate pests, and poor milling.
Fruits and vegetables: bruising, rotting, senescence, and wilting.
Roots and tubers: sprouting, rotting, senescence, and wilting, insects.
Meat, milk, poultry, and fish: growth of microbes.
Dry fish: fungi and insects.

He concludes by noting that protecting food against these sources of loss is generally more difficult in the tropics (Bourne, 2014:344, 348):

High storage temperatures accelerate most deteriorative changes that occur in stored food. Even stable foods such as cereal grains deteriorate more quickly as the temperature increases. Hence, tropical zones have a more difficult problem in maintaining food quality than do temperate zones…The rapid growth of plants in the humid tropics is matched by a rapid rate of decay.

Shelf life for all foods is shorter in the tropics than in the temperate zones. One cannot expect to get as long a shelf life in a hot climate as in a cold climate…All foods, even stable foods, should be stored in the coolest place possible-…Exposing foods or storage structures to direct sunlight increases the temperature and the activity of all spoilage vectors. In temperate climates, most foods are stored out of the sun. In contrast, in tropical climates one often sees food products exposed to the sun.

Fruits and vegetables have a high-moisture content, typically 80–95%…They are vulnerable to spoilage by growth of many microorganisms. It is the skin that protects them from microbial spoilage. Therefore, maintaining intact skin to prevent invasion by microflora is necessary for preservation of horticultural crops.

Naturally occurring deteriorative chemical reactions [in foods] are accelerated in the tropics. In addition, the year-round high temperatures of tropical countries allow pests such as rodents and insects to feed and multiply throughout the year, whereas the cold winters of temperate zones slow down or stop reproduction and reduce the feeding activities of these pests. The low temperatures in very cold climates kill many of the pests.

The value chain procedure described in this book aims to identify the types of quality threats that affect agricultural products, along with the sources of loss and degradation of products that hinder realization of their inherent competitiveness and thus block farmers from receiving the potential rewards associated with efforts to produce them.

Value chains often defy neat categorization, and the actors participating in them can be quite diverse. Therefore diagnosing problems in value chains and seeking solutions for them can require examination of the roles of a wide array of actors. Fig. 6.2 does not display the full range of institutions that can be involved in the activities and transactions of value chains. Table 6.3 gives a fuller indication of the kinds of institutions that can be involved in value chains, including governmental and international entities as well as private companies and farmers. Their participation varies widely over types of product and also over time. The complexity of value chains can be seen in the table. It also suggests that coordination, which depends in turn on value chain governance, is one of the keys to the efficiency and survivability of a value chain. This is the topic of the next section.

TABLE 6.3 Institutional Characteristics of Value Chain Components

Entity, Concept	Organizational and Institutional Nature
Resources	
• Land	Assets under legislated and traditional land tenure regimes
• Labor	Family labor; employees
• Water	Resource managed by associations of water users
• Absorptive capacity of environment	Environmental laws, regulations
Technology and Knowledge	
• Plant breeding programs	Public and private research
• Plant propagation programs	Seed multiplication schemes, private companies
• Livestock breeding programs	Public and private research
• Input development programs	Private research
• Crop and animal management	Farmers, coops and associations, companies
Primary Production, Postharvest Management	
• Smallholder family farms, ponds	Individually owned and leased assets

TABLE 6.3 Institutional Characteristics of Value Chain Components—cont'd

Entity, Concept	Organizational and Institutional Nature
• Commercial farms, ponds	Privately owned and leased assets
Processing Entities	
• Artisanal processors	Individuals, cooperatives and associations
• Cooperatives, producer associations	Activities of cooperatives and associations
• Processing companies	Public and private companies
Marketing Entities	
• Local intermediaries	Individuals, private companies
• Wholesalers	Private companies
• Retail agents	Private companies
• Exporters	Private companies, cooperatives and associations
Financial Entities	
• Community savings and lending groups	Informal entities
• Rural savings, credit associations	Formal entities supported by bank supervision
• Microfinance entities	Donor-supported formal entities
• Input suppliers	Private companies
• Processors, marketing agents	Private companies
• Banks	Private and public entities
• Agricultural innovation funds	Public and donor-supported entities
Technology Dissemination Agents	
• Public extension services	Government entities
• NGOs, development projects	Programs of international and local donors
• Exporters, wholesalers, retailers	Activities of contract farming
• Processors	Activities of contract farming
• Cooperatives, associations	Activities of cooperatives, associations
• Agricultural advisory services	Private entities

NGOs, nongovernmental organizations.

6.2 BUYER–SELLER LINKAGES AND VALUE CHAIN GOVERNANCE

Most value chains do not have a formal, overarching institutional structure. Rather, they are a set of business relationships forged by parties that see mutual benefit in sharing information and working together in other ways. Various institutional arrangements, often piecemeal, help protect the interests of the participants. "Value chain actors safeguard against risk of opportunism through joint investment, monitoring systems and specific organizational arrangements such as contracts" (Trienekens, 2011, p. 58). The glue that holds value chains together can sometimes go beyond purely financial motives: "According to network theory, relationships are not only shaped by economic considerations; other concepts like trust, reputation and power also have a key impact on the structure and duration of inter-company relationships" (Trienekens, 2011, p. 59).

Other conceptual foundations of value chains, in addition to network theory, are explored by Prowse (2012:27–40).

Following Dunn (2005), Webber and Labaste (2010:20) have described four types of relationships that define the governing structure of value chains:

- *Market relationship*: Arms-length transactions in which there are many buyers and many suppliers. Repeat transactions are possible, but little information is exchanged between firms, interactions are limited, and no technical assistance is provided.
- *Balanced relationship*: Both buyers and suppliers have alternatives, that is, a supplier has various buyers. There are extensive information flows in both directions, with the buyer often defining the product (that is, design and technical specifications). Both sides have capabilities that are hard to substitute, and both are committed to solving problems through negotiation rather than threat or exit.
- *Direct relationship*: Main buyer takes a large percentage of supplier's output, defines the product (that is, design and technical specifications), and monitors the supplier's performance. The buyer provides technical assistance and knows more about the costs and capabilities of the supplier than the supplier does about the buyer. The supplier's exit options are more restricted than those of the buyer.
- *Hierarchical relationship*: Vertical integration of value-added functions within a single firm. The supplier is owned by the buyer, or vice versa, with the junior firm having limited autonomy to make decisions at the local level.

As mentioned, the *market relationship* is the least likely to be sustainable because of its univariate character. For smallholders, the need to scale up output to meet the quantity requirements of profitable markets plus the nature of the opportunities can dictate which of these relationships the farmer enters into. In the absence of alternative forms of organization of small farmers, marketing intermediaries—local and regional wholesalers within a country—are responsible for aggregating harvests to the scale required by markets. These intermediaries often are inefficient and significantly increase the cost of marketing (Ffafchamps, 2004). An extreme example is the marketing chain for cardamom in Guatemala, where the product passes through as many as six levels of intermediaries between a small producer and an exporter.

The traditional alternative for scaling up is for producers to organize marketing cooperatives. Sometimes they work well and succeed in building social capital, and they give smallholders the best chance to enter into a *balanced relationship* for governance of a value chain. Denmark, the United States, and the Basque Country of Spain, among many other places, have shining examples of cooperatives. The leading dairy producer in Central America is a Costa Rican cooperative.

Making cooperatives in low-income countries effective in business management and marketing can require serious and sustained support. Experience shows that developing a cooperative's capacities can take several years. In an African case, the International Fund for Agricultural Development pointed out that "In the coffee sector, only 21% of growers are members of the 167 coffee cooperatives, because of widespread cooperative mismanagement, poor governance and high indebtedness, which result in little returns to members." This is not to deny the longer run importance of a program for strengthening cooperatives. However, a faster, more sure way to scale up harvests would be helpful for the small farmers of the world.

Cooperatives require expertise in production and postharvest technologies, administration and financial management, as well as marketing, and the learning processes in these areas can be quite long. There are alternatives that supply those needs to farmers. One is contract farming, which corresponds to the *direct relationship* model of governance. A marketing agent, exporter, or agroprocessor purchases harvests from numerous small farmers (outgrowers). Prices are agreed in advance, and that eliminates a large source of uncertainty for farmers; the partnership includes technical assistance and input supply from the buyer's side.

For export vegetables the model of contract farming is used in the Dominican Republic and Kenya among other countries. The experience of the purchase of sorghum for beer in Uganda is another example. In that case, a local NGO trained the farmers and the buyer helped organize them into groups and provided them with inputs on credit. The natural colorant annatto (*achiote*), a valuable crop when the right variety is planted, is grown in Panama under contract with a buyer that provides the seeds and technical assistance for 20 years. In Rwanda Minimex, the country's largest corn mill developed a joint venture for production of high-quality corn with 500 outgrowers. The elements of that scheme include delivering high-yielding varieties to farmers and renting them farm machinery, as well as giving training on postharvest management and guaranteeing purchase of the crop.

Sometimes exporters will guarantee producers a base price and include a price escalation clause if the export price proves to be higher than initially envisaged. This kind of stipulation, other than providing technical assistance, helps cement partnerships between buyers and producers and provides additional encouragement for cooperation in identifying problems and working out solutions. However the relationship is structured, it is no small challenge to keep farmers in a contract farming arrangement. Root Capital, a leader in providing finance to smallholders who have market linkages, has commented (2015:25):

> *Agricultural enterprises typically work in competitive environments, in which suppliers have the option to sell to other buyers. Enterprises also generally lack the bargaining power to sanction suppliers who fail to comply with sourcing agreements or standards. As a result, they run the risk that farmers will side-sell product to other buyers despite formal or informal agreements. When side-selling is extensive, enterprises investments end up supporting farmers who might not remain in their supply chain. Extension becomes a public good, rather than a business investment.*

On the other hand, farmers sometimes have legitimate complaints about buyers that encouraged them to produce a value product and then failed to pick it up or, as in the case of some tomato growers in Rwanda, about a buyer that consistently collected the boxes of tomatoes late, after they had sat in the sun on the roadside, and discounted prices for the condition of the tomatoes. In the Dominican Republic producers of vegetables for supermarkets suffer from the practice of the supermarkets of making payment only with a 3-month delay.[3] A common complaint of beef producers in Ethiopia was delays in payments, and low payments because their production was not aligned with market preferences (GebreMariam, Amare, Baker, Solomon, & Davies, 2013).

There are risks on both sides. Risk reduction mechanisms can include use of cooperatives or village leaders to identify farmers who can supply products in accordance with buyer requirements, stable price guarantees, and even modest insurance coverage. In a poultry value chain in India, the purchasing firm offers farmers insurance that covers up to 5% mortality for the chicks and a price tied to a stable industry price index rather than to volatile prices, plus a bonus if the spot price rises substantially above the industry price index (Prowse, 2012, citing Ramaswami, Singh Birthal, & Joshi, 2005). Prowse offers a detailed list of issues for farmers to be aware of regarding contracts before they enter into sales agreements, along with other provisions to reduce risks and help ensure the sustainability and success of value chain arrangements.

Satellite farming is a variant of contract farming under which a larger farm (nucleus farm) produces a high-value product and serves as a demonstration in production methods for the surrounding smaller farms. They agree to plant the same crop, follow the same cultivation procedures, and sell to the same buyer.

Another promising approach is consolidated land rental, or joint investments on agricultural landholdings. For entrepreneurs it is a chance to till larger expanses of land while it achieves the scaling up aim for small farmers. An entrepreneur reaches a multiyear lease agreement with a number of farmers whose lands border each other. At the outset the agreement offers participating farmers at least the income they earned before. Then, in subsequent years, the income payments are ratcheted up. In addition, these farmers and their families may be the first to be hired on the consolidated farm, although, on occasion, the investor prefers to bring in farm laborers with whom he or she has worked before. This approach is not yet widely explored, but it is being tried out on a small scale in China, in the agrarian reform cooperatives of Honduras, in Rwanda, and elsewhere. Although consolidated land rental can offer small farmers higher incomes, it does break the tie between them and their land.

Cooperatives can consolidate land for the purpose of cultivating larger areas of the same crop if all members agree to plant the same crop at the same time. In Guatemala, for example, small producers in a cooperative have joined their tiny plots on a steep hillside to produce blackberries for export. However, the cooperative model usually does not deliver improved technology to farmers the way the other models do. Nor do most cooperatives have the market linkages that are inherent in the other approaches, so training is required in that area as well. Efforts are being made to improve assistance to cooperatives in these respects. Also in Guatemala, the nongovernmental organizations CRECER and Grameen Foundation (2015) have been effective in strengthening a coffee cooperative of 1300 farmer members.

One lesson that stands out is that cooperatives have to arise out of farmers' own willingness to work together. If they are imposed on farmers by a higher authority, the glue that holds them together will be weak. This is a lesson that some governments in the tropics appear not to have learned. In addition, eventually they have to be in a position to provide, and pay for, their own extension services. A related conclusion is that cooperatives have to become profit-making, in other words true entrepreneurial entities. Often the aim is to break even, but that does not provide the continuing impetus needed to achieve the significant technological and marketing improvements that usually are necessary, or to generate a positive return on the investment of their own resources.

Independently of the mode of organization of farmers, Prowse (2012, pp. 21–23) offers a detailed and cogent summary of the benefits of contract farming for both buyers and smallholders in developing countries:

> *For firms [buyers], the opportunities provided by contract farming are clear and convincing:*
>
> *(1) increased reliability in supply quantity and quality (reducing screening and selection costs); (2) the off-loading of production risk onto farmers, in many cases; (3) greater control over the production process and crop attributes, to meet standards and credence factors; (4) reduced coordination costs, as a more regular and stable supply permits greater coordination with wider activities; (5) greater flexibility in expanding or reducing production (since there are fewer fixed assets, especially compared to full vertical integration); (6) economies of scale in procurement, via the provision and packaging of inputs. In addition, lower direct-production risk can improve a firm's credit rating, and also allow a firm to maintain intellectual property protection (for example, for new germ-plasm or genetically modified crops).*
>
> *There are also less tangible potential benefits. Contract farming can provide greater confidentiality in pricing levels (so that that competitors are less able to access this information). It can also provide status and reputational benefits, through involvement in national development programs or projects that have state involvement. On a broader note, and especially where access to land is highly politicized, it can overcome land constraints. For example, firms may find it hard to obtain land, or may run the risk of expropriation if they do own it. Overall, contract farming can increase profits from, and improve governance of, the value chain.*
>
> *Contract farming also offers numerous opportunities for farms: it can allow access to a reliable market; it can provide guaranteed and stable pricing structures; and most importantly, it can provide access to credit, inputs, pro-duction and marketing services (seed, fertilizer, training, extension, transport, and even land preparation). On a wider note, contract farming can open doors to new markets for a farm's produce, stimulate technology and skill transfer (particularly for higher-risk crops, which resource-poor farmers might typically avoid), and it can support farmers in meeting vital sanitary and phyto-sanitary standards.*
>
> *For farms, the main opportunity from contract farming is the promise of higher incomes. But, while important, this is not the sole criterion:. . . stability and technical knowledge [are] inter alia. . . important reasons why farmers join contract-farming initiatives. . . . Contract farming can also provide many additional benefits and opportunities: it can increase on-farm diversification;*

technical assistance and knowledge transfer can spill over onto adjacent fields and into nearby villages; by-products from contract farming can be used for other farming activities; it can simplify marketing decisions, thus improving efficiency; it can stimulate the broader commercialization of smallholder farming; and, finally, contracts can be used as a form of collateral for credit.

...a clear rationale for contracting smallholders can be found in the literature on the relative merits of small versus large farm production in sub-Saharan Africa...Small farms are frequently the most efficient agricultural producers, and have advantages over large farms in terms of labor-related transaction costs, in particular supervision and motivation. However, small farms often suffer from capital constraints, and a lack of capacity to adopt technological innovations. Moreover...smallholders often lack the ability to meet exacting standards from actors further down the value chain. Contract farming can overcome these limitations: it can deliver the scale benefits typically associated with large-farm production systems. Economies of scale through the firm decrease the cost of inputs and transport. In addition, firms have a comparative advantage in marketing and technical knowledge, and product traceability and quality.

In terms of poverty reduction, contracting with smallholders can reap large dividends: small farms are generally owned and operated by the poor, often using locally-hired labor, and often spend income within nearby locales, creating multipliers...Overall, there are good reasons why contract farming with smallholders can succeed.

Prowse points out that variants of contract farming can include the use of intermediaries as buyers, especially when the processing facility is located at a considerable distance from the producers (a model popular in Thailand and Indonesia) and public—private partnerships in which a local or national government entity is party to the arrangement. This variant is common in China, but runs the risk of politicization of the contract terms. He also confirms that value chain relationships based on verbal agreements are less likely to endure than those based on formal contracts.

There have not been many attempts to evaluate the effects of interventions designed to improve value chains for smallholders, mainly because of the complexity of the chains. One such evaluation was conducted for a US Agency for International Development—supported project for value chain development in Liberia by Rutherford, Burke, Cheung, and Field (2016:76, 77). Some of its conclusions were:

By aggregating produce, [farmers] reportedly arranged more lucrative sales agreements with buyers. No comparison groups reported produce aggregation and sales or cooperation beyond the typical labor groups...Focus group discussions...found that treatment farmers had stronger linkages with buyers in the capital (the major outlet) than did comparison farmers.

Moreover, based on the [focus group discussions, FGDs], treatment farmers had more outlets for their crops. They also cited the benefit of what they call "bucket sales" —aggregating produce among farmers...

[they] described the need to have a good, long-term relationship with a buyer. This was stronger for treatment groups (six of eight FGDs) than comparison groups (three of seven FGDs). As some treatment group farmers described the relationship: "We market before we produce. Now we know the time when a particular good is scarce and will make money."

Regarding overall benefits of the value chain development program, Rutherford et al. commented that (2016:78):

Food insecurity decreased over time for treatment households, but remained relatively stable among comparison households —a difference that was statistically significant...This was supported by the qualitative findings, [project] staff reports, and monitoring of rice harvests over time.

Overall, we concluded from the...FGDs that treatment farmers were more able to provide for their families in terms of food, housing, and sustainability, as exemplified by this statement from a treatment group participant: "Through my garden [proceeds], I send children to school and bought zinc to build a house. At first I didn't have a plan, now I have a plan."

However, in conclusion the study also reported that (Rutherford et al., 2016, p. 79):

The three educational outcomes examined were enrollment, attendance, and expenditures —all three of which improved for all children. Though none of the changes in individual outcomes were statistically significant between children of treatment and comparison households, all of the outcomes trended in a positive direction for the children of treatment farmers.

The authors pointed out that this result, of lack of difference between the treatment and comparison groups, may have resulted from the happenstance that the comparison groups received more benefits from child feeding and health programs and also said that more research is needed on intrahousehold dynamics.

In an analysis of contracting with producers in the Indian poultry sector, Ramaswami et al. (2005) found that contract farming is more efficient than the noncontract alternative, owing primarily to its lower feed—conversion ratio, that it gives higher incomes to producers, and that it shifts more of the risk from producers to processors.

From the viewpoint of enhancing competitiveness, value chains need to be *efficient* and *stable*. Efficiency means acceptable levels of productivity on the part of all the links in the chain, and it also means that each link has to be profitable for the entity or entities representing that link. In addition, efficiency in a value chain implies avoidance of damage to product quality or the ability to overcome threats to quality when they emerge. Stability does not imply a static character, for continuous innovation is needed, but rather it means durability or sustainability over time. To meet this requirement, efficiency in the sense of profitability is required, and the governance structure of the chain also needs to be able to involve all the actors in diagnosing problems as they arise and generating consensuses on how to solve them. Value chains that are not stable cannot sustain their presence in the marketplace, and if they dissolve their replacements may not be as efficient.

Price levels that are not consistent with producer profits have been a recurring issue in some value chains for industrial crops that have monopsonistic or oligopolistic buyer structures, i.e,. one or few processing facilities that buy from farmers. These experiences have occurred with cotton, oil palm, rubber, sugarcane, and other industrial crops. As these relations are not sustainable over the long run, pricing structures have tended to move in an acceptable direction from the viewpoint of farmers, but the issue still persists in some cases. Poulton and Tschirley (2009) have described five cotton value chain structures in Africa:

- Market based and competitive—many buyers.
- Market based and concentrated—few buyers.
- Regulated with a national monopoly.
- Regulated with local monopolies.
- Regulated in hybrid arrangements incorporated elements of the other structures.

The last four of these kinds of value chain correspond to the *direct relationship* structure mentioned earlier, with elements of the *hierarchical relationship* in the cases of government monopolies and hybrid arrangements.

In the same well-researched volume, the chapter by Baffes, Tschirley, and Gergely (2009) points out that over time producer prices for cotton lint have improved in Benin, Burkina Faso, Cameroon, Mali, Mozambique, and Tanzania, in part owing to a restructuring of the value chains, whereas they have fluctuated or declined in Uganda (slight decline), Zambia, and Zimbabwe (2009:70). The chapter by Poulton, Labaste, and Boughton (2009) documents differences in productivity among cotton farmers and farming systems, and concludes that "The stark finding…is that between 25% (Burkina Faso) and 75% (Mozambique, Uganda, and Zambia) or more of cotton producing households would be better off hiring out their labor than applying it to their own cotton plots" (2009:129). Another pervasive issue is quality of raw cotton, and the main concerns are contamination of the cotton lint with extraneous material and insufficiently long fiber staples (Estur, Poulton, & Tschirley, 2009). These findings raise questions about the long-run sustainability of some of the cotton production in those countries.

For the processing stage of the cotton value chains, the chapter by Gergely (2009) finds that cotton gins are not competitive—are money losers—under the national regulated monopoly arrangement in Mali and under the local monopolies in Burkina Faso (2009:146), in good measure because of high overhead costs and high levels of inventories. In these two countries, the cotton gins are kept afloat by subsidies from the national treasuries, and their value chains do not meet the efficiency criterion at the stage of the processing industries.

6.3 VALUE CHAINS AND INNOVATION

Continuous innovation over time is becoming a requirement for remaining competitive in today's international context. Agricultural value chains of all types are increasingly seen as sources of innovation for producers and stimuli for other entities to create and offer innovations. They can encourage and even develop innovations in areas that traditional agricultural research programs would not have considered. Examples include improved forms of packaging, simple crop dryers and coolers for smallholders and agricultural cooperatives, mechanization adapted to the needs of rural women, information and management systems for value chain coordination, small-scale water capture and irrigation systems, environment-friendly coffee mills, and soil management techniques for moisture conservation. Innovations in the forms of association and collaboration also are essential the for the development and competitiveness of modern value chains: "the most important innovations are organizational: coordinating production by large numbers of farmers of products of consistently high quality (frequently highly perishable) and delivering them to numerous distant retail sales points (increasingly abroad)" (Ekboir, 2012, p. 53).

Sometimes value chains disseminate known technologies to smallholders in what is more properly considered an extension effort, but they become innovations at the level of farmers. Innovations are sometimes required to resolve issues found in value chains through the kind of analysis of barriers to competitiveness that is described in the case studies of this book. A review of agricultural innovation systems in Africa has commented as follows on the role of value chains:

> Innovations can arise at any point of the value chain as a result of mediated or coordinated interactions among different actors. Thus, the appearance of innovation does not necessarily depend on any specific government role or action...The entire value chain is critical: The need to maintain grades and standards within the value chain, not only in export markets but also in evolving domestic and urban markets, drives innovation in agribusiness.

> Coordination is a key to success: Quality assurance, driven by the search for competitive advantage within domestic markets or access to export markets, in turn requires coordination along the value chain...[It] is necessary to assure

continuity of supply as well as efficiency in the assembly and bulking process of some commodities...Technical innovation is usually associated with complementary adaptive organizational innovation (Kim, Larsen, & Theus, 2009: 8, 10).

As an example of this dynamic process, innovation has been vigorous in the value chain for cassava in Tanzania:

[For farmers] innovations included the use of high-yielding and disease-tolerant varieties. . .and in one case, farmers are participating in fertilizer trials with researchers...However the level of testing new ideas and knowledge was considerably higher among cassava processors; at least 75% reported such testing. Processors used new dehullers [and] widespread organizational innovation to enable them to access new markets...they reported accessing new processing and packaging technologies...Some processors innovated in blending cassava flour with corn and packaging it into smaller packages of one-, two- and five-kilogram bags (Mpagalile, Ishengoma, & Gillah, 2009: 143).

In Colombia, associations of producers and processors for a few industrial and high-value products have succeeded in innovating through world-class research entities and effective extension services for their value chains. The innovations include development of new varieties, better cultivation techniques, and new processed products. Oil palm, sugarcane, and coffee are the leading products in the country in this regard.[4]

In, 2009 In Madagascar, technical assistance provided by a value chain for French beans, led by the major European company Lecofruit, generated compost-making skills among participating farmers and led them to apply compost to other crops with considerable success. For farmers in this vegetable value chain, rice yields increased to 64% above those of nonparticipating farmers. One of the ingredients of the successful extension was visits of once a week or more to farmers by company advisors, plus the use of assistant extension agents (Minten, Randrianarison, & Swinnen, 2009).

In Kenya, corn and tomato growing have seen a large number of innovations along the value chain:

New seed varieties, fertilizer blends, soil fertility analysis technologies, bio-pesticides, and [plastic greenhouses]; in processing, the use of moisture meters, diversification into new products, and proliferation of posho mills [simple mechanical apparatuses for grinding grains into flour]; [in marketing] intense outreach innovations like packaging into smaller sizes and branding campaigns that make products acceptable to a wide range of consumers; organizational innovations like farmer clustering, training for agrodealers...information and communication technology (ICT) usage, implementation of the warehouse receipt system [and] service diversification (Odame, Musyoka, & Kere, 2009:89).

Formalization of value chain arrangements and effective coordination along the value chain are elements that favor innovation. As pointed out by Lynam and Theus (2009, pp. 20−21):

> Comparing the cases of Rwanda [coffee] and Ghana [cassava] supports the basic argument that successful innovation within a commodity subsector has a higher probability of success if the market structure has the characteristics of a formal marketing chain....Success is more likely because margins tend to be higher in formal market chains, and coordination across the value chain is more easily implemented...maintaining quality required significant coordination at all stages of the Rwandan coffee value chain from farmer to processing to exportation...through the interventions of specialized NGOs supported by government and donors...[that] promoted a range of technical and organizational innovations, including new varieties and changes in harvest routines at farm level; timely bulking, sorting, and transport processes; aggregating floating, washing and depulping functions in new wet processing stations; cupping to grade the coffee and control quality at the processing station; forming farmer cooperatives to develop financial and management skills; and developing links to premium buyers in the United States and Europe. A principal point in the Rwandan case is that no one innovation would achieve the quality objective; all were necessary.

In general, export-oriented value chains promote formalization of those chains, making them more tightly integrated and more likely to involve contract agriculture. This tendency brings with it stimulus for innovation based on external market preferences. Trienekens and Willems studied the pineapple value chain in Ghana and the grapes value chain in South Africa and arrived at the following conclusions (2007:57−58):

> The analysis of the two chains shows that innovation follows international market demand. In Ghana we see business investments by large producer-exporters in quality control, tractors to improve production processes, transportation and pack houses. In South Africa respondents reported business investments in cold stores (by producers and producer associations), transportation (transportation companies) and infrastructure in general. Especially in the field of quality and safety of produce...important changes have taken place in production systems and the use of technology...Respondents in both chains reported a strong relationship between Western standards like Eurep-Gap and these developments. It is important to note that investments in both chains focused on infrastructure (e.g.,trucks) and product-related improvements (e.g., more environment-friendly pesticides).

It is evident that the role of value chains goes well beyond moving products from the farm to the market. The following section and the case studies illustrate approaches to identifying value chain issues that hinder realization of competitive potential, as a guide for orienting resources and collaborative efforts for their solution.

6.4 THE QUALITY DIMENSION OF COMPETITIVENESS

As pointed out, farm products may be lost or damaged for a wide variety of reasons. When the appropriateness of varieties is taken into account along with vulnerability to crop and animal pathogens and other types of value chain degradation, the threats to quality become numerous, and correspondingly the threats to competitiveness. For that reason, a product and its value chain have to pass multiple tests, or pass through a sequence filters, to be judged competitive. At the farm level, the quantitative measures of the first track can be used to assess cost-price competitiveness in production. The second track consists of qualitative evaluations of the status of quality in the various stages of the value chain. These evaluations may also include occasional judgments concerning the efficiency of processing and marketing agents of the chain because DRC or LRC calculations are usually not feasible for those stages since cost data in processing and trading companies usually are proprietary.

Assessments of quality necessarily have to be based on interviews or extended relationships in which the evaluator has relevant experience and key agents in the value chain are interviewed. Although attempts can be made to structure the interviews, as was done for the information reported in Section 10.6, the conversations inevitability go off on tangents and the digressions can be as informative as the responses to planned questions. Flexibility is also needed regarding the ways in which the ensuing information is organized. Although qualitative evaluations are inherently judgmental and often defy preconceived structures, the aim here is to construct a framework that identifies problem areas with as much precision as possible to facilitate solutions without imposing undue rigidity.

Alternative avenues for quality evaluations include: (1) sequentially along *the stages of the value chain*, i.e., from seed selection to cultivation to harvesting to postharvest drying and sorting, to transport and storage, to processing, to further transport, and to marketing; or (2) via a checklist based on *the nature of issues* or threats to quality described in Section 2.4.7. Evaluating quality issues by stage of the value chain is important to develop solutions; for example, an artisanal drying process that leaves excess humidity poses a different problem than a storage process that allows humidity to accumulate on a crop. Also, threats to quality may differ by the market of destination: for a landlocked country, transportation networks may be sufficient for supplying the domestic market but transportation by air may be too costly for high-value export products. Equally, a crop variety that is acceptable in the domestic market may not be the one that is preferred by consumers in export markets, and a crop that does not have an acceptable appearance for the fresh product market may be acceptable for the processing market. At each stage in a value chain, identification of the issues opens the door to efforts to find solutions.

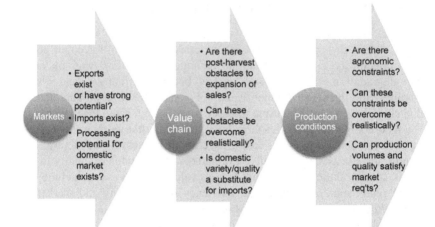

FIGURE 6.4   Filters for selection of most promising crops.

For Colombia, Norton and Argüello (2007) and for Rwanda, Turner and Norton (2009) reviewed quality and productivity issues for high-value tropical crops by stages of each value chain, with the aim of identifying the most serious constraints to the realization of a crop's competitive potential. They applied a series of filters corresponding to value chain stages and determined if the crop passed each filter or not, and if not, why. Turner and Norton summarized the procedure in the diagram similar to the one in Fig. 6.4.

Norton and Argüello (2007) followed a comparable procedure in which the sequential filters were stated as genetic material and production, postharvest and processing; and marketing.

These studies evaluated product quality (and productivity) from three perspectives: (1) stages of the value chain, (2) markets of destination, and (3) the nature of the threats to quality. The procedures can be strengthened and formalized through a finer disaggregation of each value chain and through a more systematic review of the threats to quality, based on the principal quality attributes mentioned in Section 2.4.7. The method used in this study is to incorporate all three perspectives: value chain stages, markets, and quality threats. This is done by applying quality assessors (perspective 3) at each stage of the value chain (perspective 1), and to do it separately for each final market (perspective 2). For this purpose, the basic stages of the value chain can be defined as:

1. Seeds and breeds
2. Production
3. Harvesting
4. Postharvest

5. Storage
6. Transportation
7. Processing
8. Market development

In applications of the framework, the definition of the stages may have to vary, and that is part of the framework's flexibility. For example, different transportation conditions may prevail from farm to processing industry when compared with the stage of processing industry to consumers, so the transportation stage may have to be repeated. The same applies for storage. Quality assessors are applied to all of the stages and separately for each market of destination. The definition of the types of markets also will differ by the country and product and may include, for example, domestic fresh, domestic processed, export fresh, export processed, and export organic. The quality assessors indicate areas in which issues may arise that disqualify a product from entering a market or that result in the product receiving a lower price. Following the discussion in Section 2.4.7, those assessors are the following, along with their abbreviations that are used in this assessment framework:

GE: Appropriateness, cleanliness, and uniformity of seed varieties and appropriateness of breeds
HY: Product hygiene, e.g., absence of bacterial contamination
CH: Management of agrochemicals and veterinary medicines and supplements, to avoid residues and other contamination of the final product
CM: Other aspects of crop management
TA: Taste, appearance, uniformity, ripeness
PD: Pests and diseases that affect quality and yields
HH: Excess heat or humidity that causes product deterioration
IM: Impurities mixed in with the product
SP: Product spoilage
PK: Packaging
PR: Lack of facilities for adequate processing
CE: Certifications required for some markets
LC: Lack of coordination and vertical integration in the value chain
NM: New market development required

In addition, perceived weaknesses in overall value chain governance will be indicated by the symbol VG and adverse external and policy issues by the symbol EX and PI, respectively. This list can vary in the same manner that value chain stages and markets may be defined in different ways for each application. The analyst may wish to introduce additional quality assessors based on further quality attributes in addition to the ones listed, or limit the analysis to fewer assessors.

The purposes of these assessments are (1) to assess product competitiveness from a quality viewpoint, in which quality includes the ability to supply requisite quantities and reliability in fulfillment of supply commitments and (2) to identify barriers to the realization of competitive potential. The quality dimension alone is not sufficient to declare a product competitive; for that the price-cost dimension has to be taken into account.

To systematize the findings, an assessment matrix is developed for each value chain. The rows are stages of the chain, and the columns are the markets of destination. An entry in the matrix signifies a problem that corresponds to a particular threat to quality, indicated by one of the aforementioned symbols for a quality assessor. Equally, a matrix entry indicates an opportunity to more fully realize the inherent competitiveness by resolving a problem. A matrix with no entries would indicate complete absence of quality issues that might affect competitiveness. This matrix is the core of Track 2 competitiveness evaluations. It represents the organizing framework for assessing quality barriers to the fulfillment of competitive potential.

Once a matrix is complete, the logical next challenge would be to estimate the cost of overcoming each of the quality issues that represent impediments to competitiveness. However, that is a separate exercise and, unfortunately, it usually is not carried out. If done, it would establish a basis for comparing the cost across products of interventions designed to facilitate the realization of the potential for competitiveness, i.e., the cost of moving to a higher market indifference curve in Fig. 4.4.

The evaluation of the quality dimension of competitiveness thus consists of filling out a template in the form of a Value Chain Quality Assessment (VQA) matrix that looks similar to the matrix shown in Table 6.4a and attaching comments that describe the nature of the issues. An example in Table 6.4b

TABLE 6.4A Value Chain Quality Assessment Matrix

Value Chain Stage	Domestic Markets		Export Markets	
	Fresh	Processed	Fresh	Processed
1. Seeds and breeds				
2. Production				
3. Harvesting				
4. Postharvest				
5. Storage				
6. Transportation				
7. Processing				
8. Market development				
9. Value chain governance				
10. External, policy factors				

presents the quality issues for tree tomato (tamarillo) in Rwanda, which is described further in the case study in Chapter 8. It can be seen from the table that tree tomato passes the seed (genetics) filter and the harvesting filter but that it encounters problems in all the other filters.

Geometrically, all the facets of quality cannot be put on a single axis because there is not even an ordinal ranking of them, but if the *number* of quality issues were considered as points on an axis in the assessment space, then the rows and columns of the matrix in Tables 6.4a and 6.4b would be the two other axes of value chain stages and markets, whereas this third "quality axis," of quality assessors, would be represented by the numbers of entries in the cells.

A refinement of the VQA matrix can be made by classifying the quality issues into two groups: those that completely obstruct access to a market and those that simply lower the price received in the market. Issues in the first group could be assigned a value (multiplier) of 2.0 and the less serious issues in the second group could be assigned a value of 1.0. Then the numerical value of the cells in the matrix could be added up to gain a very rough idea of the number and magnitude of issues related to quality, for comparison across products and over time. The higher the score, or position on the "third axis," the more numerous and/or onerous the quality issues. For tree tomato in Rwanda, this procedure would result in a matrix like that shown in Table 6.4b,

TABLE 6.4B Illustrative Numerical Value Chain Quality Assessment Matrix (Tree Tomato in Rwanda)[a]

Value Chain Stage	Domestic Markets		Export Markets	
	Fresh	Processed	Fresh	Processed
1. Seeds and breeds				
2. Production	CH	CH	2CH, PD	2CH, PD
3. Harvesting				
4. Postharvest	HY	HY	2HY	HY
5. Storage	SP	SP	SP	SP
6. Transportation	SP	SP	SP, 2 PK	SP, 2 PK
7. Processing		PR		2 PR
8. Market development			2CE, NM	
9. Value chain governance			2VG	2VG
Sum of scores	4	5	14	12

Total score on quality issues: 35.
[a]*Based on information in Turner and Norton (2009).*

where the quality score is a high 35. (The entries in this particular matrix are explained in Chapter 8.) This template can be used for different purposes. It can serve as a checklist of issues to be addressed for projects aimed at developing value chains. It can serve to help assign priorities for agricultural research entities or extension programs that wish to enhance the potential of high-value products. The template can also be used to fill gaps in business plans. And before the effort to fill it in begins, the template in Table 6.4a can orient efforts to assess obstacles to competitiveness, serving as a checklist of issues to review.

In addition to enabling comparisons of competitiveness issues over crops, making these kinds of indicative calculations can facilitate comparison of the level and nature of effort required to secure a product's place in different markets. In the illustrative case of tree tomato in Rwanda, evidently the requirements of penetrating export markets are more demanding than for domestic markets, but the returns also may be considerably higher. Also, they are a little less demanding for processed export products than for fresh export products. The difference would be even greater if processing capabilities were established in Rwanda, perhaps by modifying existing fruit processing facilities.

When compared with some other crops, this score of 35 would be high, but as Turner and Norton (2009) point out, at the time of their study Rwanda was inexperienced in tree tomato exports and the problems associated with the crop were not insuperable.

These matrices are qualitative tools, but when developed they provide a systematic way of looking at quality issues in regard to their effects on competitiveness. They can provide guidance regarding those issues and their seriousness and how they vary by product and to what extent they are shared among products. For example, cold chain deficiencies that result in spoilage of several fruits and vegetables will appear in all the corresponding VQA matrices, and that would constitute an argument for establishing regional cold storage and transportation facilities throughout the country to serve multiple products. However, to seek definitive solutions to quality problems, it is important to return to the original, textual information derived from the value chain interviews, which always provides a more complete picture of issues and potentials. The VQA matrices simply provide guidance on where to dig deeper and which crops, markets, and value chain links merit priority consideration.

This approach to the assessment of quality issues for competitiveness is applied to crops in Rwanda and Colombia in Chapters 8 and 10, respectively, with further commentary on the method as it is applied.

ENDNOTES

1. Prowse (2012, p. 49) surveyed 28 successful value chains in developing countries and found that only two had no upstream flows. In three cases only extension was provided to farmers by the chain, and in all other cases the chain provided varying combinations of seeds, agrochemicals, credit, and extension.

2. These threats are described in Ministry of Agriculture and Animal Resources, Rwanda (2011).
3. Communication from Carlos Rivas Almonte based on his experiences.
4. In Ecuador and Colombia significant increases in smallholder oil palm yields were achieved through an extension project developed by the Latin American Fund for Oil Palm Innovations (FLIPA) and the International Center for Tropical Agriculture (CIAT), in collaboration with Asociación de Palma Aceitera de Ecuador (ANCUPA) and the Federación Nacional de Culti- vadores de Palma de Aceite de Colombia (FEDEPALMA). See MDF Training and Consultancy (2014).

Part III

Case Studies in the Competitiveness of Tropical Agriculture

Chapter 7

Colombia: A Strategic Assessment of National Crop Competitiveness*

7.1 SCOPE AND METHODOLOGY OF THE STUDY

This case study, supported jointly by The World Bank, the Food and Agricultural Organization of the United Nations, and the US Agency for International Development arose out of a concern for the faltering economic performance of Colombian agriculture at the time and the resulting effects on the rest of the economy and rural poverty. Adapted from Norton and Balcázar (2003) it uses Track 1 methodologies and develops an extensive discussion of policies related to competitiveness. It also identifies important quality issues in agricultural value chains. The context was widespread drug cultivation and armed civil conflict that especially ravaged rural areas. A weakening economy seriously inhibited efforts to reduce poverty, which still affected more than 60% of the rural population. By the same token, the study commented that the Colombian rural sector had large economic potentials that were still untapped and could be exploited productively under favorable conditions. It presented scenarios with different crop emphases in the sector's growth, based on the competitiveness analysis.

The stated purposes of the study were to assess: (1) the sector's current and potential competitiveness, (2) the factors that determine or inhibit its competitiveness, (3) the extent to which the smallholder (campesino) economy was competitive and could become more so, and (4) in a long-run perspective, the role of policies in determining the sector's competitiveness. The study presented the outlines of strategic options that could strengthen the sector's competitiveness, with emphasis on the smallholder subsector and development of transition measures to ease the path to adoption of new orientations. Given the great variety of ecosystems in Colombia, a spatial dimension was incorporated into the analysis.

---

* This case study is adapted from Norton and Balcázar (2003). Valuable contributions were made by many colleagues, particularly Carlos Federico Espinal and Henry Samacá.

The Competitiveness of Tropical Agriculture. http://dx.doi.org/10.1016/B978-0-12-805312-6.00007-6

The main methodological pillars of this study were export performance and the cost-price track (Track 1). The analysis was based on calculations of long-run competitiveness (LRC) coefficients through use of a nation-wide database on production costs (farm budgets) from the Ministry of Agriculture that was not available previously, plus a review of trends in Colombia's international trade. Parts of the study ventured into Track 2 territory by identifying key constraints for each product in areas such as genetic material, production technologies, marketing, and producer organization. This was especially relevant for potatoes, cotton, and cassava, with lessons for products that are applicable in other countries. In addition, links and trade-offs between competitiveness of the primary producing sector and the processing sector are brought out, especially for industrial crops, including tobacco and cotton. These kinds of observations underscore the fact that competitiveness evaluations cannot be limited to the quantitative dimension alone.

Long-run competitiveness coefficients (LRCs) were calculated by department in Colombia, since the farm budgets were intended to reflect the most representative production practices in each department. The new database on production costs at the farm level had to be cleaned of obvious errors before being used. It supported calculations of LRCs for 18 of the 32 departments in the country, and for an average of seven crops per department. It permitted the assessment of LRC in livestock activities for five departments. In total, 27 crops and 6 livestock products were analyzed plus 12 commercial forestry species. However, the range of agricultural products produced in Colombia is much wider than the coverage of the study.

Food safety and quality issues for export markets were reviewed as they pertain to Colombian products, and in some cases they turned out to be crucial for realizing Colombia's agricultural competitive advantage. Also, illustrative scenarios were constructed in which existing agricultural subsidies in industrial countries were reduced or removed, and calculations were made of how the sector's competitiveness would have looked in those cases.

The study said that for Colombia's competitive potential to be fully realized, appropriate marketing links and input supply channels have to be established, inputs have to be financed, and uniformity of product quality has to be assured, among other requirements. For this reason, the calculations of comparative advantage by product and region within Colombia were supported by reviews of the main factors that determine competitiveness, including agroclimatic conditions, connectivity, agricultural research and extension efforts, the agricultural financial system, and levels of rural education. The construction and management of irrigation systems, including issues of their ownership, were reviewed in light of the importance of irrigation for agricultural productivity and product quality since irrigation accounted for about 20% of the area in crops and about 40% of crop output.

An assessment of institutional considerations accompanied these analyses, particularly for the agricultural technology system and the framework of food safety and quality control, to identify institutional bottlenecks to enhanced competitiveness of Colombian agriculture and possible ways to overcome them. Other issues discussed included the effects of the continuing violence in rural areas and the role of farmer organizations in competitiveness.

Competitiveness was also evaluated via LRCs under different hypothetical policy scenarios to illustrate factors that affect it. An illustrative sector-wide scenario, covering all major crops, was developed that provided estimates of net effects of alternative policy orientations on total sector employment and income. Although it appeared clear that new strategic emphases could improve the sector's performance in the long run, it was noted that any shift toward a different strategic framework would have to take full account of the existing situation and offer producers incentives and assistance for making a transition.

In addition to the calculations of LRCs, revealed comparative advantage (RCA) calculations were made at the level of 4-digit trade data to provide a dynamic perspective on competitiveness. Use of RCAs allowed the study to examine a significant number of agroindustrial products for which LRCs could not be calculated owing to lack public information on cost structures.

The study is presented with adaptations in the remainder of this chapter. It should be borne in mind that some conditions in Colombia have changed since the study was carried out, but its principal thrusts remain valid. As well as illustrating the application of competitiveness assessment methodologies, and although the calculations in the study are now dated, the study provides an un-usually complete example of analysis of the effects of fiscal and trade policies, and of the construction of policy dialogs and strategic options for rural areas on the basis of competitiveness analyses for the agricultural and forestry sectors.

7.2 GRAINS AND OILSEEDS

Calculation of RCA coefficients (Eq. 3.2) for Colombia suggests a future potential in some processed forms of grains and short-cycle oilseeds, although not in all. As Table 7.1 illustrates, most of these processed products showed an increasing relative importance in Colombia's basket of export goods over time. Many of these exports go to the Andean regional market. Agroindustrial exports had previously benefited from a duty drawback (*certificado de reembolso tributario* or CERT) of 2.5%, but nevertheless, the trend in these exports represents another indication that Colombia's competitive advantage in the area of grains and short-cycle edible oils lies more in processed products than in primary products. A policy implication is that an export-oriented strategy should lean toward lower tariffs on the imports of raw grains and oils, to make the processed products more competitive, instead of trying to protect the grain and oilseed production sectors.

The LRC calculations for grains and soybeans are presented in Table 7.2 (using Eq. 5.3). The results confirm the aforementioned indications, that this is not a subsector in which Colombia has LRC. The only crop/department in

TABLE 7.1 Colombia: Revealed Comparative Advantage for Processed Products of Grains, Oilseeds (Export Values in Million US$)

Product	FAO Code	Tariff Code	Export Value, 2001	Product Chain	Revealed Comparative Advantage				
					1997	1998	1999	2000	2001
Pearl barley	46	1104	0.132	Cereals for human cons.	0.64	2.84	10.19	14.89	12.94
Bread, cookies, wafers	22	1905	33.548	Cereals for human cons.	2.02	2.39	2.04	2.06	2.30
Corn flour	58	1102	0.891	Cereals for human cons.	0.64	0.66	0.87	2.75	3.67
Baby food	109	1901	6.845	Cereals for human cons.	1.57	1.15	0.57	1.47	2.38
Soybean cake	238	2304	11.242	Cereals, poultry, pork	0.02	0.02	0.26	0.62	0.70
Food for domestic animals	843	2309	3.491	Cereals, poultry, pork	0.05	0.05	0.10	0.13	0.19
Breakfast cereals	41	1904	5.448	Cereals for human cons.	0.36	0.33	0.53	0.67	1.54
Malt extract, other food preparations	115	1901	3.892	Cereals for human cons.	0.51	0.56	0.53	0.54	1.04
Ground oats	76	1104	0.471	Cereals for human cons.	0.09	0.04	0.77	0.33	1.91
Dough for bakeries, pastries	114	1901	1.102	Cereals for human cons.	0.03	0.04	0.01	0.07	0.58
Soybean oil	237	1507	3.069	Oilseeds, fats and oils	0.01	0.03	0.07	0.67	0.53
Nuts, sesame, & palm seed	289	1207	0.176	Oilseeds, fats and oils	0.00	0.00	0.50	0.09	0.04
Soya	236	1201	–	Cereals, poultry, pork	0.00	0.01	–	0.00	–
Cereal flours excluding wheat	111	1102	0.904	Cereals for human cons.	0.09	1.15	0.85	0.28	0.52
Corn, sesame oils	60	1515	0.633	Oilseeds, fats and oils	1.02	0.60	0.19	0.12	0.61

Cons., consumption; *FAO*, Food and Agricultural Organization of the United Nations. The first row section of the table includes products with an RCA >1.0 (on average) and generally increasing; the second section, products with an RCA <1.0 and generally increasing; and the third, those with an RCA generally decreasing. RCAs for some other categories of edible oils were not included here because they may include oils from perennial crops. The tariff codes are those applied in the Andean Common Market.

TABLE 7.2 Colombia: Long-Run Competitiveness Indicators for Grains and Soybeans, 2001

Department	Yellow Corn	White Corn[a]	Sorghum	Rainfed Rice	Irrigated Rice	Soybeans	Wheat	Barley
Valle del Cauca	1.80		1.49		1.72	3.84		
Santander		2.31						
Tolima		1.59	2.12		1.75	3.12		
Córdoba	2.26	1.38	2.18	2.01	1.75			
Magdalena	1.80	2.19			1.80			
Cesar	1.30	1.80	1.81		2.40			
Meta	1.37	2.59	1.90	7.97	5.63	1.79		
Nariño		3.34					2.88	
Sucre	1.90			9.34				
Bolívar	2.23	2.30	1.26	2.01	2.52			
Boyacá		4.98					2.07	8.98
Antioquia					2.46			
Huila	1.82	1.78	0.96		3.41			
Norte de Santander		4.61			2.24		2.11	
Quindío	2.41							
Cundinamarca	1.30	2.76	1.20	2.01	1.69	1.79	1.44	1.68
Minimum value	1.30	1.38	0.96	2.01	1.69	1.79	1.44	1.68
Maximum value	2.41	4.98	2.18	9.34	5.63	3.84	2.88	8.98
National average	1.88	2.64	1.61	5.33	2.49	2.92	2.13	5.33

The national average for each crop is weighted by acreage in the corresponding departments in that crop. For white corn, the production technology is the traditional, least mechanized one.

[a] Traditional technology of production.

which marginal competitiveness is indicated is sorghum in Huila (see map in Fig. 7.1), and there appears to be a possibility that sorghum could become competitive in Cundinamarca with reasonable increases in productivity and decreases in costs, but the general conclusion is that it will be quite a struggle for grains to acquire inherent competitiveness in Colombia. In the case of yellow corn, for example, each dollar saved by producing instead of importing implies an opportunity cost of domestic resources that ranges from 1.3 dollars in Cesar to 2.4 dollars in Quindío. Zones that possess relatively high levels of productivity in corn, for example, the Valley of the Cauca, Córdoba, and Tolima, do not escape the general competitive disadvantage of Colombia in corn and other grains.

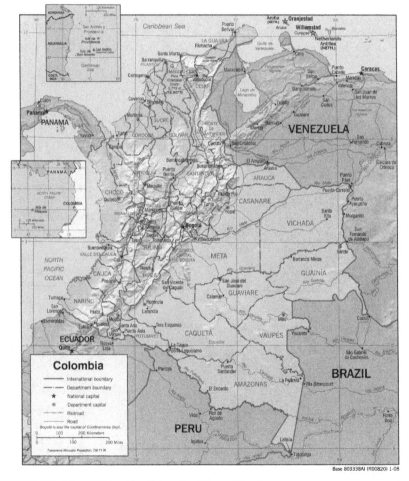

Base 803338AI (R00820) 1-08

FIGURE 7.1   Map of Colombia. *University of Texas Libraries.*

Production of feedgrains in Colombia evidently is sustained by the high tariff levels (at the time of this study). In turn, these tariffs reduce the competitiveness of poultry and pork as well as processed grain products.

White corn has an even more marked competitive disadvantage than yellow corn. Its LRC ranges from 1.4 to almost 5.0. This situation reflects the relatively low productivity of labor in white corn. It is a crop relatively intensive in the use of labor, but the opportunity cost of labor is too high in relation to its low productivity in white corn and the low world prices of that crop. The calculations reported here are made for corn as a grain. White corn sold and consumed as a vegetable may be an exception to this study's conclusions because its higher international transport costs, relative to value, confer a greater advantage to domestic producers. This advantage would be strongest in rural areas. Under current conditions, the weakest of Colombian grains in international competitiveness is rice. None of its departmental LRCs fall below 2.0, and some of them are extremely high.

It may be asked, to what extent is the lack of Colombian competitiveness in grains attributable to the fact that these products are subsidized in major exporting countries such as the United States and the European Union? The extent of those subsidies is well documented and is a major, recurring topic on the agenda of international trade negotiations. The US Department of Agriculture, among other agencies, has made estimates of how world market prices would be affected by the removal of those subsidies and other impediments to international trade in agriculture. It has been estimated that, as of 2000, a complete liberalization of international trade and elimination of agricultural subsidies would result in the following percentage price increases on world markets for agricultural products: wheat, 18.1; rice, 10.1; other grains, 15.2; fats and oils; 11.2; sugar, 16.4; and meat and milk, 22.3.[1]

It is valid to inquire why a developing country should transmit the distortive effects of international subsidies to its own agriculture. The question acquires more force in light of the pervasiveness of rural poverty in most developing countries, and also the quasi-irreversibility of labor flows between agriculture and the rest of the economy. If families are driven out of agriculture in part because of low prices on world commodity markets caused by policy decisions in richer countries, then if those decisions are changed someday it would be virtually impossible to induce those families to return to rural areas and participate in reviving agriculture. Thus a long-term structural shift in lower income economies would be taking place on the basis of policy decisions made elsewhere.

To analyze this issue in the Colombian context, the LRC coefficients were recalculated with international prices 20% higher for grains and soybeans. Such an increase could arise from changes in world market conditions or from changes in subsidy policies in Organisation for Economic Co-operation and Development countries. The results are shown in Table 7.3. It can be seen that

TABLE 7.3 Colombia: Long-Run Competitiveness Indicators for Grains and Soybeans, 2001, With a 20% Increase in International Prices

Department	Yellow Corn	White Corn[a]	Sorghum	Rainfed Rice	Irrigated Rice	Soybeans	Wheat	Barley
Valle del Cauca	1.26		1.06		1.21	2.50		
Santander		1.92				1.53		
Tolima		1.33	1.27		1.27			
Córdoba	1.51	1.13	1.33	1.48	1.24			
Magdalena	1.30	1.83			1.32			
Cesar	1.00	1.50	1.25		1.56	1.28		
Meta	0.96	1.82	1.20	2.79	2.55			
Nariño		2.65					1.88	
Sucre	1.41			3.37				
Bolívar	1.59	1.87	0.91	1.48	1.75			
Boyacá		3.82					1.52	3.92
Antioquia					1.55			
Huila	1.14	1.37	0.69		2.01			
Norte de Santander		3.07			1.55		1.72	
Quindío	1.84							
Cundinamarca		2.24	0.84		1.19		1.11	1.32
Minimum value								
Maximum value	2.41	4.98	2.18	9.34	5.63	3.84	2.88	8.98
National average	1.88	2.64	1.61	5.33	2.49	2.92	2.13	5.33

The national average for each crop is weighted by acreage in corresponding departments in that crop. For white corn, the production technology is the traditional, least mechanized one.
[a]*Traditional technology of production.*

on the whole Colombian grain production still would not be competitive even under an optimistic scenario about liberalization of world agricultural policy.

Under this scenario, sorghum would strengthen its competitiveness in Huila and would move into the competitive category in Bolívar and yellow corn would attain marginal competitiveness in Meta, whereas the Andean cordillera falls away to the vast eastern plains, the Orinoquia (*Altillanura del Oriente*). Nevertheless, these products in these regions would be vulnerable to fluctuations in the exchange rate, and they would not be competitive in most regions of the country. Colombia's competitive disadvantage in rice, wheat, barley, and soybeans would not be reversed by a 20% increase in international prices.

The conclusions of these analyses are clear: that, in general, Colombia's strengths in agriculture do not extend to grains and short-cycle oilseeds—with the possible exception of corn in Meta—but that the country may have potential for increasing its exports of processed grain products. For those producers whose productivity/cost relationship is better than the average, the best hope lies in continuing to increase productivity. In the case of corn as a feedgrain, sorghum, and soybeans, the sparsely populated high plateau of the eastern plains of the country, the *Altillanura del Oriente*, offers intriguing possibilities for obtaining high yields. Meta already is the most competitive area in Colombia in soybeans, as well as being one of the two most competitive areas in yellow corn. However, cultivation of the *Altillanura* would have to take place on a large scale, in plots of thousands of hectares, to take advantage of the economies of scale offered by mechanization, and large initial investments would have to be made in improving the soil quality in light of the relatively high quantities of aluminum content of the soils of that region and their high acidity. In addition, a substantial, long-term commitment would have to be made for agricultural research on those crops under those conditions, and the feasibility of such a project might not be known until 10 years or more of experience accumulate. Misión Paz (2001, p. 179) is optimistic about the potential of the Orinoquia for corn, soybeans, and cassava and states that recently hybrid varieties have become available that are resistant to aluminum and acidity in the soil. However, private sector field trials in the region have given disappointing results so far. Thus, an important potential may exist in that region, but its economics are as yet uncertain.[2]

In the case of corn, it may be argued that its importance lies in the fact that it is a subsistence crop for many rural families. About one-fourth of Colombia's corn is grown under mechanized conditions with high doses of fertilizers and pesticides. The other three-fourths of the crop is planted mainly on hillsides with traditional techniques of cultivation, utilizing mainly family labor with very little application of agrochemicals. However, between 1991 and 2001 the area planted in corn under traditional cultivation techniques fell by about one-third. It is unclear how much of this decline represented a move to more modern cultivation techniques and how much was attributable to families abandoning the land because of the violence. The area of corn under more modern technologies did increase rapidly during this period, and average yields increased from about 2.8 t per ha to about 3.7 t per ha, but still they fall short of competitive levels in relation to costs.

To the extent that corn is used as a subsistence crop for home consumption on small farms, that role would not be affected by a reduction in tariff protection. In Colombia, the range of crops used for home consumption on small farms is wide and it includes plantain, cassava, beans, and some fruit and vegetables.

Along with sugar and milk, rice is one of the three products with the largest monetary value of tariff protection in Colombia. It also benefits from "absorption agreements"[3] and receives substantial fiscal subsidies including large subsidies for storage costs. Therefore its competitive disadvantage is even greater than indicated in Tables 7.2 and 7.3. It may be argued that the position of rice could be strengthened through productivity improvements, but they would have to be very large. During the period 1991—2001, rice yields grew by only 2.0% per year. That is a decent rate of increase for many crops but not nearly enough to overcome rice's competitive disadvantage (on average) in the foreseeable future. For the more efficient rice producers who wish to pursue its cultivation in spite of its strong competitive disadvantage in general, the most important avenues to pursue include widening the range of varieties available, improving the efficiency of irrigation, improving phytosanitary controls and reducing the use of agrochemicals, and increasing the use of certified seeds.

For many crops, rates of irrigation water application often are far in excess of what is needed and contribute to degradation of soil quality in addition to higher costs of production. It has been estimated that for rice irrigation applications could be reduced by as much as 30% (CONSULTPLAN, 1998). The best zones for irrigated rice are currently more competitive than those producing rainfed rice, and a rationalization of irrigation practices would increase that advantage. However, rainfed yields could be improved in Córdoba, for example, through use of certified seed and the construction of small dams to capture surface runoff from rains.

With regard to the use of agrochemicals, pesticides represent 35% of the total cost of production of rainfed rice and 27% of the total cost of production of irrigated rice, for Colombia as a whole (data provided by Tomás Sicard, 2003). These percentages have almost doubled since 1981, which is an indication that the pesticide strategy in rice is unsustainable. The high humidity in Meta means that zone is an especially heavy user of pesticides for rice. Hence, if rice in Colombia is to become internationally competitive, fundamental changes will have to be made in the way it is produced. An advantage of this subsector is that the milling and drying technology is relatively advanced when compared with that of rice producers such as Brazil, Uruguay, Venezuela, and the United States.

The presence of international subsidies is a powerful argument for not eliminating all tariff protection on grains. However, from the viewpoint of long-term national development policy, the wisdom of continuing to provide the sector's strongest incentives for the cultivation of grains is put into doubt not only by the aforementioned results above but also by the fact that those crops generate relatively little employment per hectare. Table 7.4 shows the average amount of employment, in person-days, generated by 27 major crops. In this table it can be seen that the grains rank among the lowest crops in terms of capacity to create employment.

From a perspective of value chains, some of the most competitive components of the cereal chains appear to be breads and cookies. Their exports increased from US$6 million in 1991 to US$35 million in 2001. Their principal markets are in Venezuela and Ecuador, which means their export strength may partly depend on the arrangements of the Andean Common Market. It is noteworthy that the main raw material for these products, wheat, is entirely imported, except for very small quantities of specialty wheat grown at high altitudes in Nariño.

The competitive potential of processed flour products is shown by the increases over time in their RCA value in Table 7.1. LRC calculations, which are the most solid indicators of underlying competitiveness, could not be made for processed foods because their information on costs is proprietary.

TABLE 7.4 Colombia: National Average Coefficients of Labor Use by Crop (in Person-Days/ha)

Crop	Labor Use	Crop	Labor Use	Crop	Labor Use
Isabella grapes	520.0	*Lulo*	116.4	White corn	39.8
Green asparagus	339.4	Potatoes	104.6	Oil palm	35.2
Coffee	249.7	Plantain	97.4	Irrigated rice	33.4
Granadilla	220.8	Citrus	96.9	Yellow corn	23.3
Light tobacco	214.2	Industrial cassava	87.9	Rainfed rice	22.0
Dragon fruit	208.9	Mango	84.4	Wheat	20.6
Dark tobacco	196.3	Beans	75.9	Soybeans	18.0
Passion fruit	166.2	Sugarcane	65.1	Barley	15.0
Cacao	121.7	Cotton	49.9	Sorghum	13.4
Papaya	121.1				

Data on Isabella grapes are from Misión Paz (2001).

7.3 FRUIT AND VEGETABLES

Misión Paz (2001, p. 160) pointed out "the increasing demand for products that are fresh and rich in vitamins and other nutrients, and that do not contain either fats or calories, has caused the market for fruit and vegetables to grow at an accelerated pace in the world.... In the United States, imports of vegetables and fresh roots and tubers grew at 9.1% annually from 1995 to 1999."

Fruit and vegetables have been important in Colombia's external trade balance, and they also figure among the crops that generate the most employment per hectare. Bananas alone generate 34,543 direct jobs. They are Colombia's third largest agricultural export, after coffee and flowers. Fruit production is concentrated mainly in the central Andean Departments of Santander, Cundinamarca, Valle del Cauca, and Tolima. Santander leads in the production of pineapple, mandarin oranges, and guava (*guayaba*); Valle del Cauca leads in bananas for the domestic market, *granadilla*, dragon fruit, and grapes; Tolima leads in soursop (*guanábana*), lemons, and mango; and Cundinamarca leads in citrus, strawberries, other berries, and Cape gooseberry (*uchuva*).

Cape gooseberry exports have grown rapidly, mainly to Europe (principally to Germany), and in spite of its production and marketing being handled in an artisanal manner, this crop is now Colombia's third most important fruit export, after bananas and plantains. Like several other fruits it is mainly a smallholder crop. *Granadilla* (related to passion fruit) is another crop whose main export market is Europe. *Lulo* (or *naranjilla*, a tomatolike fruit), passion fruit, berries, papaya, and other fruit crops have shown potential as well, although it has not been possible to analyze the comparative advantage of all of them in the context of this study. *Lulo* and oranges have a considerable domestic market for juices. The importance of fruit as an input to agroindustry can be seen in the fact that in 1998 two-thirds (65.3%) of the country's 1680 food processing enterprises required fruit as an input into their processes.[4] Fruit is used in products as diverse as processed dairy products, starches, breakfast cereals, cookies, soft drinks, mineral waters, preserves, baby foods, and other prepared meals. Industries using fruit as an input account for about 17% of the total industrial employment.

Bananas are cultivated mainly in lowlands, in the northern zone of Antioquian Urabá, and to a lesser extent in the northern part of the Department of Magdalena. These are areas especially affected by the violence, and banana production has been notable for sustaining its levels in spite of the lack of security. There has been a slight downward trend in production, more in Magdalena than in Urabá, because of phytosanitary problems. Banana exports depend in good measure on preferential access to Europe, and in the long run this dependency implies a certain fragility of the export position of bananas.

Plantain exports have increased considerably since the early 1990s. Its production is more widespread throughout the country. The largest producing areas are the Andean zones and the inter-Andean valleys, and the main producer for export is the Antioquian Urabá. Plantain also has spawned an incipient processing industry in Bogotá and the Cauca Valley. A key to plantain's competitiveness is the provision of genetic material of good quality. In this case, materials obtained in vitro are not acceptable because they are easily contaminated with viruses and are difficult to manage in the field. In areas such as Caldas and Quindío, for example, plantain has good access to research centers and high-quality genetic material, as well as good soils. Where plantain is not very successful it is due to poor soils, a humid climate, weak access to healthy genetic materials, and the need to apply greater amounts of fertilizer and pesticides because of the soils and climate.

The acreage planted in fruit other than bananas and plantains more than doubled from 1991 to 2001, increasing from 93,072 to 190,197 ha in that period. Table 7.5 presents data on exports in recent years of fruits other than bananas and plantains. Export growth of these products is very strong, at nearly 15% per year. It would have been even stronger had not a large part of the passion fruit processing industry collapsed for reasons unrelated to the product's inherent competitiveness.

Colombia has negotiated preferential access for some fruit exports in parts of Latin America. For example, *bananito* (and bananas) do not pay tariffs when they are exported to Central America and Chile, and fruits such as dragon fruit (*pitahaya*) and *uchuva* can enter Mexico under a 3.9% tariff. Fruit exports have received little incentives from the government, mainly duty drawbacks of 2.5−3.0% and, for some, access to credits from the export bank Bancoldex. On the import side, these categories of fruit enjoy tariff protection levels that are standard for Colombia, of 15% and 20% according to the degree of processing.

Citrus production expanded at more than 8% per year during the 1990s, and it is estimated that citrus trees covered 41,555 ha in 1998. The main producing area is the center of the country (Santander, Cundinamarca, and Tolima, as mentioned, and also the mountainous Central Department of Boyacá) but the most rapid growth of production has occurred in the Orinoquia, to the east, and the Atlantic Coast. Citrus fruit does not participate much in international trade, neither in exports nor in imports, and the same has been true of mango. It would appear that there is considerable scope for further growth of this market since, for example, Colombian levels of consumption of orange juice are still well below those of other countries: 5 L/person/year versus 12 L in Venezuela, 20−30 L in most European countries, 44 L in the United States, and 60 L in Germany. Citrus production receives tariff protection of 15% and 20%, and citrus exports can enter many Latin American countries with low tariff barriers. Based on export performance and informed opinion, lemons and limes appear to be the most competitive of the citrus fruits for fresh sales, and oranges have demonstrated potential for juice on the domestic market.

TABLE 7.5 Colombia: Exports of Fresh and Processed Fruit Other Than Bananas and Plantains, 1997–2001 (000 US$ fob)

Type of Fruit and Processed Fruit	1997	1998	1999	2000	2001	% Growth Rate/ year
Coconut, cashews, and Brazil nuts	3	7	19	34	64	114.9
Almonds, walnuts, hazelnuts, and pistachios	6	0	28	230	210	143.3
Figs, pineapple, avocadoes, guayaba, and mangos	485	996	1,707	1,527	1,995	42.5
Citrus	28	2,093	1,459	212	496	105.2
Fresh grapes, raisins	138	50	77	10	12	−45.7
Fresh cantaloupe, watermelon, and papaya	8	143	31	38	264	140.0
Fresh apples and pears	8	0	66	254	128	100.0
Fresh peaches, plums, and cherries	11	6	28	61	72	60.0
Fresh Cape gooseberries, dragon fruit, *granadilla*, passion fruit, and others	7,956	9,507	9,854	11,290	13,525	14.2

Frozen fruit (strawberries, berries, etc.)	591	595	397	357	386	−10.1
Fruit preserves (strawberry, peach, etc.)	3	5	4	10	1	−24.0
Dried fruit	120	6	42	73	124	3.3
Sweets of vegetables and fruit and fruit skins	57	69	80	59	27	−17.0
Jams, jellies, purées, and fruit pastes	480	791	574	827	1,327	28.9
Other forms of preserved fruit	3,159	3,831	3,362	3,103	8,007	26.2
Nonfermented fruit juice	4,021	4,289	4,323	2,365	2,686	−9.6
Subtotal, fresh fruit	9,357	13,408	13,712	14,096	17,277	16.6
Subtotal, processed fruit	7,717	8,980	8,339	6,354	12,047	11.8
Total	17,074	22,388	22,051	20,450	29,324	14.5

Ministerio de Agricultura y Desarrollo Rural, *Estadísticas Agropecuarias*, 2002.

Trade data suggest the possibility of a comparative advantage for most fruit and vegetables in Colombia. In the case of citrus, it is worth repeating that to date the traded quantities are small, and hence they may not be indicative of true potentials. Table 7.6 shows the RCA coefficients for principal fruit and vegetable line items in the customs classifications.

Isabella grapes are not shown in the table, but Misión Paz (2001, p. 164) points out that producers in the Cauca Valley plan to increase production from 11,000 to 32,000 t per year, for export to Ecuador and Venezuela during the counterseason when Chilean grapes are not available, and for the domestic market. Thus there are a number of indications of potential competitiveness in the area of fruit and vegetables, but the existence of a degree of tariff protection in Colombia for these products, albeit less than for grains, means that the trade data alone are insufficient to determine whether they have a comparative advantage. Therefore, LRC calculations were carried out for the study to determine how solid their competitiveness might be. These calculations are summarized for selected products in Table 7.7.

The LRC coefficients confirm the results of the RCA calculations and indicate that Colombia has notable LRC in fruit and vegetable crops. In addition, their high rate of employment creation per hectare means they have significant benefits in terms of poverty reduction and social equity in rural areas. If labor's opportunity cost were set at a lower level than the prevailing rural wage, as some analysts prefer to do, then the competitiveness of fruit and vegetables would be even greater, and the competitive disadvantage of grains would be even stronger. One of the hurdles that fruit and vegetables face for further expansion is inadequate production technologies. In its National Plan for Technology Transfer, the Colombian Agricultural Institute (ICA)[5] has pointed out that 92.4% of fruit farms use little or no improved technology, 5.3% use some improved technology, and only 2.3% use improved technology intensively.

Another hurdle is a high risk of plant diseases, owing in part to the use of traditional technologies. For example, in 1998, Antioquia was the principal producing area of *granadilla*, but then its plantings of that crop were devastated by an infestation (of the disease caused by *Nectria haematococca Berk.*), and now the Cauca Valley is the main producer of that crop. Bananas and plantains struggle against *sigatoka*. Another risk faced by producers of these crops is the volatile prices and a long-term downward trend of prices in world markets. As more producers enter these profitable international markets, prices fall. For example, international prices *declined* at the following annual rates between 1995 and 2000 for this sample of products, all of which experienced rapid increases in world export volumes: mangoes, 5.2%; pineapples, 1.9%; lemons and limes, 4.7%; cantaloupes and other melons, 5.0%; canned mushrooms, 7.3%; fresh mushrooms, 9.1%; fresh onions and shallots, 8.4%; eggplant, 5.6%; papayas, 1.3%; figs, 6%; and asparagus, 5.6%.[6] These negative price trends can be expected to continue for most fruit and vegetable crops.

TABLE 7.6 Colombia: Revealed Comparative Advantage for Fruit and Vegetables (Export Values in Million US$)

Product	FAO Code	Tariff Code	Export Value, 2001	Product Chain	Revealed Comparative Advantage				
					1997	1998	1999	2000	2001
Bananas	486	0803	364.868	Bananas					
Figs	569	0804	0.376	Fruit	5.38	6.43	6.32	7.00	8.11
Mangoes	571	0804	1.339	Fruit	0.15	0.76	1.64	1.35	1.61
Uchuva and fresh fruit n.e.c.	610	0810	13.528	Fruit	7.30	13.24	10.2	11.61	13.12
Onions	403	0703	1.218	Vegetables	0.56	0.01	0.84	5.60	0.87
Other vegetables including temporarily preserved mixtures	474	0711	1.045	Vegetables	1.24	1.23	2.45	1.12	1.54
Dry beans	176	0713	7.004	Vegetables	0.20	1.41	5.23	2.91	2.82
Peppers, whole and ground	689	0904	2.598	Vegetables	1.10	2.75	3.29	1.56	3.56
Canned mushrooms	451	2003	2.429	Vegetables	1.32	1.34	2.13	2.23	2.06
Fruit juice n.e.c.	622	2009	2.355	Fruit	1.81	1.97	1.86	0.98	1.04
Asparagus	367	0709	1.930	Vegetables	4.53	5.48	3.09	1.85	2.32
Lemons and limes	497	0805	0.374	Citrus	0.00	0.08	0.00	0.09	0.26
Fresh pineapple	574	0804	0.267	Fruit	0.17	0.18	0.15	0.24	0.27
Fruit preparations	623	0811	0.386	Fruit	0.45	0.55	0.46	0.47	1.10

Continued

TABLE 7.6 Colombia: Revealed Comparative Advantage for Fruit and Vegetables (Export Values in Million US$)—cont'd

Product	FAO Code	Tariff Code	Export Value, 2001	Product Chain	Revealed Comparative Advantage				
					1997	1998	1999	2000	2001
Fruit with rinds n.e.c.	234	0802	0.210	Fruit	0.01	–	0.04	0.35	0.34
Other dry fruit	620	0813	0.124	Fruit	0.18	0.01	0.08	0.11	0.20
Processed peanuts	246	2008	0.134	Fruit	0.02	0.09	0.12	0.11	0.16
Vegetables in vinegar	471	2001	0.390	Vegetables	0.07	0.15	0.30	0.14	0.28
Fresh mushrooms	449	0709	0.472	Vegetables	0.00	–	0.12	0.28	0.31
Garlic	406	0703	0.149	Vegetables	–	0.01	–	0.47	0.14
Carrots	426	0706	0.162	Vegetables	0.04	0.18	0.01	0.07	0.19
Lettuce	372	0705	0.165	Vegetables	0.00	–	0.02	0.16	0.08
Processed tomatoes	391	2002	0.116	Vegetables	0.04	0.03	0.04	0.04	0.05
Oranges	490	0805	0.121	Citrus	0.00	0.47	0.38	0.02	0.03
Jellies, marmalades, etc.	626	2007	1.320	Fruit	0.21	0.35	0.04	0.05	0.08
Tomatoes	388	0702	0.217	Vegetables	0.01	0.22	0.06	0.17	0.03

The first section of the table includes products with an RCA >1.0 (on average) and generally increasing; the second section, products with an RCA >1.0 and generally decreasing; the third, products with an RCA <1.0 and generally increasing; and the fourth, those with an RCA <1.0 and generally decreasing. The tariff codes are those applied in the Andean Common Market. *n.e.c.*, not elsewhere classified.
Based on official trade statistics.

TABLE 7.7 Colombia: Long-Run Competitiveness Indicators for Fruit and Vegetables, 2001

Department	Granadilla	Citrus	Passion Fruit	Lulo	Dragon Fruit	Papaya	Plantain	Mango	Green Asparagus
Valle del Cauca	0.33	0.26	0.18		0.76		0.26		
Tolima		0.29					0.22	0.30	
Córdoba			0.19			0.15			
Meta		0.21	0.19			0.20			
Risaralda		0.28				0.19			
Antioquia	0.72	0.22			0.88	0.15	0.36	0.29	0.73
Norte de Santander				0.23					
Caldas		0.26	0.14			0.17	0.34		
Cesar							0.32		
Quindío	0.85	0.27	0.14	0.46		0.20	0.26	0.29	0.73
Minimum value	0.33	0.21	0.14	0.23	0.76	0.15	0.22	0.29	0.73
Maximum value	0.85	0.29	0.19	0.46	0.88	0.20	0.36	0.30	0.73
National average	0.64	0.26	0.17	0.35	0.82	0.18	0.29	0.29	0.73

These calculations take into account transportation and marketing margins and rates of product spoilage. The national average for each crop is weighted by acreage in corresponding departments in that crop.

Nevertheless, other countries in Latin America have overcome these obstacles and have expanded their exports of fruit and vegetables more than Colombia has. Today, Peru earns annually US$15 million only in mango exports; Costa Rica, about US$100 million in pineapple exports; Argentina, US$83 million in lemon and lime exports; and Mexico, US$124 million in mangoes and more than US$50 million in avocadoes. The wide market potential for fresh fruit is illustrated by the experience of berries, of which there are 40 varieties native to Colombia. In 1992, only 189 mt were exported, but they went to external markets as diverse as Germany, Finland, the Antilles, Aruba, Curacao, Belgium, Spain, Australia, Canada, the United States, and the United Arab Emirates.

Brazil has successfully based its transformation of a smallholder sector in part of the Northeast on the irrigated production of fruit and vegetables. In the district of Petrolina-Juazeiro, in the north of Minas Gerais, 83% of the cropping pattern of smallholder participants comprises fruits and vegetables. The leading crops include mangos (27% of the area), grapes (19%), industrial tomatoes (13%), onions (13%), and plantain (11%). The governmental development agency for the district successfully stimulated agricultural research and the involvement of processing industries, and with these ingredients, fruit and vegetable crops became a major vehicle for raising the incomes of the rural poor (Damiani, 2001).

Colombia's variety of climates and growing seasons means that over the long run it is in a favorable position to take advantage of windows of opportunity in the huge market of the United States. For example, normally cantaloupe exports to the United States face a tariff of 29.7%, but during the period August 1–September 15 their tariff is reduced to 12.8%. In any case, for the time being Colombia has the additional benefit of being able to export many fruit and vegetable products to the United States without paying any tariffs, through a bilateral agreement.

Some observers dismiss the importance of fruit and vegetables on the grounds that each occupies a small amount of acreage but taken together they are quite important for generating both income and employment. According to official statistics of the Ministry of Agriculture and Rural Development, in 2001, at constant 1994 prices, fruit and vegetables excepting plantains and bananas accounted for 23% of the value of total crop output without coffee. Bananas and plantains added another 19%, so that all together fruit and vegetables represented more than 40% of the value of agricultural output. Their proportionate contribution to agricultural employment is considerably higher.

The principal reasons for Colombia's lag in promoting fruit and vegetable exports to take advantage of their inherent competitiveness appear to be the following:

1. Colombia has attained certification of admissibility to the US market on phytosanitary grounds for very few products, and this means that the

market for fresh produce there is almost closed to Colombian exporters. During the past 10 years, such certification has been obtained for only one product (although it is close to being obtained for another one). In a much shorter period Costa Rica obtained phytosanitary certification for three products. Costa Rica and New Zealand now export *uchuva* to the United States, although it is a product native to Colombia. Three years ago Argentina gained admissibility to the US market for fresh citrus, something Colombia has not achieved. Although the violence in rural areas may complicate the process of obtaining additional certifications, what is most needed is an appropriate institutional response in the phytosanitary area and public sector leadership in meeting certification standards.

2. Although these products have a comparative advantage, they lack a system of market intelligence that would facilitate the establishment of links between producers, exporters, and buyers in other countries.

3. Technological development is weak in many fruits and vegetables, reflecting the fact that the commitment to agricultural research in this area is insufficient. As has been pointed out for Latin America in general, "Research in tropical countries focused more on traditional food crops that offer fewer competitive advantages in national and international trade, neglecting products such as fruits and vegetables, where the region enjoys clear advantages" (Kondo, 2000).

4. There is insufficient awareness among producers of the importance of product quality in all senses: absence of chemical residues, adequate flavor, uniformity of appearance, and timeliness of delivery. Achieving the desired quality in some cases may require supplementary irrigation. However, consensus building among value chains has helped promote a greater awareness among all producers of the importance of product quality for future growth prospects.

5. The deficiencies in the area of product quality are in part a result of the weakness of producer organizations for fruit and vegetables. It is noteworthy that among all the parafiscal funds—which are funded jointly by producers and government and are used to support research and technology transfer, marketing, phytosanitary improvement, infrastructure development, and other forms of production support—the parafiscal fund for fruit and vegetables has experienced the highest rate of producer evasion of their responsibilities for making contributions to the fund, at 95%. In contrast, the evasion rate among producers is 0 for poultry, 4% for oil palm, 5% for cacao, 6% for beef cattle, and 20% for rice (Contraloría General de la República, Colombia, 2001, p. 15). In part, this situation is attributable to the spatial dispersion of producers of fruit and vegetables and the large number of intermediaries who participate in marketing those products and who are legally responsible for collection of the parafiscal fund contributions (set by law at 1% of the sales value in the case of fruit and vegetables).

6. In the past, a high priority has not been assigned in the country's international trade negotiations to opening up markets for Colombian exports in this area and to providing high-quality marketing assistance. Rather, emphasis has been placed on maintaining tariffs for import substitutes and there has not been a coordinated strategy for export of fruit and vegetables, or indeed an agroexport strategy in general.

Overcoming these drawbacks is a considerable challenge, but it is one that promises benefits throughout rural society if it can be fulfilled.

7.4 COFFEE, SUGAR, OIL PALM, AND CACAO

This category of crops includes two of the four largest agricultural exports, coffee and sugar, and one of the fastest growing exports, palm oil. Exports of raw cacao products have decreased sharply since 1994, but exports of chocolates have increased by a larger amount. All four crops are linked to important agroprocessing industries in Colombia. The sugar and oil palm subsectors also have been leaders in technological and institutional development in the past 10 years, and coffee has long had a strong institutional base. Palm oil production has been increasing, whereas cottonseed oil and soybean oil production have been declining.

Nominal tariff protection for palm oil, cacao, and sugar has fluctuated markedly over the past 15 years. In the case of palm oil, this is due in part to the functioning of the Andean price band system. Nominal protection was less than 5% in 1994 and 1995, jumped to about 20% in 1996, then fell to −30.5% in 1997, rose to about 2% in 1998, and since then has fluctuated in the range of 45−77%. Domestic prices have varied by smaller amounts since the price bands offset part of international price fluctuations.

The cacao product chain has tariff levels of 10%, 15%, and 20%, depending on the degree of processing of the product, but market forces often have depressed domestic price levels below what the tariffs would indicate. Nominal protection was negative in 1990−94 and 1996, and slightly positive in 1995, 1997, and 1998. Then it rose to about 23% and 35% in 1999 and 2000, respectively, before falling to negative levels again in 2001 and 2002. On average, it has had little or no nominal protection.

The sugar chain has tariff rates between 5% and 20%, but its tariffs often move out of that range because it belongs to the Andean price band system. Its domestic price is managed through a stabilization fund, taking as reference prices the level indicated by the price bands and the export prices. The net result is a highly protected subsector. In the 5 years 1998−2002, the nominal protection rate for white sugar ranged from 124% to 180%, and for raw sugar, it ranged from 68% to 148%. The only gaps in this pattern of protection for these crops are preferences negotiated in bilateral agreements for imports of vegetable oils and sugar from Mexico and Chile, and also for imports from countries of Latin American Integration Association.

Imports of coffee are trivial, and imports of sugar are less than 10% of sugar exports. For vegetable oils all taken together, including soybean oil, imports are about four times larger than exports, but the latter have been growing much faster since the early 1990s owing to the dynamism of the oil palm subsector (Table 7.8, showing the calculations of Eq. 3.1). In the case of cacao, exports have not increased but, as mentioned, their composition has shifted toward processed products and imports have increased rapidly, so the relative trade balance (RTB) has deteriorated (Table 7.9).

RCA coefficients for these products are presented in Table 7.10. The picture that emerges is a mixed one. First, obviously some of these products have a proportionately strong presence in export markets, coffee, and sugar in particular. Second, palm oils have rapidly growing exports and are becoming important in value. Third, most of the products with increasing RCA coefficients have a significant processing component. Even raw sugar passes through an industrial process before being marketed in that form.

TABLE 7.8 Colombia: RTB of Vegetable Oils (Trade in Millions of Dollars, Period Averages)

Item	1991/92	1993/95	1996/98	1999/2001
Imports	125,492	204,460	329,082	303,583
Exports	6,572	10,535	40,385	75,933
Trade balance	−118,920	−193,925	−288,697	−227,651
Total trade	132,064	214,995	369,468	379,516
RTB	−0.90	−0.90	−0.78	−0.60

RTB, relative trade balance.

TABLE 7.9 Colombia: RTB for The Cacao Chain (Trade in Millions of Dollars, Period Averages)

Item	1991/92	1993/95	1996/98	1999/2001
Imports	1,287	6,824	9,841	15,779
Exports	17,484	21,153	22,261	18,824
Trade balance	16,197	14,329	12,420	3,045
Total trade	18,772	27,977	32,102	34,603
RTB	0.86	0.51	0.39	0.09

RTB, relative trade balance.

TABLE 7.10 Colombia: Revealed Comparative Advantage for Other Long-Cycle Crops and Their Products (Export Values in Million US$)

Product	FAO Code	Tariff Code	Export Value, 2001	Product Chain	Revealed Comparative Advantage					
					1997	1998	1999	2000	2001
Raw sugar and *panela*	162	1701	150.745	Sugar	9.39	15.30	10.33	14.12	14.10
Sweets from sugar	168	1704	117.296	Sugar	7.80	10.84	12.33	10.87	13.82
Glucose syrup etc.	172	1702	5.729	Sugar	1.89	3.42	4.42	4.83	8.14
Palm oil	257	1511	26.228	Oilseeds, oils, fats	2.31	3.13	3.26	3.49	2.92
Palm kernel oil	1275	1513	5.860	Oilseeds, oils, fats	2.44	2.60	4.91	7.16	6.87
Margarine, edible oil mixtures	1242	1517	29.313	Oilseeds, oils, fats	1.43	1.07	6.41	17.29	16.41
Cardamom etc	702	0908	1.534	Others	4.15	1.32	1.83	3.90	3.27
Sugarcane	156	1212	0.604	Sugar	117.70	90.81	141.04	85.41	57.39
Refined sugar	164	1701	66.245	Sugar	8.75	8.87	7.31	7.47	6.27
Molasses	165	1703	2.970	Sugar	3.47	10.07	5.93	7.20	3.29

Cacao butter and oils	664	1804	5.519	Cacao	4.06	3.02	3.37	0.90	2.62
Green coffee	656	09011	768.573	Coffee	83.86	79.39	66.28	63.20	68.86
Coffee extracts	659	2101	89.587	Coffee	34.39	36.41	26.16	29.86	23.00
Palm kernels	257	12071	0.138	Oilseeds, oils, fats	23.47	17.41	7.24	2.84	14.30
Glucose, sugar syrups n.e.c.	167	1702	5.729	Sugar	0.02	0.16	0.49	0.63	0.51
Chocolate, other preparations	666	1806	21.582	Cacao	0.46	0.49	0.46	0.57	1.45
Cacao paste and powder	665	1805	4.701	Cacao	0.12	0.09	0.08	0.01	0.17
Thyme and bay leaves	723	0910	0.324	Others	0.32	0.32	0.44	0.55	0.49
Cacao paste	662	1803	0.410	Cacao	0.50	0.01	0.03	–	0.42
Raw cacao	661	1801	0.683	Cacao	0.23	0.23	0.07	0.11	0.17
Roasted coffee	657	09012	0.800	Coffee	0.84	0.34	0.18	0.15	0.31

FAO, Food and Agricultural Organization of the United Nations. The first row section of the table includes products with an RCA >1.0 (on average) and generally increasing; the second section, products with an RCA >1.0 and generally decreasing; the third, those with an RCA <1.0 and generally increasing; and the fourth, those with an RCA <1.0 and generally decreasing. The tariff codes are those applied in the Andean Common Market.

Exports of sweets from sugar and chocolate preparations are increasing quickly in importance. In the case of chocolate, they go principally to the Andean Community and the United States.

These results tend to confirm the hypotheses that Colombia's principal comparative advantage lies in long-cycle crops and agroindustrial products. Industrial efficiency requires an organizational ability in which Colombia excels when compared with most other countries of its income level. A major question that remains is the extent to which sugar would remain competitive if it did not enjoy the benefits of managed prices on the domestic market. The same question applies to palm oil in light of its high degree of border protection in recent years. It is a beneficiary of the protection policy that is applied, for example, to soybean oil.

Answers to these questions are sought through the LRC calculations presented in Table 7.11. It can be seen that all four categories of long-cycle crops do possess considerable LRC. Sugarcane, thanks to its very high yields in the Cauca Valley (among the highest in the world), and oil palm emerge as the

TABLE 7.11 Colombia: Long-Run Competitiveness Indicators for Other Long-Cycle Crops, 2001

Department	Cacao	Sugarcane	Palm Oil	Coffee
Valle del Cauca	0.64	0.55		
Santander	0.70			
Tolima	0.54			0.49
Magdalena			0.66	
Cesar			0.53	
Meta			0.49	
Nariño			0.59	
Bolívar			0.54	
Antioquia	0.39			0.64
Huila	1.01			
Quindío				0.68
Minimum value	0.39	0.55	0.49	0.49
Maximum value	1.01	0.55	0.66	0.68
National average	0.66	0.55	0.56	0.60

The national average for each crop is weighted by acreage in corresponding departments in that crop.

most competitive of the four crops by a small margin—in spite of the strength of Malaysia and Indonesia as palm oil producers. However, when the spatial dimension is taken into account, the most competitive localized crop turns out to be cacao in the northwestern Department of Antioquia. The only case of doubtful inherent competitiveness is cacao in Huila in the south, which is one of the larger producers of that crop.

For oil palm, Meta to the east is the region with greatest competitive advantage, although Cesar and Bolívar in the north are close behind, and for coffee it is Tolima.

All these crops are highly intensive in labor use and therefore have considerable capacity for creating employment. Data on their costs of production indicate that more than 30% of costs are represented by the employment of skilled and unskilled labor. In addition, employment is relatively stable and well remunerated in most of these crops over the medium term and therefore it is employment of high quality. The same is true of bananas and plantains. In sugarcane, oil palm, and coffee, producers also are well organized. These are crops grown by large numbers of smallholders (especially coffee, cacao and oil palm). The presence of smallholders in combination with the economies of scale in processing and marketing provides strong incentives for social cooperation and organization in the rural areas where the crops are dominant. Coffee is especially important in all these respects. In 1999, it created 406,794 direct jobs, and 95% of coffee producers held less than 5 ha of land. Therefore it is of overwhelming importance for rural society as well as for the national economy.

Coffee's LRC has slipped in recent years with the entry of producers such as Vietnam and the consequent decline in international coffee prices. This points to the need to improve productivity (control costs) and place greater emphasis on high-quality coffees, to differentiate Colombia's product in world markets. However, whereas some producers are going out of coffee, those that have been able to control costs remain competitive internationally. They tend to be the smaller producers. Before the coffee crisis, smallholders cultivated 445,000 ha of coffee, representing 45% of the total area, and after the crisis set in fully, they were cultivating 670,000 ha of coffee, which was 77% of the total (data from the coffee census put together by Jaime Forero). One of their strategies for economic survival has been to increase the interplanting of plantain with coffee.

The importance of quality cannot be sufficiently emphasized. Now 40% of coffee imported into the United States, by value, is specialty coffees (Comisión de Ajuste de la Institucionalidad Cafetera, 2002, p. 14). Although the quality of Brazilian coffee is improving, neither Brazil nor Vietnam is as well endowed as Colombia with natural conditions for producing high-quality coffee. However, when compared with some Central American countries, Colombia has been slow to improve its coffee grades and develop new, high-quality brands, although historically quality has been a strong selling point for

Colombian coffee. Thus the opportunities for Colombia in coffee are still there, but seizing them will require a concerted effort in research, extension, and marketing, all aimed at positioning Colombia at the high end of the evolving market.

In spite of cacao's LRC, its production and exports have declined abruptly in the past 3 years from levels recorded in the early and mid-1990s, and now chocolate producers import a share of their cacao requirements. Colombia is the world's ninth largest producer of cacao but only the 41st largest in its exports (Ministerio de Agricultura y Desarrollo Rural, Colombia, 2002). Smallholder cacao cultivation systems often include several other crops, including coffee, plantain, cassava, fruit, and trees grown for timber. The average area planted in cacao by these smallholders is about 3 ha. Perhaps in part because of the small scale of cacao farms and their geographical dispersion, an effective phytosanitary campaign has not been mounted for this crop, and plant diseases are a major reason for its lack of dynamism. Equally, it is weak in the area of agricultural research. In short, technical problems at the farm level need to be resolved before cacao can realize its apparently high potential. Santander is by far the largest producing area for cacao, and therefore it would be a logical place to concentrate an intensified research and extension effort for the crop, and to the extent that productivity gains could be achieved there, the program could be broadened to other areas. The competitiveness of cacao would imply that a strengthened technology development effort is warranted for this crop.

A favorable circumstance for palm oil is that world demand is expected to grow at least in the range of 3.3–3.9% annually for the next 20 years. These are projections of the palm oil federation FEDEPALMA, but it has been pointed out that they are conservative when compared with other international projections, especially when the possibilities of the incipient oleochemical industry (substituting vegetable oils for petroleum products) are taken into account (Misión Paz, 2001, p. 131). Palm oil is the tropical counterpart to the temperate soybean oil and can be directed to many of the same uses. Another advantage for the crop in Colombia is that, beyond areas presently planted in palm, there are literally millions of hectares of additional lands with appropriate soils for the crop.

For calculating the LRC of sugar, the reference price was not the free world market price, which is acknowledged to be distorted by subsidies in developed countries, but rather the average price on three markets: the free world market, the international quota export market, and the domestic market. Thus the reference price was 10¢ per lb. rather than 5¢ per lb., which often prevails on the free market. At the latter price, Colombian sugar would not be competitive, but neither would sugar from other exporting countries, because that price is the result of surplus production arising from subsidies in those countries.

Regarding domestic pricing arrangements and tariff protection for sugar, to a degree they are justified so that Colombia could be well positioned to eventually export on a wider world market if international subsidies were reduced. For palm oil, the LRC calculations indicate that the crop can be competitive on world markets without the support provided through tariffs and pricing arrangements. The issue is one of balancing growth stimulus to the sector with the interest of domestic consumer welfare. Additionally, given the apparent competitiveness of processed sugar products (and some livestock products, as shown below), more competitive domestic markets for sugar and palm oil could enhance Colombia's export prospects for products with a higher degree of processing content.

7.5 TUBERS, TOBACCO, AND COTTON

The Andes are the region of origin of the potato plant. Today potatoes are the 10th most important crop in Colombia in area planted and the 6th most important in value of production. They are the most important crop in the higher, colder regions of the country. Close to 90% of potato producers have less than 5 ha together they produce 35% of the potato harvest. Medium- and large-scale producers generally have substantially lower unit production costs because their yields are so much higher, although some of the smaller producers are much more productive than others.[7]

The potato processing industry is extensive; 70 processing facilities existed in 1997, although 15 medium-sized and large companies control 95% of the market for processed potatoes. Exports of potatoes and other tubers earned more than US$20 million in 2001. Potatoes and their products have import tariffs of 10% and 20%, respectively, with preferences granted to a number of Latin American countries.

Cassava is both consumed fresh and used as an animal feed, in milled form, and work is underway to exploit its potential for making starch and industrial glues. In human consumption it is a basic subsistence crop in some regions of the country and is also marketed through the complete range of retail outlets. Cassava is the tropical alternative to corn, in the same sense that palm oil is the tropical counterpart to soybean oil. Physiologically corn yields are favored by the long hours of summer sunlight at latitudes far from the equator, a factor that would make it difficult for tropical countries ever to compete effectively in world markets for corn. Cassava, however, prospers in the conditions of light at latitudes closer to the equator, and very high yields are obtained in countries like Thailand and even in southern Africa.

Two kinds of tobacco are grown in Colombia, light and dark. The tobacco product chain includes extensive processing facilities and a substantial export industry. However, in recent years tobacco imports have increased more rapidly than exports. At the farm level, tobacco is one of the crops most intensive in labor use, and it often is grown on very small holdings. In

Santander, Boyacá, North Santander, and the Atlantic Coast, the average size of tobacco farm is 1.5 ha or smaller.

Cotton occupied about 262,000 ha in 1991, but in 2001, it was sown on only 53,000 ha. Its production fell dramatically in 1994 and then dropped further in 1998–2000. Since then it has recovered very slightly. The nominal rate of protection was negative in 1989–91, rose to 21% and 41% in 1992 and 1993, respectively, fell to 5% in 1994, and remained at slightly positive or negative levels until 2001, when it rose abruptly to 22%, only to drop again to 3% in 2002. More than through the low average protection rates, its profitability has been undermined by falling world prices, the rising real value of the Colombia peso in the 1990s, and the rising costs of plant protection at the farm level.

The cotton value chain includes important subsectors producing cotton thread and yarn, fabric, and clothing. Colombia is a net importer of thread and yarn as well as fabric, but it also exports them and is a significant exporter of clothing. Denim and serge fabrics are among the larger exports of fabric. Exports of cotton products have more than compensated the fall in raw cotton exports, and total exports of all forms of raw and processed cotton fiber are now in the neighborhood of US$600 million per year, which makes the subsector's role in the balance of payments comparable to that of flowers.

The RTB coefficients for the potato, cotton, and tobacco chains are shown in Tables 7.12–7.14. On the surface, Colombia would appear to be losing competitiveness in all three chains but, as indicated, there have been important developments at the industrial end of all of them. This is another indication of Colombia's relative strength in agroprocessing industries.

On average, over this span of years, the RTB coefficients suggest that potatoes have the strongest export potential of the crops in this group. The more detailed information provided by the RCA coefficients in Table 7.15 confirms Colombia's growing potential in fresh and processed potatoes, of a

TABLE 7.12 Colombia: RTB for the Potato Chain (Trade in Millions of Dollars, Period Averages)

Item	1991/92	1993/95	1996/98	1999/2001
Imports	110	1,211	7,258	6,461
Exports	8,790	11,186	6,574	15,042
Trade balance	8,680	9,975	−684	8,581
Total trade	8,900	12,396	13,832	21,502
RTB	0.98	0.80	−0.05	0.40

RTB, relative trade balance.

TABLE 7.13 Colombia: RTB for the Cotton Chain (Trade in Millions of Dollars, Period Averages)

Item	1991/92	1993/95	1996/98	1999/2001
Imports	61,983	189,458	253,464	258,978
Exports	425,044	388,516	341,354	331,869
Trade balance	363,060	199,057	87,890	72,892
Total trade	487,027	577,974	594,817	590,847
RTB	0.75	0.34	0.15	0.12

RTB, relative trade balance.

TABLE 7.14 Colombia: RTB for the Tobacco Chain (Trade in Millions of Dollars, Period Averages)

Item	1991/92	1993/95	1996/98	1999/2001
Imports	4,952	6,045	26,811	49,847
Exports	51,799	28,155	24,725	33,093
Trade balance	46,847	22,110	−2,086	−16,753
Total trade	56,751	34,201	51,536	82,940
RTB	0.83	0.65	−0.04	−0.20

RTB, relative trade balance.

variety of types, and in processed tobacco. Leaf tobacco is a not-insignificant export too, but it appears to be losing ground in relative importance. This table does not reveal Colombia's strength in cotton textiles and other kinds of textiles because the coefficients were not calculated for subsectors considered to be part of industry, but the export values reported in Table 7.13 confirm the robustness of the textile sector, in spite of the decline of cotton cultivation.

It should be noted that cotton is not an especially employment-intensive crop. It ranks above grains in employment per hectare but below sugarcane and well below fruit, vegetables, and tobacco (Table 7.4). Therefore, the loss in employment owing to the decline of cotton cultivation can be made up through expansion of the area under other kinds of crops as well as through growth of the textile industry (based on imports of cotton and other materials).

The LRC calculations for these crops are shown in Table 7.16. These LRC results confirm the competitiveness, at the farm level, of potatoes and tobaccos in Colombia. Dark tobacco is the most competitive of these crops, especially

TABLE 7.15 Colombia: Revealed Comparative Advantage for Tubers, Tobacco, and Cotton (Export Values in Million US$)

Product	FAO Code	Tariff Code	Export Value, 2001	Product Chain	Revealed Comparative Advantage					
					1997	1998	1999	2000	2001	
Fresh and refrigerated potatoes	116	0701	8.604	Potatoes	1.19	0.76	1.43	5.10	2.91	
Other potato and root products	149	07149	11.963	Vegetables	5.29	8.46	47.7	53.66	68.74	
Cotton lint	770	14042	0.149[a]	Cotton	3.23	1.06	1.60	0.20	0.62	
Leaf tobacco	826	2401	12.608	Tobacco	1.64	1.69	1.84	1.53	1.04	
Prepared and preserved potatoes	475	20041	0.351	Potatoes	–	0.01	0.20	0.53	0.79	
Other preparations, preserved potatoes	472	20052	1.704	Potatoes	0.43	0.93	1.22	1.50	0.37	
Frozen potatoes	118	07101	0.480	Potatoes	0.06	0.01	0.04	0.08	0.20	
Cigarettes	828	2402	21.695	Tobacco	0.03	0.07	0.27	0.61	0.90	
Cotton fiber, uncarded, uncombed	767	5201	0.165	Cotton	0.06	0.03	0.02	0.01	0.01	

FAO, Food and Agriculture Organization of the United Nations. The first section of the table includes products with an RCA >1.0 (on average) and generally increasing; the second section, products with an RCA >1.0 and generally decreasing; the third, products with an RCA <1.0 and generally increasing; and the fourth, those with an RCA <1.0 and generally decreasing. The tariff codes are those applied in the Andean Common Market.
[a]Exports of cotton fibers, threads, yarn, and weavings (tariff codes 5201–5212) were US$33.010 million in 2001.

TABLE 7.16 Colombia: Long-Run Competitiveness Indicators for Tubers, Tobacco and Cotton, 2001

Department	Potatoes	Industrial Cassava	Dark Tobacco	Light Tobacco	Cotton
Valle del Cauca	0.55	2.07		0.33	0.67
Santander			0.27	0.44	
Tolima					1.08
Córdoba					0.91
Magdalena					1.69
Cesar		1.75			1.55
Meta		0.78			0.70
Nariño	0.49				
Sucre			0.21		1.01
Bolívar			0.31		
Boyacá	0.47			0.34	
Antioquia	0.53	2.53			
Huila				0.53	0.57
Norte de Santander	0.47			0.33	
Quindío	0.43	1.76			
Minimum value	0.43	0.78	0.21	0.33	0.57
Maximum value	0.55	2.53	0.31	0.53	1.69
National average	0.49	1.78	0.27	0.39	1.02

The national average for each crop is weighted by acreage in corresponding departments in that crop.

in Sucre and Santander. The competitiveness of these crops is almost as strong as that of fruit and vegetables and represents a solid base for their long-run development.

Potatoes also have a comparative advantage when produced by small-holders, as discussed later in this report, in spite of their generally lower yields. A major hurdle for potato growers is to reduce their dependency on agro-chemicals. Chemical residues on the crops are a source of concern from the viewpoints of consumer health, farmworker health, export prospects, and costs of production. They are a significant quality issue. It is widely known that potato growers regularly use excessive quantities of fungicides, herbicides, insecticides, and antibacterial compounds. In the case of the smallholders of the *Sabana de Bogotá*, these applications of chemicals represent around 31% of the cost of production. On larger farms, they represent about 27% of the cost of production (data from Tomás León). In the future, increasing numbers of export markets will be closed to products that do not meet food safety standards.

Potato exports also face phytosanitary barriers in the case of Ecuador and Peru, and Venezuela is the only significant export market for Colombian fresh potatoes (they go to processing plants in Venezuela). Colombia's National Potato Council currently is working on a project to market French fries to the Andean Community, Panamá, and Trinidad and Tobago. The aim is to export a high-quality product, since it is thought that Colombia cannot compete with some other countries in lower quality potatoes. Overcoming the phytosanitary barriers to fresh potato exports to Ecuador and Peru will require greater pro-ducer organization and close coordination with research and extension agencies.

Viewed in terms of competitiveness, cotton presents mixed results. It demonstrates a competitive advantage in four departments (Córdoba, Valle del Cauca, Huila, and Meta), although only marginally so in Cordóba. It has a competitive disadvantage in Cesar, Sucre, Magdalena, and Tolima, although it is close to being competitive in the first two of these departments. These regional differences spring from differences in cost structures, in particular the costs of pest control. Those costs in turn are related to crop management and rotation practices. In Huila, cotton rotates with sorghum, rice, and corn. Also, the least competitive regions have experienced degradation of soils over the long term through inadequate soil management. Tolima in particular has suffered compaction and salinization of soils in cotton-growing areas, and Cesar also has suffered degradation of its soils.

A key to cotton's success in Meta has been addition of nutrients to the soils, leveling of the fields, and provision of adequate drainage. In Meta, only 22% of the costs of cotton are devoted to pest control, whereas the corresponding figure in Cesar is 32%. Traditionally, the Atlantic coast zone has experienced the greatest problems with pest infestation in cotton, including *Anthonomus grandis*, the pink worm *Heliothis virescens*, and the borer *Spodoptera frugiperda*. These pests have not tended to appear in the *Altillanura Oriental* because cotton is grown there in dispersed form, along river floodplains. However, it is important to be vigilant because in 2001 *A. grandis* was detected in Meta. The Ministry of Agriculture and its local agencies have reported significant increases in cotton yields and profitability through the use of organic fertilizers and by plowing with a vibrating "chisel" instead of discs. Cotton is another crop that requires change in techniques of cultivation to be competitive and sustainable over the long run. Given its susceptibility to pests, it also is a crop that clearly can benefit from biotechnology, provided the appropriate institutional structures for biotechnology research and controls can be put into place. Worldwide, cotton is one of the four crops that have received most of the attention in transgenetic research to date (the other three being corn, canola, and soybeans), and this provides grounds for optimism regarding the potential for that kind of research to improve the competitiveness of Colombian cotton.

Cotton's profitability undoubtedly has been affected by subsidies in industrialized nations. To test this hypothesis, the LRCs for cotton were recalculated with a 15% increase in international prices. The results are shown in Table 7.17.

It can be seen that if it were possible to consistently offset the effects of international subsidies on prices (say through tariff policy) cotton would become viable in more regions of the country, essentially everywhere except in Magdalena and Cesar. However, in Sucre and Tolima the competitive advantage would not be very robust and profitability of the crop would be threatened by fluctuations in the real exchange rate over time. If more border protection were provided to cotton, it would be necessary to ask about the consequences for the textile industry, which would pay a higher price for a basic raw material.

Cassava has a competitive disadvantage everywhere except in Meta. When sensitivity analysis was done for cassava as well, with an assumed 20% increase in international prices, the crop still remained noncompetitive, except again in Meta. Nevertheless, cassava is the one crop with a competitive disadvantage that may have more potential over the long run and it warrants closer attention for this reason. The basis for its potential is that its productivity ceiling is very high in relation to actual performance, higher than for any other crop. Current national average yields are about 10 t per ha, yet a number of producers, in a wide area from the Department of Atlántico to the Cauca Valley, are obtaining 35 t or

TABLE 7.17 Colombia: Modified Long-Run Competitiveness Indicators for Cotton, 2001 (With 15% Increase in International Prices)

Valle	Tolima	Córdoba	Magdalena	Cesar	Meta	Sucre	Huila
0.56	0.86	0.72	1.05	1.10	0.58	0.83	0.47

more per ha. A group of producers in South Africa is regularly obtaining yields of about 50 t/ha, and in Colombia in the Department of Córdoba, the clonal variety SM 1433-4 has yielded 84 t/ha on a plot of 10 ha.[8] Although the average yields in Thailand are only 15 t/ha, it is considered that the practical genetic ceiling is in the range of 80 t/ha.

Although cassava is an important component of the diet of many families, especially in rural areas along the coasts, its most productive uses appear to be in animal feeds and in diverse industrial applications, especially as starch and glue. This flexibility permits a direct calculation of the parameters necessary for it to be competitive with corn. The technical equivalence between the two crops is 0.7: a ton of cassava flour is equivalent to 0.7 t of corn. This means that if the reference price of corn were about US$110/t, then the equivalent price of fresh cassava would have to be no more than US$34/t.[9] At that price, a combination of cost and yield that would give cassava a comparative advantage would be a cost of 2.6 million Colombian pesos (at prices of 2002) and a yield of 37 t/ha. That combination and even better results have been attained in various regions of the country, according, for example, to reports of the International Center for Tropical Agriculture (*Centro Internacional de Agricultura Tropical,* CIAT) about the Atlantic Coast and of the Corporación Colombia Internacional (CCI) about the zone of San Juan de Arama in Meta.

In spite of cassava's competitive disadvantage, Colombia already is its world's third largest producer, mainly for supplying the domestic market. The country's favorable conditions for cassava in the long run are (1) the international research center CIAT has the greatest knowledge of cassava and the largest gene bank for the crop in the world; (2) cassava is readily adaptable to a variety of soils, including poor and acid soils; and (3) Colombia already has advanced relatively far down the road of developing and testing commercial varieties and mechanization of some cultivation tasks (Misión Paz, 2001, p. 144). The adaptability of cassava to acid soils is one reason for the interest in promoting its trials, in rotation with corn and/ or sorghum or soybeans, in the *Altillanura.*

Cassava's main hurdles include susceptibility to pests (especially the white fly); the high water content of the root (65%), which makes it expensive to transport; and the lack of processing facilities that would purchase the crop on a significant scale. The problem of the white fly has been managed successfully in the northern part of the Cauca Valley through techniques of integrated pest management (CLAYUCA n.d.:2). The water content problem is being addressed through the promotion of small-scale, inexpensive drying of plants in rural areas. The question of developing an industrial market on a substantial scale is more difficult to resolve, because it requires a committed investor who is convinced that the farm-level problems can be resolved. However, the banana enterprises are experimenting with a cassava processing plant in the Urabá that would produce glue for banana boxes, and interest has been shown in another plant that would produce cassava fries to compete with French fries. A seed capital fund for related areas could be a useful catalyst.

In the case of cassava, trying to change relative prices through trade policy does not appear to be the answer, but technological development could be. From the viewpoint of public policy, more rapid exploration and development of cassava's potential has been held back by the lack of commitment of sufficient research resources to the crop. Although the outcome of such a commitment cannot be predicted with certainty, all indications are that this is a promising area.

7.6 LIVESTOCK PRODUCTS

Livestock products reviewed in this study include poultry and eggs, pork, and beef and milk. Over the past 20 years, poultry and milk have been the fastest growing products by a wide margin. Poultry production increased by almost two and a half times from 1982 to 2001. Milk production almost tripled from 1979 to 2001, and pork production increased by about 10% between 1991 and 2000. Beef production fluctuated around the same level from 1982 to 1997 (of slaughter of 3—3.8 million head per year) and then dropped abruptly in 1998—2001 to about 2 million head. By carcass weight, the domestic availability of beef was 24% lower in 2001 than in 1982. It is possible that the decline in cattle operations is attributable to the rural violence. Cattle ranchers report a significant higher frequency of problems of physical insecurity than crop farmers do, and many of those who stay in the business manage to do so by paying "protection" that amounts to about 10% of the value of their output. Although beef is no longer a dynamic product, its volume of production remains high. Colombia is the ninth largest producer of beef in the world and the largest in the Andean region.

Corresponding to production trends, beef exports have trended downward in the past 10 years, although with significant fluctuations, but exports of dairy products have grown very rapidly, especially in the past few years (Tables 7.18 and 7.19). A small amount of beef exports go to Venezuela; at present they are prohibited from entering the US market because of foot and mouth disease, but there are hopes that parts of Colombia can be declared free of this disease in the foreseeable future. Dairy exports are almost entirely milk, and their principal markets are Ecuador and Venezuela. Because of the presence of international subsidies on milk and the efficiency of dairy producers such as New Zealand and Denmark, it would not be realistic for Colombia to aspire to export dairy to the wider international market. However, continuing to export to the Andean region, in the framework of the Andean Community, appears to be viable. The RTB value for dairy products indicates that Colombia has been a net importer but that this products' trade balance is moving toward net exports.

In poultry subsector, eggs are the principal export, on a relatively small scale. Poultry trade is hampered by the lack of sanitary clearance for poultry meat to enter the United States and by Colombia's controls over import of chicken parts.

TABLE 7.18 Colombia: RTB for Beef and Beef Products (Trade in Millions of Dollars, Period Averages)

Item	1991/92	1993/95	1996/98	1999/2001
Imports	621	11,722	14,071	5,546
Exports	51,792	8,288	28,884	10,359
Trade balance	51,171	−3,434	14,814	4,813
Total trade	52,413	20,010	42,955	15,905
RTB	0.98	−0.17	0.34	0.30

RTB, relative trade balance.

TABLE 7.19 Colombia: RTB for Dairy Products (Trade in Millions of Dollars, Period Average)

Item	1991/92	1993/95	1996/98	1999/2001
Imports	12,757	20,984	52,503	43,203
Exports	1,015	4,018	9,653	42,795
Trade balance	−11,742	−16,965	−42,850	−408
Total trade	13,772	25,002	62,157	85,998
RTB	−0.85	−0.68	−0.69	0.00

RTB, relative trade balance.

The study of Misión Paz indicates that extensive beef production is more efficient than fattening in feedlots, because of lower costs and fewer problems for the environment and for animal health. It is suggested that a few basic improvements in managing pastures and herds, such as incorporation of legumes in pastures, better attention to animal health, and use of supplemental feeds in the dry season, can increase the productivity of extensive beef operations significantly. In the same spirit, it is also recommended that installation of cooling tanks for milk can permit double milking each day, increasing milk yields significantly, and to date only 35% of the country's milk is produced in this way (Misión Paz, 2001, pp. 150–152). Thus the technological frontier for beef and milk production could be expanded with proven techniques of management. Milk plays a role in incomes of smallholders. It is estimated that 70% of dairy farmers produce less than 70 liters of milk per day (Jaime Forero).

The LRC coefficients, based on Eq. (5.8) in this case (Table 7.20A), show LRC for all livestock products in Colombia. It is strongest for poultry and

TABLE 7.20A Colombia: Long-Run Competitiveness Indicators for Livestock Products, 2001

Department	Poultry	Milk	Beef Fattening	Eggs	Pork Fattening	Pork, Complete Cycle
Valle del Cauca	0.51	0.70	0.91	0.52	0.75	0.65
Santander	0.52			0.34		0.81
Antioquia	0.42	0.69	0.74	0.56	0.84	0.88
Quindío	0.62	0.64		0.47	0.96	
Cundinamarca	0.54			0.31		0.72
Minimum value	0.42	0.64	0.74	0.31	0.75	0.65
Maximum value	0.62	0.70	0.91	0.56	0.96	0.88
National average	0.52	0.68	0.82	0.44	0.85	0.77

The national average for each product is weighted by production value in corresponding departments in that crop. (there are inconsistencies in the data for pork.)

eggs, followed by milk. However, when the calculations are made with prevailing rates of tariff protection on imports of corn, sorghum, and soya included in the border prices, pork loses its competitiveness and poultry's competitiveness becomes only marginal. Milk's competitiveness also diminishes somewhat. Thus the protection policy for feedgrains is an important factor in determining the international competitiveness of livestock products, especially pork and poultry.

In Colombia, poultry and pork production differ significantly in their scale of operations and the degree of vertical integration of the enterprises. Unit costs decline markedly as the scale increases in these industries, because of (1) technologies that are viable only on a large scale, (2) greater efficiency in marketing and purchasing of inputs, and (3) the possibilities for vertical integration in production of feed mixes. The difference in competitiveness between the poultry and pork subsectors may be explained by their differences in scale of operations, poultry being more advanced in that regard. A characteristic of pork production is that it requires fairly large expanses of territory that is lightly populated because of the need to dispose of large quantities of waste. In principle, Colombia possesses such areas in abundance, but to date they have been little exploited for purposes of large-scale pork operations.

Concerns have been raised about the competitiveness of Colombia's poultry industry in the face of competition from Brazil. However, two factors attenuate this competition. One is the high cost of shipping poultry from Brazilian ports to Colombia. The other has been the recent Brazilian currency revaluation. The LRCs reported for poultry in Table 7.20A refer to comparative advantage vis-à-vis the US market. To address the question about Brazil, the LRCs were recalculated with Brazilian prices as of June 2003. The results of those calculations are shown in Table 7.20B. It can be seen that poultry has a marginally lower competitive advantage when confronted with competition from Brazil, but it still retains an advantage.

In the case of cattle, analysis of 17 representative livestock operations in various regions of the country revealed that the competitive advantage of this subsector is strongest for the most intensive operations, contrary to the conclusion of Misión Paz. In fact, for extensive cattle raising, the LRC indicator is approximately unity, which indicates that its competitive advantage is nil. Because many of these ranchers do not take into account the opportunity cost of their land, nor of the financial capital they put into the operation, they perceive a profit in cashflow terms (financial terms) when in fact they may be losing in economic terms, that is, when compared with alternative forms of investing the resources. However, the competitive advantage for intensive operations, both for beef and for dual-purpose operations of beef plus milk, is real and substantial.

The analysis for all livestock products provides another confirmation of Colombia's comparative advantage in subsectors that embody more product processing. Because of particular structural barriers in world markets—the issue of sanitary certification for beef and poultry, plus the heavy international subsidies on milk—Colombia's competitive advantage in these products may not have an opportunity to be expressed through more exports in the near future. However, should the sanitary hurdles be overcome, the potential for export growth in these products is very considerable. In the meantime, until better international arrangements can be negotiated, the distorted nature of international trade in poultry products would appear to justify continuation of a significant import tariff on chicken parts.

TABLE 7.20B Colombia: Long-Run Competitiveness Indicators for Poultry vis-à-vis Brazil

Department	Valle del Cauca	Santander	Antioquia	Quindío	Cundinamarca
LRC value	0.72	0.68	0.57	0.83	0.75

7.7 FORESTRY

The forestry sector includes products that range from logs to boards, plywood, and wood manufactures. It is estimated that in Colombia about 25 million ha of land are appropriate for forestry (Misión Paz, 2001, p. 120). Currently, the country has about 8 million ha of natural forests and forest plantations, of which only a little over 100,000 ha are plantations. The timber industry is poorly regulated with a significant component of illegal, unregistered activity. Frequently the wood felled in Colombia is characterized by low yields and poor quality, and it often is not delivered in a timely fashion to sawmills. In spite of these deficiencies, very recently exports of wood products have begun to increase at an accelerated pace (Table 7.21). In species like pine and eucalyptus, the natural growth of wood in cubic meters per year is higher than in major exporting countries such as the United States, Argentina, and Chile, so Colombia possesses physical basis for competitiveness in the sector.

In international markets, wood products have been increasing at a substantial pace for the past 20 years, and it is expected that growth will continue to be strong. In South America, Brazil and Chile already are large-scale exporters of forest products, and Argentina, Uruguay, and Paraguay are increasing their efforts in this area. Among the benefits of sustainable forestry are considerably greater employment per hectare than in cattle raising and environmental benefits through control of soil erosion and protection of watersheds. Watershed benefits, normally considered to be economic externalities, can be internalized in financial terms in the forestry sector. In Honduras, a county (*municipio*) on the Caribbean coast pays *campesinos* in the upper reaches of a watershed for forest protection services, and an international project has been effecting compensation to *campesinos* for the same purpose in the watershed of the Cajón dam. In Panama, a pilot

TABLE 7.21 Colombia: RTB for Forestry (Trade in Millions of Dollars, Period Averages)

Item	1991/92	1993/95	1996/98	1999/2001
Imports	7,382	28,219	28,336	21,932
Exports	13,360	9,808	13,470	26,658
Trade balance	5,979	−18,411	−14,867	3726
Total trade	20,742	38,028	41,806	47,590
RTB	0.20	−0.48	−0.36	0.08

RTB, relative trade balance.

project of the Inter-American Development Bank pays *campesinos* in Darien US $250–US$300 annually for specified tasks in forest protection, with provision for quarterly monitoring of their fulfillment of duties (Misión Paz, 2001, p. 169). In Costa Rica there are even more examples of markets for environmental services.

Forests can also provide the setting for the provision of environmental services that have a market value, including sequestration of greenhouse gases and bioprospecting. As of mid-1999, 46 countries had projects for the capture of greenhouse gases, of which 14 are in Latin America, including Colombia. The potential market for the findings of bioprospecting is also very large; international sales of medicinal plants amounts to about US$14 billion. Costa Rica has been a leader in both carbon capture and organized bioprospecting. Therefore, there is a wide variety of potential economic benefits from adequate forest management, but Colombia has progressed relatively slowly in this area. Increasingly, success in exporting forest products will require certification of sustainable forest management, and that in turn requires appropriate institutional development to support the sector.

For the present study, LRC calculations were made for 12 forest species using Eq. (5.8), with the results shown in Table 7.22. Most of the species analyzed have a clear competitive advantage, especially so in broad-leaf species. Economic analysis thus points to strengths in the forestry area, provided that guarantees can be provided that the forest exploitation is sustainable. In light of these results, it would be worth rethinking the incentives for reforestation and also for proper management of natural forests. At the moment the fiscal incentive for reforestation is tied to the availability of funding to back it, and there have been reports of viable reforestation projects being put on hold in the expectation that eventually sufficient funding will become available to allow the project to access the incentive provision.

TABLE 7.22 Colombia: Long-Run Competitiveness Indicators for Forest Species

Specie	LRC	Specie	LRC
Ceiba	0.44	Caribbean pine	0.72
Oak	0.47	*Eucalyptus tereticornis*	0.82
Gmelina arbórea	0.48	*Pino oocarpa*	0.84
Alder	0.55	*Pino pátula*	0.90
Teak	0.55	Cypress	0.90
Walnut	0.58	*Eucalyptus grandis*	1.39

7.8 EXPORTS, COMPETITIVENESS, AND RISK

Exporting represents an important avenue for increasing production and income. This is especially true for agriculture. By Engel's Law, aggregate domestic demand for food products will grow by only about two-thirds of the growth of incomes. Therefore accelerated growth of production can occur only through exports or import substitution. In the case of Colombia, it has been found that the principal crops that compete with imports do not have a competitive advantage, so expansion of their output would have to come at the expense of higher domestic prices to consumers or greater fiscal outlays, or both. On the other hand, most export crops do have a long-run competitive advantage, and therefore increases in their output and sales can occur without fiscal incentives.

Exports also play a role in increasing the quality of domestic production, since markets in industrialized countries tend to be more demanding of product quality than the domestic market in developing countries. Export markets transmit information on standards of product quality in the various dimensions mentioned earlier in this book. The need to keep abreast of quality requirements in diverse markets is one reason, along with the need to improve technologies of production, why export agriculture is increasingly a knowledge-intensive industry. Largely because of the role of exports, during the late 1980s and early 1990s, agriculture was the fastest growing sector in Chile and the sector that created the most new jobs in scientific, technical, managerial, and administrative areas. In order for Colombia to reap full benefits from its comparative advantage in export crops, its agriculture will have to undergo a similar transformation in acquisition of skills and knowledge.

Table 7.23 shows that Colombia has a solid agricultural export sector that has been able to increase the volume and value of its exports in spite of the collapse of world coffee prices[10] and declines in some other agricultural prices. Exporting, of course, brings with it just this risk, of fluctuating or falling prices on world markets. Risk can be attenuated in part through use of fixed price or futures contracts, but such arrangements are not available for very many agricultural products. Another response over the longer run is to continually increase productivity, so that profit margins are not affected as much by reductions in prices. This is being done already in some exporting sectors such as oil palm.

A third response to price risk on international markets is diversification of the basket of export goods. At the farmer level, especially for smallholders, diversification is very important, especially to strengthen the resilience in light of the increased climate variability associated with global climate change. This diversification can include apiculture, small ruminants, and other farm animals as well as crops for home consumption and the domestic market. It also is an avenue for creating income sources in the months between harvests of major

TABLE 7.23 Colombian Agricultural and Agroindustrial Exports

Year	All Agriculture			Coffee			Agriculture Without Coffee		
	Value	Volume (mt)	Implicit Price	Value	Volume (mt)	Implicit Price	Value	Volume (mt)	Implicit Price
1991	2,725,273	2,931,527	930	1,336,506	734,048	1,821	1,388,767	2,197,479	632
1992	2,735,496	3,411,010	802	1,261,280	968,243	1,303	1,474,216	2,442,767	604
1993	2,568,930	3,570,342	720	1,152,339	788,036	1,462	1,416,591	2,782,306	509
1994	3,738,380	3,738,679	1.000	1,998,664	674,969	2,961	1,739,716	3,063,710	568
1995	3,696,077	3,277,315	1.128	1,840,985	559,630	3,290	1,855,092	2,717,685	683
1996	3,431,972	3,346,211	1.026	1,581,294	601,024	2,631	1,850,678	2,745,187	674
1997	4,288,902	3,672,325	1.168	2,264,670	618,216	3,663	2,024,232	3,054,109	663
1998	4,044,727	4,053,068	998	1,895,530	637,197	2,975	2,149,197	3,415,871	629
1999	3,409,659	4,347,583	784	1,348,537	568,697	2,371	2,061,122	3,778,886	545
2000	3,151,651	4,356,743	723	1,069,823	508,619	2,103	2,081,828	3,848,124	541
2001	2,915,900	4,021,635	725	769,386	560,248	1,373	2,146,514	3,461,387	620
2002	2,947,726	4,448,738	663	782,180	579,083	1,351	2,165,546	3,869,655	560
Annual Growth Rates (%):									
1991/02	1.0	3.1	-2.1	-4.3	-3.9	-0.5	4.3	4.7	-0.4
2000/02	-3.3	1.0	-4.4	-15.7	6.5	-22.1	2.0	0.3	1.7

Agricultural exports include agroindustrial exports. Values are in thousand US$.

crops such as coffee and fruit trees. At the enterprise level, FEDEPALMA provides an example in marketing palm kernel oil as well as palm oil itself, and in working to develop new products based on processing of African palm. At the national level promoting continuing diversification should be an integral part of an agricultural export strategy. It implies providing temporary support and/or incentives for small products as well as large ones, encouraging export of a wider variety of fruits and vegetables and nuts both fresh and processed, organic products, and other specialized products. Since these products tend to be highly labor intensive, a national diversification strategy goes hand in hand with employment creation and poverty reduction.

The degree of diversification of Colombia's agricultural exports over the past 10 years is impressive. The number of products (as defined by eight-digit customs codes) with an export value of more than US$5 million was 32 in 1991. In 2001, this number was 64 (excluding flowers and counting bananas and plantains as one item). Virtually all the increased diversification has arisen in the agroindustrial product lines, with very few exceptions such as eggs and root crops. On the one hand, this is another indication of Colombia's relative strength in agricultural processing, but on the other hand, it is another confirmation that the country is not exploiting fully its competitive advantage in fruits and vegetables—and its probable competitive advantage in some nuts and spices.

The other side of diversification concerns markets for the destination of export products. Here also the trend has been toward less concentration. Traditionally, Colombia's principal markets have been the United States, the EU 15, Venezuela, Central America and the Caribbean, and Peru. In the past 10 years, however, although the US share of Colombia's exports has grown slightly, the exports have expanded rapidly to countries such as Chile, Mexico, Ecuador, and Canada. In 1991, only five destinations each accounted for more than 1% of Colombia's total agricultural and agroindustrial exports. In 2002, nine countries each received more than 1% of Colombia's exports. Between 1991 and 2002, exports to seven countries of destination increased by more than 10% per year. This is a healthy trend for the long run.

7.9 POLICY IMPLICATIONS OF COMPETITIVENESS FINDINGS IN COLOMBIA

7.9.1 Overall Findings

In the original study a number of policy comments were made in relation to specific products in the foregoing, and the study followed the product analyses with an extended policy chapter covering a wide range of issues. This section presents a few excerpts from that chapter. Some of the institutional details are particular to Colombia and some Colombian policies have changed since the time of this case study, but the issues addressed have emerged to one degree or another in many countries, so the discussion here may provide helpful perspectives. The original report addressed other topics such as policies for trade, land tenure, irrigation, agricultural extension, agricultural finance, exchange rate policy, and other aspects of fiscal policy in addition to those discussed here.

The general conclusion of the foregoing analysis is that Colombia has an inherent competitive advantage, at times very strong, in the following categories of production, some of which overlap: (1) perennial crops, as opposed to annual crops; (2) export products, as opposed to import substitution crops; (3) livestock products; and (4) many agroindustrial products. These products offer Colombian agriculture the greatest possibilities for growth of income and employment over the medium and long term. Almost all of them, with the exception of extensive livestock operations, generate more employment per hectare than the products without competitive advantage, and they have the best perspectives for future expansion of output levels.

Each product or subsector faces its own set of issues to be resolved to realize its potential for growth. Some of them require better and stronger strategies for pest management, some products require an intensification of the agricultural research effort, some require a strengthening of producer organization, some require better marketing information and strategies, and so forth. However, all products with competitive advantage have clear potential for substantial growth. Smallholder producers are very much involved in quite a few of them—in crops such as coffee, oil palm, cacao, bananas, plantains, other fruit, and potatoes—so these products offer significant opportunities for low-income rural families to improve their standard of living.

Therefore realizing the sector's growth potential is important both for poverty alleviation in a sustainable way—more than just alleviating the short-run symptoms of poverty—and for improving incomes throughout the economy.

The results of other parts of this study and other investigations support the following conclusions about net incentives provided through fiscal and trade policies: (1) policies prevailing for the past decade and more do not support the products with greatest long-run competitive advantage, and therefore they do not represent the best strategy for promoting sector growth; (2) as a corollary of the first conclusion, net support (or net effective protection) is very unequal among products, which in itself discourages efficiency in resource allocation and therefore undermines growth prospects; (3) the country's agroexport strategy is only partially developed; (4) pricing policies for sugar and palm oil are special cases in light of their market structures and they warrant close review; and (5) policy incentives mostly go to the larger producers, and very little if at all to *campesino* producers.

7.9.2 Sector-Wide Effects of Policy Shifts Toward Competitive Products

A numerical simulation has been developed to test the importance of these biases in policy incentives. Utilizing information on approximate supply response elasticities of major crops, plus the information on costs and employment that was compiled for the competitive advantage analysis, a scenario was created to project the approximate effects of hypothetical changes in policy incentives. The scenario, covering the noncoffee cropping sector, consists of assuming that trade and fiscal policies are modified in the direction of a 10% increase in the prices of the products with a competitive advantage, *relative* to the crops without a competitive advantage.

In reality, given the prevailing orders of magnitude of economic protection for some crops in Colombia, this scenario would represent a modest change—a 5% decrease in the prices of the highly protected crops and a 5% increase in the prices of those that display competitive advantage. Nevertheless, a shift of this nature would result in significant benefits for the sector in terms of increased income and employment. As Table 7.24 shows, the number of jobs in the sector would increase by about 40,000, and sectoral income would increase by 3.6%, in spite of the decline for grains and oilseeds.

While corn, rice, sorghum, wheat, and barley would decrease in acreage and production, and therefore in the income they generate, annual crops with competitive advantage such as potatoes and cotton would experience increases. The largest increase would be experienced by perennial crops, above all the exportable perennials.

If Colombia's tariff regime were obliged to undergo a restructuring through negotiations with the World Trade Organization, which is quite possible, then the percentage changes in relative incentives could be much greater than shown in this scenario. Only a 20% change in *relative* incentives between the two classes of crops would create about 80,000 new direct jobs and raise sector income by about 7.2%. These effects are of course approximations, but given that actual data on Colombian agriculture are used, they are unlikely to be substantially in error. The simulated increase in production of crops does take into account market limitations, in the form of downward-sloping demand curves.

Changes in incentives policies also would affect the livestock sector. As noted earlier, tariffs on imported corn and soybeans influence the degree of competitive advantage of livestock products, particularly poultry and pork. If those tariffs were reduced, costs and prices of livestock production would fall and meat consumption would increase. This kind of effect can be predicted with some confidence, at least in approximate magnitudes, because these products have very little participation in international trade, and therefore prices do vary according to supply and demand considerations in the domestic market within a reasonable range of values.

Table 7.25 shows the approximate percentage effects on these livestock products of a hypothetical reduction of 10% in the tariffs on corn and soybeans. Not only would the livestock production sector gain but also consumers would benefit. They would pay lower prices and accordingly would tend to consume more of these products, an effect which I have taken into account in the calculations.

TABLE 7.24 Colombia: Effects on Employment and Income of a Shift in Incentives Toward Products with Competitive Advantage (Shift of 10% in Relative Prices toward Crops with Competitive Advantage)

Product Group	Production, 2001 (000 mt)	Changes in: Production (000 mt)	Area Cultivated (000 ha)	Employment (000 d)	Jobs	Gross Income Million pesos[a]
Annual crops	8,222.1	54.1	−97.2	420.0	2,386.6	1,576.5
Grains	3,799.2	−413.5	−123.4	−3,667.9	−20,840.6	−86,590.3
Oilseeds	309.5	−3.3	−6.8	−95.9	−544.7	−7,560.1
Others	4,113.5	471.5	33.0	4,183.9	23,771.9	95,726.9
Perennials excluding coffee	14,035.0	526.9	44.5	6,601.3	37,507.1	174,121.6
Exportables	4,533.1	459.2	47.0	5,487.8	31,180.5	128,697.6
Others	9,501.9	67.7	−2.5	1,113.5	6,326.6	45,423.9
Agriculture except coffee	22,257.1	581.1	−52.7	7,021.3	39,893.8	175,698.1
Percentage change		2.6%			2.5%	3.6%

[a]In millions of constant 1994 pesos.

TABLE 7.25 Colombia: Effects on the Livestock Sector of Variations in Tariffs on Corn and Soybeans (Simulated Effects of a 10% Decrease in Those Tariffs)

Product	Percentage Effect on:			% Share in Production Cost		Price Elasticities Used	
	Cost	Price	Consumption	Corn	Soybeans	Supply	Demand
Poultry	−4.37	−1.86	3.14	44.8	20.3	1.25	−1.69
Eggs	−4.14	−2.43	1.92	43.0	18.6	1.12	−0.79
Pork	−4.39	−1.49	2.54	44.1	21.2	0.88	−1.71
Milk	−0.63	−0.26	0.26	5.7	3.7	0.69	−0.98

Given the considerable benefits to the nation that would arise from a change in incentives policies along the lines suggested, logically questions arise about how a transition to a new policy framework could be carried out without causing undue disruption and economic harm to some groups of producers in the short run. If it were desired to compensate temporarily those producers who were most affected by the transition, recourse would have to be made to fiscal channels, such as mechanisms of payments per hectare like those used in Europe and Mexico.

But the policy framework should make it clear that the support is transitory with a specified time limit and is intended to encourage changes in cropping patterns.

7.9.3 Export and Pricing Policies

An integral agricultural export policy would consist of much more than trade negotiations (which the original study discussed extensively). It would contain the following elements:

1. An aggressive program to ensure that the country's exports satisfy food safety requirements in importing countries. This issue will become very important in the future when new, tough restrictions on food safety for imports go into effect in the European Union. Such a program would require extensive training of producers, laboratories to certify compliance with food safety standards, international agreements on certification procedures, and enhanced research into integrated pest management techniques to enable producers to reduce their reliance on agrochemicals, the residues of which are a major concern for food safety.
2. An accelerated effort in the area of obtaining certification of freedom from plant diseases, along the lines of what Costa Rica has accomplished for its fresh fruit exports.
3. Strengthened international marketing assistance: first, through more effective consular support abroad (selection of attaché personnel on the

basis of proven technical expertise in agricultural marketing); second, through fiscal support for producer participation in trade fairs and for trial shipments of new products and to new markets; and third, through an extensive, up-to-date market information system that reaches all areas of the country in real time.

4. Support for strengthening producer associations in fruit and vegetables, where they remain weak when compared with the associations for some other products.

5. Strengthened agricultural research for export products with significant potential, especially for fruit, vegetables, cassava, and cacao.

6. Effective, if modest and transitional, fiscal incentives for export products, such as funding for market exploration and for contracting with specialized expertise in production and product handling.

Some of the elements of a wider agroexport strategy are already in place, but when compared with other Latin American countries the effort has not yet been strong. Colombian producers who have attempted to export new products complain of lack of market information and support from the public sector. In terms of present institutions in the sector, the CCI performs some of these functions but is not sufficiently funded to carry out its responsibilities on the scale required. Nor is the definition of its mission entirely clear and consistent. The export promotion agency also participates in market studies, and through its agreement with the national science agency Colciencias it implements programs for management training for the export sector, for obtaining international certification, and for research on improved technologies. The bank Bancoldex provides some export credits. In recognition of the vital importance of product quality for success in exporting, a national program for quality assurance provides incentives to enterprises to obtain certification according to the norms of ISO 9000, ISO 14000, QS 9000, and the seal of Good Manufacturing Practices HACCP.

In the phytosanitary area, the ICA has well-qualified staff but is also seriously underfunded. In current dollar terms, its funding dropped by more than half from 1993 to 2002. In addition, its original orientation was narrower than what is now required: its objective then was to protect Colombian growers and livestock raisers from imported diseases. It needs an additional orientation, in the direction of guaranteeing the safety of products exported, and not only in terms of plant diseases but also in regard to chemical and bacterial residues. The division of responsibilities with national health authorities needs to be worked out in this area, and appropriate new legislation enacted.

In other words, mounting an effective agroexport strategy, to better exploit Colombia's competitive advantage, requires a recasting of relevant aspects of the institutional framework in the sector more than subsidies, although selective and temporary incentives for mounting export platforms also can be effective.

In the area of pricing policy, the principal products for which agricultural prices still are manipulated are sugar and palm oil. In both products, high domestic prices are used to compensate for lower prices in export markets, a form of cross-subsidy in which domestic consumers, in effect, subsidize exports. Producers also subsidize their own exports to a degree through agreements among themselves. These mechanisms include export quotas assigned to enterprises by the industries themselves and "stabilization funds," into which enterprises may make contributions instead of taking up their full export quota.

In the case of sugar, in recognition of the importance of agroindustries based on sugar, the raw material is sold to them by sugar mills at the import-equivalent price. It is household purchasers of sugar who pay the highest price in Colombia. This price management scheme was motivated in part by the fact that the international price of sugar is distorted downward by international subsidies, and Colombian sugar would not have a competitive advantage vis-à-vis the price on the free or uncontrolled world market, as noted earlier. The referenced study of Burfisher (2001) estimates that the international price of sugar would rise by 16% under a scenario of complete elimination of international subsidies.

Thus paradoxically sugar is a highly competitive product, a bulwark of Colombian agriculture, which, nonetheless, receives substantial domestic support through the price mechanism. As a consequence, its scale of production is overdimensioned for current world market conditions, although it might be an appropriate scale if international price distortions were removed. The government and the sugar industry recognize this situation and are working toward a plan under which a significant part of the sugarcane acreage would be devoted to producing for the purpose of generating ethanol that would be combined with gasoline in automobiles, under the terms of a new law that will require the use of such a mixture. Implementation of this plan should permit a reduction in the price of sugar charged to domestic consumers, along with a reduction of the amount of Colombian sugar exports on the uncontrolled world market.

In the case of palm oil, the subsector is competitive at prevailing world market prices and the degree of international price distortion is less than in the case of sugar (11% according to Burfisher's study). Hence the justification for maintaining the price to domestic consumers at high levels is less compelling than in the case of sugar. In addition, although it is always important to ensure that incentives for agricultural production are adequate, beyond a certain point, the interests of domestic consumers should be taken into account, and edible oils account for a significant portion of household expenditures for low-income families. For these reasons, it is suggested that consideration be given to a program for gradually reducing the domestic price distortion for this product, to bring it more in line with the prevailing international price. Import tariffs on vegetable oils would decline as part of such a program, but they would not drop below the level corresponding to the distortion created by international subsidies. This kind of reform would improve the competitiveness of the livestock sector and oil-using agroindustries as well as be beneficial to consumers.

7.9.4 Agricultural Fiscal Policy for Competitiveness and Inclusiveness

In addition to broad categories of government expenditure for agriculture that are found in all countries, there are expenditures on special incentives or for specially targeted groups in the rural population. In Colombia, many of these special incentives are negotiated and coordinated through the mechanism of value chains. In light of the deterioration of international agricultural prices in recent years, two of the more important fiscal programs have been designed to fund the purchase of uncollectable farm loans from financial intermediaries: the National Program for Agricultural Reactivation and the National Program for Coffee Reactivation. Other than these programs, two of the largest fiscal outlays for agriculture in recent years have been the Incentive for Rural Capitalization (ICR) and the incentive for product storage, provided not only to rice farmers but also for corn and cotton. The ICR is used to reduce the investment cost (reduce the outstanding loan balance) of farmers who borrow for investing in farm modernization. The loans have to be rediscounted through the second-story financial institution FINAGRO (*Fondo para el Financiamiento del Sector Agropecuario*) and all participants to date have been medium- or large-scale farmers, although the areas eligible for this support include artisanal fisheries. As much as 40% of a loan balance is canceled with contributions from the ICR.

The largest part of the Ministry of Agriculture's budget increase during recent years has been attributable to producer subsidies, principally the programs for erasing accumulated farmer debts, the ICR, the incentive for reforestation, and the subsidy for crop storage. In recent years, support to the coffee sector has increased. It now includes payments per unit of coffee produced, payments for renovation of coffee plantings, funding research that formerly was financed solely by the coffee federation, purchase of bad loans made to coffee producers, and investments in infrastructure in coffee areas. Other programs supported by fiscal outlays include associative credits in the sector, a credit guarantee fund, the program for rural microenterprises, the duty-free export zone, the program for reactivating cotton, the program to support expansion of the area of higher technology corn, and many others in addition to research and extension, most of them relatively small in terms of funding.

The ICR subsidy is emblematic of expenditures on producer incentives in agriculture. Along with other producer subsidies it undoubtedly assists some agricultural entrepreneurs to better compete in markets. Nevertheless, there are three concerns about this type of instrument from the viewpoint of economic strategies: (1) by the nature of its requirements, it has effectively been targeted on the nonpoor, whereas it is the poor who most need assistance in becoming more productive; (2) it encourages capital-intensive modes of production when Colombia's comparative advantage—and social need—lies in encouraging labor-intensive production; and (3) although statistics on its allocation by crop or livestock product are not available, it seems clear through discussions with value chain participants that most of the ICR funding goes for import substitution crops, again contrary to Colombia's competitive advantage. In terms of the incidence of the ICR by farm size stratum, it has been estimated that in the 2 years, 1999–2000, about 2% of its disbursements reached smallholders (Jaime Forero).[11] Colombia is not alone in this kind of concern about the incidence of fiscal support.

The issues surrounding the ICR are illustrative of some issues concerning the role of the value chains in general. Their purpose has been to promote greater cooperation among the links in the chains—higher levels of social capital in the sector—and better identification of priority needs of each chain. They are autonomous, flexible, nonbureaucratic organizations that try to create a shared vision of the future. In some cases, they are becoming learning organizations, which is a valuable contribution because sustaining productivity growth in the long run requires learning how to learn.

However, in practice some value chains tend to become organized lobbies for government support, particularly in subsectors with a competitive disadvantage, and to the extent they are successful they therefore may distort public sector support away from the most efficient channels from a growth perspective. Also, there has been very little evaluation of the effects of the support provided through the chains, and there is no link between productivity-enhancing performance and amounts of subsidies received. Continuing to receive subsidies is not made conditional on demonstrating productivity improvements achieved with earlier subsidies. (The Government does appear to be moving to change this, starting with support for cotton.) Forging such linkages in fiscal support policy was one of the keys to the decades of rapid growth in East Asia (The World Bank, 1993). Absence of a relation between subsidies and performance is a serious defect. Another concern is that in practice the incentives gained through negotiations by the value chains tend to be dominated by agroindustrial interests so that primary agriculture does not have an effective voice in many of the chains. Most of the participants in the chains are unaware of the variety of governmental programs and benefits that may be obtained through concerted action by the chain. Given these concerns, it would appear appropriate to review the role of value chains as mechanisms for obtaining fiscal support for the sector, in the context of using sector growth objectives as guides for fiscal action.

The inclusiveness issue is equally relevant to other fiscal subsidy instruments. For example, although a recent law to provide tax exemptions for investments in long-cycle crops is a positive step toward aligning fiscal support with Colombia's competitive advantage, by definition it excludes small-scale producers, whose incomes are so low that they pay no taxes. This problem is not particular to Colombia. Costa Rica, for example, found that it pervaded a program of tax exemptions for reforestation (mainly in teak and mahogany), to the extent that some investors used the program as a tax shelter without having a real commitment to forest activities.

An example of direct support not linked to output is found in the program PROCAMPO in Mexico, which provides direct incentives to participating producers to the amount of US$100/ha. (It is not linked to current production levels.) The pioneer program of this nature was the McSharry Plan in the European Union. In the late 1990s, a variant of it was adopted in Estonia, specifically as an alternative to price and trade controls for reactivating agriculture. A major advantage of the system is that export crops, including small-scale fruits and vegetables, all would receive some incentive. Also, import-competing crops would receive benefits as well, and that could assist those farmers in making a transition made necessary by lowered import tariffs, if such a decision were to be made in tariff policy.

Another advantage of the system is that it would represent a replacement for the subsidized credit offered by the *Caja Agraria* and subsequently the *Banco Agrario*. Since state-owned development banks generally have not proven to be sustainable, in part because loan arrearage inevitably reaches unmanageable levels, it may be sounder to simply provide the needed economic support directly, with emphasis on smallholders, rather than disguising it in the form of loans, many of which eventually become uncollectable.

To be feasible in implementation terms, such a program would require a strong presence of the Ministry of Agriculture (or Department-level Ministries of Agriculture) at the *municipio* level. However, in its system designed for classifying farm sizes by family agricultural units (*unidades agrícolas familiares*, or UAF), Colombia already has the basis for determining eligibility of farmers for participating in the program and the number of hectares cultivated in each case.[12] A UAF is a measure of the land necessary to support a family at subsistence, and its size in hectares varies by region and quality of land. If this kind of program were considered for pilot implementation, key considerations would be the amount of subsidy per hectare of UAF, a limit on eligible acreage per farm (expressed in UAFs), and environmental restrictions on farming practices for participants. The Mexican program does not have an upper bound on hectareage eligible for support payments, but most observers consider that there should be one. Environmental factors would argue, for example, for not subsidizing cultivation in the *páramo*, which is vital to national water supplies and which has fragile soils, and also for not subsidizing cultivation done on steep hillsides or with excessive use of agrochemicals.

A program of direct support could also be used to improve the functioning of rural land markets by encouraging titling and the use of explicit rental contracts. For example, it could be announced in advance that after the program were in operation for a given number of years, say 3 or 4 years, support payments would be provided only to those farmers who either had registered title to their land or who worked their parcels under formal rental contracts.

The other side of fiscal policy for agriculture is raising revenues. Although there are strong arguments for generalized support for agriculture, the sector cannot remain oblivious to the national fiscal constraints, and some of the additional funding will have to come from within the sector itself. Many countries have a rural real estate tax, mainly for the purpose of providing revenues to local governments, but generally the rate of collection of that tax is low, for two reasons: cadastral values of land and real estate appraisals in developing countries usually are very much out of date, and in political terms, local governments usually are not strong enough to enforce tax collection on the larger properties.

These kinds of constraints could be eased by a threefold approach. (1) Basing land taxes on area (flat rates) rather than valuations; at least 11 developing countries now have such a system. (2) Making tax collection a joint effort of national and local governments, even though most of the revenues would go to the latter. (3) Establishing with land tax revenues a local fund for infrastructure investments and training programs, with local citizens participating in the decisions about how the funds are to be allocated. This last prong follows the principle of demand-driven investment programs for rural development that have been implemented in several countries. It would be a key to increasing the willingness of rural populations to pay taxes, since one common concern is that tax revenues may not be put to good use. In its parafiscal entities, Colombia already has examples of farmer-managed funds. In 2000, these entities managed about 90 billion Colombian pesos of funds.[13] There is another precedent for an effectively administered land tax in Colombia in that (municipal) real estate taxes traditionally have been more important revenue sources than in most other countries in Latin America.

A flat rate would not be truly flat. There would be handful of classes of land—not many, or disputes would arise about the appropriate classification of each farm, but say irrigated and nonirrigated to start with. Again, the information used to establish the UAF system in each area provides data about the quality of land, and these data could be used directly to set up the system of rates. There always would be an exemption for the first x hectares (or UAFs) on a farm, so that poor rural families would not face a tax burden. For example, if the first 10 ha were exempt, then a farm of 15 ha would pay tax on 5 ha (33% of its land), and a farm of 20 ha would pay on 10 ha (50% of its land), etc., so that the tax rate in effect would be progressive. Also, the tax would be deductible from income tax obligations, so that more commercial farms would not be taxed twice.

Another advantage of an area-based tax, in addition to administrative ease, is that it would not discourage investments in farm improvements. A valuation-based property tax system constitutes a disincentive to investments in the land. A perhaps more important advantage over the longer run is that *an area-based tax would discourage leaving land idle*. Land would be more likely to be put into production without the necessity of an administrative apparatus to determine whether land is idle and if so what should be the penalty (tax or expropriation, usually, under those systems).

For these reasons, an area-based land tax would be likely to encourage redistribution of land by market mechanisms—through both rental and sales—without the need for coercive measures of agrarian reform. Of course, collection of such a tax would require sufficient political will, but it may be argued that it would be less demanding in this sense than summoning up the political decisiveness needed to impose agrarian reform, which always is an abrasive process for the rural social fabric. Therefore, in the final analysis, a flat-rate land tax turns out to be much more than just a fiscal instrument (Uribe, 2002, p. 21).

The possibility of implementing an area-based land tax in this way is not entirely hypothetical speculation, since both flat land taxes and local investment funds have been established in various developing countries. Should such a revenue system be considered in combination with the direct payment incentives, then the smaller farms would be net beneficiaries and the larger farms would be net contributors. Fig. 7.2 shows a hypothetical illustration of implementation of the two systems together. In this illustration, farms above eight UAFs become net contributors and farms below that size become net beneficiaries. Clearly, that turning point could be set at any level by adjustment of the relevant rates.

A program of direct support payments per hectare obviously would have to begin on a pilot basis, in secure areas. It could be funded by a combination of land taxes and redirecting some of the present fiscal subsidies that go to crops and agroindustries, so it would not necessarily require a net increase in fiscal outlays. Many kinds of variations on these schemes

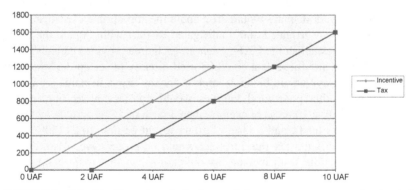

FIGURE 7.2   Colombia: Illustration of a land tax and direct payment system together (vertical scale in US$).

and their fiscal implications could be considered. The purpose here is simply to stimulate new thinking about the role of fiscal expenditures and receipts in the rural sector, in the direction of fiscal systems that have positive effects for both equity concerns (social justice) and efficiency (growth perspectives) in the sector.

7.9.5 Agricultural Research Policy for Competitiveness

The area of agricultural research illustrates the lack of alignment of public policy with the country's competitive strengths. Continuous development and adoption of improved agricultural technology is central to achieving and sustaining competitiveness. Colombia has significant accomplishments to its credit in this area. To mention only a few examples: (1) it has strong product-based research units in which producers' organizations play the principal role in defining research priorities (perhaps most successfully in coffee, oil palm, and cultivated shrimp); (2) it has been a leader in developing participatory research with smallholders; (3) through the national public agricultural research agency CORPOICA it has given a prominent role to producer groups in defining national research priorities; and (4) it has implemented a successful model of a competitive fund for allocating, in a decentralized way, resources for technology transfer (including for participatory research).

Nevertheless, on a national level yields have not increased in line with expectations, real funding for research is declining, research priorities have been questioned, there are important operational deficiencies in the extension service, and smallholders tend to receive few benefits from the technology system. In addition to the issue of declining research funding, doubts have been expressed about the appropriateness of the research agenda. In the final analysis, the relevance of research to farmers' principal problems is the main yardstick of its effectiveness. In the words of Antholt (1998, p. 360), the fundamental issue is "the importance of getting the technology right." In Colombia, an appropriate response to this challenge would involve directing significant amounts of research to the products with long-run competitive advantage, but in many cases they are the products with the least research support.

Not many years ago a workshop of the sector's business association *Sociedad de Agricultores de Colombia* was devoted to identifying research needs for 11 crops and crop groups, including coffee, sugar, oil palm, rice, and other cereals. The results were summarized in the form of yes-or-no responses to three questions. (1) Has there been a diagnosis of technologies used? (2) Have priorities been developed for research and technological development? (3) Have research goals and directions been defined? The only crops or groups for which the answer was negative for all three questions were cacao and the fruits and vegetables group, and both are lines of production with strong competitive advantage. In the case of fruits and vegetables, the answers were positive for research carried out by private investors but negative for public sector research. Almost all the answers were positive for other crops, except for a "no" for cotton (another crop with some competitive potential) for the third question, and the answer "incipient" for the potatoes (another competitive crop) for the first and third questions. Strengthening research on cacao acquires particular relevance not only because it has a competitive advantage but also because crop-level analysis indicates that the main bottleneck to realization of its potential is lack of appropriate technologies.

More recently, the mission to Colombia of the International Service for National Agricultural Research commented that the analysis of the allocation of resources for agricultural research, both for the entire system of agricultural research and for CORPOICA, suggests a high research concentration on traditional crops such as wheat, corn, and cotton, to the detriment of more promising crops in which Colombia has a natural comparative advantage such as fruits and vegetables, which, in general, have received less attention in agricultural policy. Therefore a reorientation of research priorities to the most promising crops and livestock products, and to postharvest management issues as well as production, would be very important for realizing the country's competitive potential in agriculture. Useful guidance for research priorities on these products can be gained in conversations with key participants in the respective value chains.

Crop research can also assist in facilitating a transition for crops that do not have a comparative advantage in their present form. In the case of rice, for example, Colombia's best prospects do not lie in rice as a commodity but perhaps in high-quality rice of different types. Some international markets have their own tastes and require specialized types of rice (Japan, Korea, Thailand, India, etc.). The same can be said for coffee, where high-quality varieties represent one way to overcome the present crisis in that subsector. Also, there are indications that cassava may have a potential not revealed by static competitiveness calculations, and that a strengthened research effort is essential to unlocking that potential.

ENDNOTES

1. M.E. Burfisher, ed. (2001). Estimates of this nature were first made by the Organisation for Economic Co-operation and Development in 1993 and have been made by that agency at intervals since then, and by others as well. The Burfisher estimates are cited here because they are among the most recent.
2. As this book went to press and as prospects for ending Colombia's 50 years of civil war seemed to brighten, interest was increasing in the agricultural potential of the Orinoquia for grains and oilseeds in spite of the high investment costs required.
3. These agreements required purchase of specified quantities of domestic output before imports of the same product were allowed. They have been dismantled following a ruling of the World Trade Organization.
4. Data from the manufacturing surveys of Corporación Colombia Internacional, reported in Corporación Colombia Internacional, CCI (2000).

5. Since this study the research responsibilities of ICA have been passed to the new research entity CORPOICA, and ICA now has sanitary and phytosanitary responsibilities.
6. Information in this paragraph is from the Corporación Colombia Internacional.
7. Information from Jaime Forero.
8. Consorcio Latinoamericano y del Caribe de Apoyo a la Investigación y el Desarrollo de la Yuca—CLAYUCA, n.d., and conversations with Bernardo Ospina of CLAYUCA and Hernán Ceballos of CIAT.
9. When cassava is milled, 2.6 t of fresh cassava are required to produce a ton of flour.
10. Since the time of this study coffee prices have risen and then fallen again in a pattern of permanent instability.
11. At the time of the case study, the Colombian Government was trying to limit access to the Incentive for Rural Capitalization to small- and medium-scale farmers and to perennial crops, so this distributional picture may have changed substantially after the case study.
12. Descriptions of the UAF methodology are found in (1) Dirección de Desarrollo Agrario y Dirección de Desarrollo Social, Departamento Nacional de Planeación, *UAF, Unidad Agrícola Familiar, Promedio Municipal: Manual Metodológico*, Bogotá, December, 2000; and (2) Dirección de Desarrollo Social, Departamento Nacional de Planeación, *Manual general de estratificación socioeconómica*, Bogotá, April 2002.
13. Contraloría General de la República (2002, p. 39). This publication emphasizes the importance of the parafiscal funds as rural institutions.

Chapter 8

Rwanda: Competitiveness by Quality Criteria, Track 2

8.1 INTRODUCTION

This case study applies the Track 2 methodology for evaluation of the competitiveness and of the barriers to its realization. It ties together this analysis with the design of an agricultural development project for Rwanda. This chapter adapts and reinterprets the findings of the Turner–Norton study (2009), applies filters corresponding to stages of value chains, and develops the matrices of quality assessments [value chain quality assessment matrix (VQA)] described in Section 6.4. On the basis of the VQA matrices it identifies the most pervasive, cross-cutting quality issues and derives conclusions ranging over all the crops studied. As in the case of the Colombian studies, circumstances have evolved rapidly in Rwanda since this study was carried out, so some of the comments and conclusions may not be relevant for that country today. However, it illustrates the variety of issues related to competitiveness of high-value crops and also puts the analysis in the context of regional markets and an evolving domestic market.

The principal objective of the original study was, on the basis of agricultural and market analyses, to identify appropriate crops for planting in areas developed throughout Rwanda under a new hillside irrigation project. In addition, following the kinds of value chain and quality assessments outlined earlier, the study identified critical factors and constraints for success in growing, managing in the postharvest stage, and marketing those crops so that activities designed to overcome the constraints could be included in the implementation of the project.

The study was limited to horticulture crops and tea and coffee because these are generally the highest value crops and hence would repay the substantial investment required for irrigation development on hillsides and watershed conservation.

Quality turned out to be the single most pervasive concern for selecting the crops to be recommended. This concern embraced production methods including approaches to disease control, postharvest handling procedures, the use of cold chain facilities, product and process certifications, and even the

The Competitiveness of Tropical Agriculture. http://dx.doi.org/10.1016/B978-0-12-805312-6.00008-8
Copyright © 2017 Elsevier Inc. All rights reserved.

type of packaging. It was concluded that to improve quality as well as yields at the farm level a complementary program of intensive, hands-on technical assistance would be needed for several years. Market conditions also played an important role in these analyses because of the growing East African regional market, the opportunities for import substitution in a rapidly expanding agricultural sector, and the dangers of oversupplying the domestic market. In this context, the case study illustrates an approach to analyzing domestic demand prospects for high-value products.

Apart from quantitative data on domestic demand and regional price comparisons, all the information for the study came from detailed interviews with sector experts and persons involved in the production, processing, and trade of fruits and vegetables, plus information contained in a few technical studies of agronomic issues. In the end, quality evaluations have to be based primarily on hands-on judgments of persons experienced in all stages of the relevant value chains.

As well as producing high-quality tea and specialty coffees, Rwanda produces a wide variety of fruit and vegetable crops, for both the domestic and export markets. There are severe limitations on data availability and quality of data for these products. In general, the standard agricultural data sets are compiled for what are regarded as "principal" crops (and principal livestock products), but in these compilations fruits and vegetables are always treated in aggregate form, although they are an important part of the sector. (The same is true in many other countries.) Nevertheless, some more disaggregated information has been found in specialized studies, and this case study makes extensive use of it.

Although trade in fruits and vegetables is not as well documented as for grains, legumes, and root crops, and production estimates are not available for the more minor but numerous fruits and vegetables, these crops represent a very important part of the sector. Also, they are quite labor intensive to produce, and these crops along with tea and coffee generate well over half of the employment in the sector.

On a per-hectare basis horticulture crops give high returns to farmers. The Rwanda Horticulture Development Authority (RHODA) estimated that 1 million persons are involved in horticulture, either in production or the value chain. For the farmers who grow horticultural crops, on average they gain 45% of their income from those crops while dedicating only 12% of their land area to them. Fig. 8.1 shows the value of production of principal horticulture crops in Rwandan francs as of 2008.

8.2 THE CONTEXT: DOMESTIC DEMAND FOR HIGH-VALUE CROPS

According to integrated household living conditions surveys in Rwanda [Enquête Intégrale sur les Conditions de Vie des Ménages (EICV2) in French],

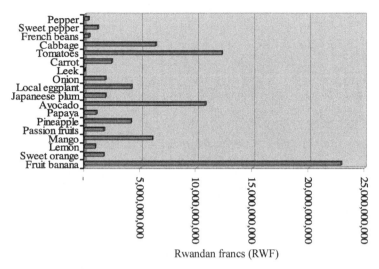

Rwandan francs (RWF)

FIGURE 8.1   Rwanda: value of production for leading horticulture crops (in Rwandan francs). *Masimbe, S., Gahizi, F., Musana, B., & Sangano, J., & G&N Consultants, Ltd., October 2008. A survey report on the status of horticulture in Rwanda. Prepared for the Rwanda Horticulture Development Authority (RHODA).*

vegetables, bananas, and other fruits account for about 17% of food expenditures for the average household and 10% of the total expenditures.[1] In rural areas, these products represent a much greater share of the average household food budget, about 35%. The difference is mainly attributable to the fact that urban consumers spend proportionately more on processed food, meat products, and rice.

As income levels rise, consumers tend to decrease the share of their household budget that is devoted to food. At the same time, they increase their consumption of some classes of food items relative to others. This is true in Rwanda and everywhere else. Although cross-sectional survey data are not strictly applicable to changes experienced over time, information from the integrated household living conditions surveys shows how food consumption patterns change as households move up to higher income strata. The products for which they tend to decrease their consumption shares by the greatest amounts include corn, sorghum, cassava and other root crops, pulses, potatoes, and sweet potatoes. The items that increase in relative terms include rice, vegetables, fruit, vegetable oils, meat products, and processed foods. Table 8.1 presents estimates of income elasticities of demand that support these trends.

These parameters confirm that the growth of domestic demand for fruits and vegetables will continue to be robust for the foreseeable future. When the parameters are calculated for individual income strata, additional insights become available. The overall income elasticity of demand for rice, for example, is high, whereas it is much lower for the upper income groups. In the

TABLE 8.1 Rwanda: Income Elasticities of Demand from the Living Conditions Surveys

	2001–02 Survey			2005–06 Survey		
	Rural	Urban	National	Rural	Urban	National
Corn	0.4	0.4	0.4	0.7	1.2	0.8
Rice	4.1	0.9	2.6	2.9	1.2	2.1
Sorghum	0.5	0.6	0.6	0.5	1.0	0.6
Wheat	4.7	1.1	2.6	0.6	1.0	0.7
Cassava	0.5	0.5	0.5	0.3	0.4	0.4
Potatoes	0.7	0.6	0.8	0.6	0.8	0.7
Sweet potatoes	0.1	−0.3	0.2	0.2	1.0	0.3
Other roots	0.3	0.5	0.4	0.3	1.0	0.4
Pulses	0.3	0.4	0.4	0.3	0.9	0.4
Bananas	1.0	1.1	1.1	0.7	1.4	0.9
Vegetables	0.7	0.7	0.8	1.1	1.0	1.1
Fruits	0.6	0.8	0.6	1.3	1.1	1.3
Oil crops	1.7	0.9	1.6	1.2	1.0	1.2
Other crops	3.4	0.9	2.3	3.1	1.1	2.1
Livestock	3.0	1.1	2.2	3.0	1.4	2.3
Processed food	1.1	1.1	1.2	1.3	1.2	1.3
Nonfood	1.5	1.1	1.2	1.3	0.9	1.1

survey EICV2 in highest of five income strata in urban areas, this parameter has a calculated value of 0.9, versus 5.8 for the lowest of the five urban income strata. In rural areas the difference was even more striking in the case of rice: 1.5 for the highest stratum versus 9.6 for the lowest stratum. This suggests that, for example, expenditures on rice consumption will respond less than proportionately to additional income perceived by the highest urban income group, and much more than proportionately for the lowest group. Hence future growth in rice consumption will be supported mainly by income increases for households that are in the lower income strata.

For vegetables and fruit, the picture is quite different. To illustrate, in EICV2 in urban areas, the income elasticity of demand for vegetables is a uniform 0.9 across all five income strata. That is, an extra 10% of spending power results in an extra 9% of expenditures on vegetables at all income levels. For fruits other than bananas, the urban income elasticities vary from 1.4 for the lowest income group to 0.9 for the highest: a difference but much less marked than the one for rice. For bananas the respective elasticities are 2.1 for the lowest urban stratum and 1.1 for the highest, again showing a smaller variation than in the case of rice.

Thus a more disaggregated look at behavioral parameters for demand confirms that the domestic market for fruit and vegetables will continue to respond strongly to national economic growth.

Unfortunately the survey data do not disaggregate consumption of fruits and vegetables by individual products. Time series data on prices are not available for the calculation of price elasticities of demand, but international experience shows that the price elasticity of demand for individual horticulture products is almost invariably greater than 1 in absolute value (less than -1.0), and therefore a basic criterion for selection of crops for the hillside irrigation areas is that they can (1) be exported, (2) substitute competitively for imports, or (3) fill unused processing capacity on the domestic market. (For the aggregate of fruits and vegetables the price elasticity of demand is typically in the range -0.6 to -0.9, but since households may substitute among fruits and among vegetables in response to price variations, the individual products are more price sensitive.)

Therefore, although domestic demand for horticultural products will grow, it will be important to avoid oversupplying the market at any point in time as the consequence could be a significant drop in the domestic price. This points to the importance of three outlets mentioned earlier for additional production. Regarding the third outlet of the processing stage of value chains, the income elasticities of demand for processed foods are high so that their market can be expected to expand more rapidly than the market for fresh products.

8.3 REGIONAL TRADE AND PRICE CRITERIA FOR COMPETITIVENESS

Coffee and tea of course have large export markets throughout the world. For fruits and vegetables the regional export markets have special importance. A considerable amount of informal trade in fruits and vegetables takes place across the borders in the region, often by boat, bicycle, and foot as well as by vehicle. Documentation is imprecise, but it is clear that these products move fairly freely between Rwanda on the one hand and Uganda, Congo, Burundi, and Tanzania on the other hand, and to and from Kenya as well. As would be expected when trade is fluid, prices tend to converge except for transport margins. However, in some cases substantial price differentials have remained, beyond what is explained by transportation costs, and these differences tend to indicate where comparative advantage or competitiveness lies.

Table 8.2 below compares supermarket prices in Nairobi, Kampala, and Kigali for selected fruits and vegetables, as of May 2009. Those products for which Rwanda's price is higher than in both the other countries are highlighted in boldface. Products for which Rwanda has a favorable growing environment stand out for having relatively low prices in the country; examples include

TABLE 8.2 Supermarket Prices of Fruits and Vegetables in East Africa US$/kg, May 2009

	Rwanda	Kenya	Uganda
Avocados	0.26	0.20	0.26
Carrots	0.86	1.05	1.29
Cabbages	0.43	0.43	0.40
Mangoes, ngowe	4.32	1.32	2.00
Green bell pepper (capsicum)	0.60	1.72	1.49
Red onions	1.04	1.45	1.47
Pineapple	0.86	0.92	0.70
Tomatoes	2.42	1.19	1.29
Courgettes	1.04	1.32	1.29
Eggplant	1.29	1.19	1.29
Lemon, local	1.24	0.92	0.60
Lettuce	0.43	0.60	1.00
Passion fruit, black[a]	1.29	1.72	1.82
Banana, long (Gros Michel variety)	1.21	0.92	0.84
Broccoli	0.78	2.52	3.09
Pumpkin, yellow	2.59	0.79	0.51

[a]*Purple passion fruit.*
Sundip Jethalal, General Mgr., Fresh An Juici, Nairobi; Nakumatt Supermarket management, Kigali.

carrots, passion fruit, red onions, green bell peppers, and broccoli. Rwanda appears to be a net exporter to the region of some of these crops according to the sketchy data available.

On the other hand, by regional standards, prices in Rwanda appear to be fairly high for some other fruits and vegetables, at least in the dry season, and Rwanda imports more of these products than it exports. The high prices tend to indicate less favorable growing conditions than in neighboring countries and hence the lack of comparative advantage. Examples include lemons, pumpkins, and mangoes. These are generalizations, and it is important to be aware that there are significant geographical variations within Rwanda. For example, carrots enjoy a suitable climate in the north but not in the east or south-central parts of the country. Mangoes, although generally a more appropriate crop for the hotter climates of Uganda, Congo, Burundi, and Tanzania, can be grown reasonably well in Bugesera, which is at a lower altitude than the rest of the country, and to some extent in other areas, although perhaps not with the same quality and yields as in other countries of the region.

Another indicator that suggests that some prices tend to be high by regional standards is the fact that many processing facilities have considerable unused capacity; in some cases they have difficulties in competing with prices paid by local traders for supplies and still operate profitably, and this has been stated explicitly by some facility managers.

Reasons for the high prices include, in addition to the climate, the scarcity of land and the low yields. Rwanda can most readily export, and compete with imports, in those crops for which the climate helps confer special quality characteristics. If yields can be raised and quality improved with irrigation and intensive technical assistance, then the range of competitive crops can be widened, as this report indicates.

The data in Table 8.2 do not imply that Rwanda cannot grow more of the higher priced items, but it does suggest that crop husbandry and yields should be improved in those items and that more production would be expected to lead to lower prices. An example could be tomatoes for processing.

If additional production from the hillside irrigation areas were to result in somewhat reduced prices for a few crops, producers could be compensated by the higher yields that would result from irrigation and improved cultivation practices, and more income and employment would be generated in the processing industries, which then would source more of their raw materials domestically. Thus for tomatoes, for example, a slight decline in prices may benefit the sector as well as improve the welfare of consumers. However, in other cases where the market is saturated and there is little or no room for import substitution, then additional production would significantly affect prices and the products would not pass the market test and therefore would not be recommended for the hillside projects.

8.4 QUALITY, THE PROCESSING INDUSTRY AND INFRASTRUCTURE

For the domestic market a pervasive issue is lack of sufficient quality in fresh fruits and vegetables. A major buyer for supermarkets stated he would purchase more products locally, rather than importing them, if quality were higher. This and other evidence indicate the need for intensive, on-farm technical assistance in association with all the agricultural development projects. It has been estimated that only 10% of horticulture producers have received technical assistance, and even that has not always been of a continuing, hands-on variety. As commented below for the case of apple bananas, even when farmers are aware that better cultivation practices will improve yields and reduce disease incidence, they do not always apply the lessons. Technical assistance should cover harvest and postharvest operations also. The postharvest phase is equally important for ensuring product quality. In Rwanda, an obvious example of the need for improvements in this area is the practice of tossing tomatoes together into rustic baskets that were designed to carry harvested potatoes. The result is an inferior product when it reaches the buyer's hands, no matter how good it was when picked from the plants.

Quality is even more important for export markets. As noted in Chapter 6, in addition to meeting consumer expectations regarding taste, appearance, consistency, packaging, and other characteristics, obtaining certifications is increasingly important for penetrating export markets. These can include certifications at both the production and processing levels. Rwanda has advantages for organic production because of the traditionally low use of agrochemical inputs. However, this advantage has not always been well exploited. For example a sample of tree tomato fruit shipped to the United Kingdom was rejected because of pesticide residues in excess of the limits.

In view of the potential role that organics can play in Rwandan agricultural exports, the government may wish to support this aspect of the hillside irrigation project by reaching a consensus with farmers and decreeing organic production zones, in which only organic production will be carried out. Clearly a farmer cannot earn organic certification if the neighboring farm is using pesticides that could be carried across farm boundaries by the wind, so usually zones that embrace multiple farms have to go the organic route together. Perhaps more importantly, the ease of setting up an Internal Control System, which is by far the most cost-effective way of certifying numerous small-scale farmers, is vastly increased by certifying all the farmers in a given geographical area.

The quality issue is more important and more widespread in fruits, vegetables, tea, and coffee than in, say, grains and legumes. Hence quality has to be given priority in the development of this subsector. In overall terms, Rwanda's climate and small farm size (high ratio of labor to land) lends it a long-term competitive advantage in fruits, vegetables, tea, and coffee, which are high-value products vis-à-vis many other kinds of crops, but to fully realize this competitive advantage, it will be necessary to pay close attention to quality concerns.

The tea and coffee processing industries are well developed in Rwanda, although improvements are being implemented as a result of the privatization program for tea factories and the technical assistance provided to coffee washing stations. New types of processed products have been developed for tea recently: bagged tea and "orthodox" (rolled leaf) tea. For fruits and vegetables there are seven major processors in the country along with more than a dozen smaller processors. They produce a range of fruit pulp, juices, concentrates, sauces, fruit wines, and jams, some of which are exported, but as yet they do not have capabilities for producing the dried and fresh frozen fruits that have good markets in developed countries. A couple of entrepreneurs have carried out experiments for drying fruit, but so far they can be only considered to be artisanal.

The aforementioned underutilization of capacity in fruit and vegetable processing is a general characteristic of the sector. Overall, the unused capacity in processing facilities was 85% recently, a very high figure. The price question for one or two products explains part of this phenomenon. A bigger reason undoubtedly was the high tax of 39% levied on sugar inputs to fruit processing. This tax rate was revised downward to 5% in the past few months (while remaining at 39% for retail consumers of sugar), and it is expected that processing activity will pick up as a consequence. In some cases, another factor causing the underutilization of capacity is the lack of a solid, long-term relationship between producers and processors—the lack of a true alliance in the value chain. This issue is central to the development of high-value crops and is discussed more fully later. In other instances, capacity simply may have been constructed without a proper assessment of the market.

In the international context, Rwanda is at a disadvantage with respect to transportation since it is landlocked with only three direct flights a week to Europe. However, the considerable volumes of cross-border trade attest to the viability of trade routes with neighboring countries, informal as they may be in some instances. There also have been regular air shipments of fresh produce to Brussels. New transport options are being opened to the Middle East by air and plans are underway for a railway to Isaka in Tanzania, both of which will facilitate export possibilities. In addition, the planned new airport for Rwanda, located outside of Kigali, may open up other direct routes to Europe. Clearly, air shipment is viable only for the highest value products in relation to their weight, but there are products in Rwanda that meet this requirement.

Transportation within the country is a frequent bottleneck in spite of the small geographical extent of Rwanda. Farm access roads are generally unpaved, and their condition takes a heavy toll on the vehicles and the perishable produce carried. They often can be impassable during parts of the rainy seasons. Therefore improving rural roads, preferably by laying down tarmac, is a priority for principal producing areas for perishable products.

The cold chain for Rwandan agriculture is still underdeveloped. A new cold store facility has been constructed at the international airport, but shippers have commented that its management is not adequate for the needs of the sector. Too often a consignment of fruits or vegetables has to sit in the sun for an hour or more until the facility is opened, and when there is an electricity outage the backup generator is not always turned on promptly. Quality suffers. Adequate separation is not always provided between organic and nonorganic products, which can jeopardize organic certifications. In brief, the management of this cold store needs to be put in private hands rather than being left with a parastatal agency, to meet product quality standards.

Some producing areas have sufficiently rapid road access to the airport that local cold store facilities are not needed. However, the study team estimates that up to six cold stores should be planned for major fruit and vegetable zones, starting with Kirehe, to the east, and Bugesera in the south-central part of the country, again for maintaining quality. (See map of Rwanda in Fig. 8.2.) The other four facilities should be planned in relation to the accessibility of the watersheds being developed. In all cases, the management of these cold stores should be private.

FIGURE 8.2 Map of Rwanda with provincial boundaries. *Reprinted with permission from the United Nations.*

In cases in which the selected crops lend themselves to fairly frequent and continuous harvesting throughout the year, an economical alternative to cold stores is refrigerated trucks of a size appropriate to the volumes harvested. The decision between fixed facilities and trucks cannot be made in the abstract but rather depends on the harvesting program that emerges from the planting decisions.

It will be equally important to facilitate, through the hillside irrigation project, the installation of evaporative coolers and other simple cooling devices at the farm level, or for groups of farms. It is important to take the "field heat" out of perishable products before they begin their journey up the value chain.

Apart from irrigation itself, which is a powerful instrument for increasing farm productivity, another important piece of infrastructure is greenhouses or, in their simplest form, plastic tunnels. Their use for vegetable cultivation is growing rapidly throughout the developing world. It is recommended that small numbers of them be installed on an experimental basis in watersheds with crops for which they can be useful, and that farmers work with technical experts in a participatory manner to develop the best cultivation practices for tunnels. When their successful use has been demonstrated, they can be installed on more farms. The potential for productivity gains is very high with these structures; the yield increases often can be from two- to fivefold or more.

Finally, as stressed throughout this report, for attaining required levels of quality the harvest and postharvest *techniques* can be equally or more important than *infrastructure*. For this reason, the report concludes that intensive and sustained technical assistance to farmers will be one of the keys to success of the watershed projects.

8.5 VARIETAL DEVELOPMENT AND VALUE CHAIN ALLIANCES

Rwandan horticultural exports are still in their infancy when compared with those in Kenya and Uganda and even in Ethiopia. Therefore it is likely that new export products will emerge in coming years, and some of those with very small export volumes at present may grow to become more significant foreign exchange earners. One of the recommendations of this study is that in each project watershed a small area be set aside, typically no more than 1−3 ha, for varietal trials of newer products and for testing different techniques of production. The farmers whose land is being used for this purpose should be compensated by the project for the foregone annual crop harvests, and local farmers, particularly women farmers, should be invited to participate in the trials.

These trials should be part of a program of participatory research and extension, in which technical experts work side by side with farmers, playing the role of facilitators of the learning process as much as sources of technical information. For small farmers, participatory approaches have been proved to be more effective ways to raise the technical level of farming and farm productivity, when compared with the traditional top-down approach of "delivering" technical messages to farmers. Agriculture is increasingly a knowledge-intensive sector with numerous areas of specialization, especially in regard to cultivating high-value products. Facilitating processes of acquiring knowledge, and incorporating farmer know-how into these processes, empowers farmers to continue learning on their own and gives them a sense of ownership over the varieties and practices that are developed.

Sometimes the solutions are relatively simple and involve harvest and postharvest practices. Many farmers in Rwanda have a tendency to harvest crops too early because they are eager for cash, which sometimes reduces the weight of the harvest and in all cases reduces quality and shelf life. This problem is common in passion fruit, sweet peppers, potatoes, tree tomatoes, and coffee, for example. If the project can develop support mechanisms to overcome this problem, it will already be contributing significantly to raising farm incomes.

The participatory approach can be fruitfully extended to the selection of crops to be grown in the newly irrigated watershed areas. Although this report suggests a number of alternative crops that can be considered, and some of the relevant factors for their success, farmers should be consulted closely in the decisions about which crops to grow in a given locality. Among other advantages, this will help ensure a greater commitment to proper cultivation of those crops on the part of the farmers. It is suggested that for each watershed a list of at six to eight crops be presented to the farmers, with the aim of their selecting four to five of them. A reasonable degree of crop diversity is important for soil management and controlling crop diseases. It also helps reduce market risk, which always is present, although this report has tried to identify crops with the most promising markets.

Another general conclusion is that every effort should be made to coordinate planting programs in the watersheds with commercially solid buyers, whether they be wholesalers for supermarkets, processors, or exporters. They are stakeholders in agriculture also. As will be illustrated later for the cases of French beans and avocadoes, buyers can indicate the number of hectares that should be planted for their needs and may be willing to enter into contracts with producers in some instances. Buyers can also indicate their standards for product quality and packing (inappropriate harvesting and postharvest handling practices, including packaging, are a widespread deficiency in horticulture production in Rwanda at present). This kind of relationship within the value chain will be very important for the economic success of the watershed development projects.

The project should attempt to facilitate the formation of alliances between groups of farmers and the next links in the value chain. As long as the relationship between the seller (the farmer) and the buyer is based only on negotiating a spot price, it will be a fragile relationship because price is the one issue for which the two sides are adversaries (as pointed out in Section 6.2). A more enduring relationship also includes provision of inputs and technical assistance that demonstrably raises farmers' incomes, and collaboration in solving quality issues. Farm gate prices also can be negotiated for the longer term, including clauses for changing them in light of changes in export prices. An umbrella organization of coffee cooperatives in Rwanda already has applied this formula successfully, providing a second payment to coffee farmers whose amount depends on the price received at the time of export of the processed coffee.

The approach of the tea company SORWATHE is an example of fostering long-term, multidimensional relationships with growers, giving a deeper stake in cooperating with the processor. The factory has set up health care clinics in growing areas and has donated sewing machines (among other things) to local women to give them another source of income. The project can take a leadership role in encouraging longer term relationships between farmers and buyers of their crops. This approach of contract farming has many variants, but all the successful ones go beyond the question of negotiating a spot price for sales.

Given the value of intensive technical assistance for these crops, coordination will be equally important with the RHODA [and its equivalent in the new structure of the Ministry of Agriculture and Livestock Resources (MIN-AGRI)] and the soon-to-be-organized umbrella organization for the private horticulture sector. RHODA had developed a sound strategy for the sector, and it provides much of the basis for continuing actions to support growers and processors.

In light of the economic importance of the horticulture sector in Rwanda, it also merits consideration for a higher priority in government and international support programs. By contrast, the coffee sector has received several times the project funding that horticulture has. Coffee is indeed a valuable sector for the economy but employs about half the people that horticulture does.[2]

8.6 ASSESSMENTS OF VALUE CHAIN ISSUES IN RWANDA

8.6.1 Tea

The Field Findings

Rwanda's evident comparative advantage in tea is due primarily to altitude and climate, aided by careful crop management and processing practices. (The country's average altitude is 9000 feet above sea level and the altitude of the hilly capital is listed at 5141 feet.) Tea is Rwandan agriculture's largest export earner after coffee, and its prices have held up better during the recent world economic downturn than those of specialty coffee. From 2004 to 2008, tea exports earned an average of about US$37 million each year. The government has been privatizing tea factories, and the output of private factories has increased substantially this decade. According to statistics of the government parastatal tea authority OCIR-Thé, the industry is a source of income for more than 60,000 households and employs more than 35,000 workers.

Rwanda produces some of the highest-quality tea in the world, and its markets continue to expand. Its tea is consistently ranked first in quality in the Mombasa auctions. Currently, 97% of the domestic production is exported and tea factories have said that they can process and export all the additional tea that can be produced in the country without affecting prices received because Rwanda's share of the world market is very small. Traditionally, Rwanda exported all its tea via the Mombasa auctions, where it regularly obtained the highest prices, but some factories are increasingly making direct exports to foreign buyers in the United Kingdom, Pakistan, Afghanistan, Yemen, Egypt, and South Africa. Efforts are also underway to open up new markets in the United States and Middle East as well as the Far East.

Tea is cultivated mainly on large plantations owned and managed by about a dozen tea "factories" that process green leaves into made tea. The factories supplement the tea cultivated on their own land with relatively small amounts of tea produced by tea cooperatives and private growers, although outgrowers provide about 75% of the leaves in the case of the tea processing company SORWATHE. For the special grade called Rwanda Mountain Tea, they provide about 30% of the leaves. The majority of tea growers (farmers) are organized in *Coopthes* and *Villageois*, brought together by a single umbrella federation (Fédération Rwandaise des Coopératives de Théiculteurs, Ferwacothe) at the national level. *Coopthes* bring together growers on scattered small plots, whereas *villageois* are relatively bigger units and possess land that was formerly owned by government. Ferwacothe is composed of 15 cooperatives of various sizes with about 30,797 members. The federation controls a total of 2972 ha of tea in marshlands (valleys) and 4845 of tea ha on hillsides. Tea plantations are concentrated in the southern, western, and northern provinces of the country, and the largest concentration is around the Nyungwe forest ring.

The marketing chain for tea is necessarily short, reflecting the extreme perishability of fresh tea leaves, which must be delivered to the factory within hours of being picked so as not to suffer a significant loss in quality. Tea leaves usually are picked early in the morning when temperatures are cool, and they are transported in baskets to local collection points, which are typically within 3 km (for Rwanda Mountain Tea) or 5 km of all growers (for OCIR-Thé). After being weighed, they are loaded onto trucks for delivery to the factory. Following a brief withering period, they are chopped, fermented, dried down, cut, and packaged for export. Tea exported from Rwanda is mostly transported by road to Mombasa for the auctions. Private factories, however, increasingly sell tea directly to niche buyers and blenders in, for example, the United States.

The hillside irrigation project presents a special opportunity for tea production because almost all of it is presently cultivated without irrigation. In the dry season the tea plants become stressed, hence yields are affected, and some of them die. It is estimated that on average irrigating tea would raise its yields by 50% over the dry season. It would be particularly important to expand the production of hill tea (mountain tea), which is the higher quality variant, and the hillside project is well suited to that purpose. Tea plantation cover, besides its ability to thrive on acid and aluminum toxic soil, can also act as an effective rehabilitation crop for exhausted land. It is an effective option for soil conservation, especially on small farms. Tea has high value added per hectare, and its productivity can be improved by mulching as well as irrigation.

Another factor that makes tea especially suitable for watershed development is that approximately one-third of the area of a tea estate has to be planted with fuelwood for the factory. There are stages in tea processing that cannot be economically powered by electricity. Tree planting is part of the design of the hillside project, but the area devoted to trees, and the species, should be coordinated with the tea factories when tea is part of the new watershed planting program.

The elements of the value chain for tea are in place, with the restriction that tea factories prefer not to purchase tea produced more than 15 km from the factory because of the freshness constraints. Hence planting decisions should be coordinated with the nearest factories if they fall within that radius. It should be noted that the higher quality mountain tea is not defined by altitude but rather by slope and other conditions that make it grow more slowly, and hence it can be produced in many locations in Rwanda. However, its volumes of production are lower than those of valley tea. For example, at present only 6% of SORWATHE's tea production is mountain tea. Nationally, not more than 10% is classified mountain tea. Cultivation of tea is a labor-intensive occupation, and higher yields result from the timing and manner of pruning and the frequency and method of fertilizer application. For organic teas rock phosphate and manure are applied instead of chemical fertilizers.

The main value-added options for tea are organic and fair trade certifications. The recent implementation of price stratification and improvement in tea transport has helped increase the quality of Rwandan tea quality. The private tea factory SORWATHE has gained Fair Trade certification for its products and now has ISO 22000 certification as well. Rwanda Mountain Tea, which owns two factories (Rubaya and Nyabihu), is in the process of obtaining ISO 22000 certification.

OCIR-Thé recently developed a bagged tea product, and the facility for producing it is running over capacity. The target markets for this product are East Africa and Dubai. This value-added operation appears quite profitable. A rough indicator of the profitability is that a kilogram of the better grade of bulk CTC[3] tea may fetch a price up to US$3−US$4, whereas the bagged tea earns US$14/kg, and the cost of bagging is estimated at US$7−US$8 per kg. However, there may be limitations to the amount of tea that can be exported in bagged form because it would begin to compete with the products of the very agents who buy bulk made tea from Rwanda, and the market links to them could be prejudiced by that competition. In another value-added initiative SORWATHE has recently invested in a facility for producing "orthodox" tea (rolled leaf tea), for which only some of the harvested leaves are suitable. This is a high-value product, but not all the tea factories can afford to make that kind of investment.

Subject to the 15-km radius limitation and coordination with tea factories, there is no upper bound on the amount of new tea that can be planted in the watersheds under outgrower schemes. The project will need to support farmers for the first 3—4 years or more after planting tea, until harvests become significant. After that it is a very worthwhile crop for smallholders and also an excellent crop for soil conservation. Care needs to be taken about mining activities in tea growing areas because they diminish water quality. In some areas such as Karenda the mining is informal but no less injurious to tea production. Another area in which the government can assist tea production is paving rural roads. Rwanda Mountain Tea, for example, has a problem with 8 km of poor dirt road linking the Rubaya factory up with main roads. They have discussed this issue with the government and are hopeful the road will be paved.

In summary, Rwandan tea's strong competitive position is clear, but there are a few steps that could be taken to take fuller advantage of it, including (1) providing more irrigation for areas planted in tea, (2) planting more mountain tea (on appropriate slopes), (3) undertaking more direct marketing rather than continuing to rely so much on the Mombasa auctions as points of international sales, (4) planting trees along with new areas of tea plantings, (5) controlling mining operations in tea areas, and (6) paving more roads in tea areas.

Application of Value Chain Filters

From these findings the following assessments may be made of issues in each stage of the value chain for Rwandan tea:

- *Seeds and breeds*: no issues.
- *Production*: (1) more irrigation needed, (2) more emphasis on planting mountain tea would be beneficial.
- *Harvesting*: no issues.
- *Postharvest*: no issues.
- *Storage*: no issues.
- *Transportation*: although transport conditions have improved, roads that are still unpaved in some areas are threats to quality because of the delays imposed on product delivery.
- *Processing*: no issues (new processed products being developed).
- *Market development*: more emphasis on direct marketing (versus sales at the Mombasa auction) would bring higher prices.
- *Value chain governance*: no issues.
- *External and policy*: uncontrolled mining operations represent a threat to the crop through their contamination of water sources.

TABLE 8.3 Rwanda: Value Chain Quality Assessment Matrix for Tea

	Domestic Markets		Export Markets	
Value Chain Stage	Fresh	Processed	Fresh	Processed
1. Seeds and breeds				
2. Production				CM, TA
3. Harvesting				
4. Postharvest				
5. Storage				
6. Transportation				SP
7. Processing				
8. Market development				NM
9. Value chain governance				
10. External and policy				EX
Sum of scores				5

Total score on quality issues: 5. Here CM refers to the need to irrigate more tea areas; TA to the opportunity to increase taste aspects of quality by planting more mountain tea; SP to the effects on the product of unpaved, slow roads; NM to the need to engage in more direct marketing; and EX to the hazard to the crop posed by uncontrolled mining operations.

The Value Chain Quality Assessment Matrix for Rwandan Tea

These conclusions result in the VQA for tea shown in Table 8.3, where again the designations of issues affecting quality are:

GE: Appropriateness, cleanliness, and uniformity of seed varieties and appropriateness of breeds

HY: Product hygiene, e.g., absence of bacterial contamination

CH: Management of agrochemicals and veterinary medicines and supplements, to avoid residues and other contamination of the final product

CM: Other aspects of crop management

TA: Taste, appearance, uniformity, ripeness

PD: Pests and diseases that affect quality and yields

HH: Excess heat or humidity that causes product deterioration

IM: Impurities mixed in with the product

SP: Product spoilage

PK: Packaging

PR: Lack of facilities for adequate processing

CE: Certifications required for some markets
NM: New market development required
VG: Value chain governance issues
EX: External issues
PI: Policy issues

The overall score in Table 8.3 signifies a competitive crop by quality criteria and shows that there are few hurdles to the continuing competitiveness of Rwanda's tea industry. However, there are a handful of things that can be done to further enhance it, and one hazard to the industry that should be dealt with.

8.6.2 Coffee

The Field Findings

During a period of unfavorable price trends for commercial-grade coffee, up to 2001, prices for higher grades (fine and specialty) increasingly diverged from the former. In light of this development in world markets, in 2002, Rwanda began to put into action a strategy for enhancing the quality of its production so that it could be sold at the higher prices. One of the keys to producing the higher grades has been fully washing the coffee (Republic of Rwanda, Ministry of Agriculture and Animal Resources, 2008a, p. 30–31).

In 2002, there were only two washing stations in the country, and a goal was established to increase their number drastically, a goal that has been largely achieved. As of this study there were 130 coffee washing stations in the country. In addition, an intensive effort was implemented to develop specialty coffees in Rwanda and market them effectively. International campaigns were launched for Rwandan coffee, millions of new trees were planted to replace aging existing trees, and programs were undertaken to increase the use of fertilizers, fungicides, and insecticides on coffee.

Progress in transforming the coffee sector has been significant but somewhat slower than expected. At the beginning of the 2000s a specialty coffee sector did not exist in Rwanda, so the development of this high-value product is a signal achievement in itself. Between 2003 and the first half of 2008 farmer prices for cherry coffee more than doubled, from 60 Rwandan francs (RWF) per kg to RWF 130/kg and higher in some instances. Then the world economic crisis caused a fall in prices of the better grades of coffee in late 2008; farmer prices for fully washed coffee in Rwanda declined, and some washing stations closed as a consequence. Nevertheless, prices of fully washed and specialty coffees are expected to recover within a year or two at most, and the long-term prospects of the sector in Rwanda are bright.

The *Rwanda National Coffee Strategy* for the period 2009–12 recognizes that although progress has been made, it has been difficult on some fronts, particularly in improving the quality of the cherries harvested (harvesting more uniform batches in terms of ripeness) and turning around the coffee washing stations (Republic of Rwanda, Ministry of Agriculture and Animal Resources, 2008a). Although export prices increased for Rwandan coffee up to mid-2008, and Rwanda gained international recognition for the quality of its coffee, world prices for specialty coffees actually increased more rapidly. The pricing structure within Rwanda has been such that most farmers have had no incentives to produce better quality harvests, and input use and husbandry of coffee trees remained below what was hoped to be achieved. (Note: some second-level cooperatives have introduced a dual pricing scheme under which, after the final sale, farmers are given a second payment whose magnitude depends on the actual export price received.)

Consequently, for the medium term the coffee strategy concentrates on the following five pillars related to improving quality at the same time that rehabilitation of the plantations can produce higher yields and hence greater volumes of coffee (Republic of Rwanda, Ministry of Agriculture and Animal Resources, 2008a, p. 33–37):

- Improve monitoring and distribution of seedlings, fertilizers, and pesticides.
- Encourage a voluntary turnaround program for the coffee washing stations.
- Improve international sales and distribution mechanisms.
- Implement a census and geographical information system (GIS) study of all coffee-producing regions.
- Implement value addition activities including toll roasting abroad and partnerships with international retailers.

The *Strategic Plan for Transformation of Agriculture, Phase II* (PSTA II) also notes the importance of shading coffee trees as part of the improvements in farm-level management, implementing a program to control coffee leaf rust, and carrying out systematic fertilizer trials to have greater certainty regarding what kinds of input mixes would be productive. It also comments that it is necessary to (Republic of Rwanda, Ministry of Agriculture and Livestock Resources, 2008b, p. 81):

Identify the cause of the "potato taste" that is a major constraint to getting greater prices for Rwandan coffees and **implement as an urgent matter a programme to correct it.** *Nearly 25% of all Rwandan specialty coffees in 2008 revealed significant levels of this defect. This percentage is higher than last year. The specialty coffee industry realizes now that Rwandan coffees will continue to be infected with this defect unless remedial actions are taken. Already, Rwanda is losing income due to this problem. Several higher end buyers have even refused to continue to buy Rwandan coffee because of it. Larger roasters selling to big buyers like Walmart and Target are also refusing to buy Rwandan coffee in large quantities for these markets due to the defect.*

The source of this odd taste in the coffee has been identified as the *antestia* bug, and it is urgent to find ways to control it because it is jeopardizing Rwanda's gains in international marketing of high-quality coffee.

In conclusion, in spite of the short-term downturn in world prices for specialty coffees, it is clear that coffee can continue to be a substantial income earner for Rwandan farmers over the medium and longer term and increase its export earnings. Therefore it passes the filter of markets. However, it is also clear that efforts to improve quality, at the levels of production, harvesting, and processing, have to take precedence over acreage expansion at the present time. In effect, there are important issues in regard to the value chain and on-farm management of the crop. It is recommended that caution be exercised in planting new coffee trees in the watershed development projects until more progress is made in resolving these issues. Specifically in the case of coffee this study recommends:

1. If coffee plantations exist in some of the watershed development areas, work with the national coffee program to help intensify the efforts to improve cultivation and harvesting practices and postharvest management for those farmers, and to accelerate the rate of improvement of the washing stations.
2. Plant new coffee trees only where improved washing stations exist that are willing to process higher volumes of cherries, and implement intensive programs of farmer training for the newly planted areas.

Coffee washing stations are widespread throughout most of the country, except for the far north and northeast. In the project areas it will be important to involve any stations that exist there in decisions concerning coffee plantings, and ascertain whether they would prefer to see the emphasis placed on improving the quality of harvested coffee in their areas. Therefore the amount of coffee planted in the new project areas could turn out to be even less than official projections.

Application of Value Chain Filters

-   *Seeds and breeds*: replanting of older, less productive trees needed.
-   *Production*: (1) urgent need to solve the potato taste issue; (2) need to improve application of agrochemicals in line with good agricultural practices.
-   *Harvesting*: need to continue progress in training farmers to collect only the berries of uniform and appropriate ripeness at each harvest.
-   *Postharvest*: no issues.
-   *Storage*: no issues.
-   *Transportation*: no issues.
-   *Processing*: additional washing stations needed along with training in their management.
-   *Market development*: new external markets including in Asia would be useful but are not critical for expansion of the sector.
-   *Value chain governance*: no issues.
-   *External and policy*: need to implement more fully pricing schemes that give premia for quality.

TABLE 8.4 Rwanda: Value Chain Quality Assessment Matrix for Coffee

Value Chain Stage	Domestic Markets		Export Markets	
	Fresh	Processed	Fresh	Processed
1. Seeds and breeds				GE
2. Production				2TA, CM
3. Harvesting				TA
4. Postharvest				
5. Storage				
6. Transportation				
7. Processing				PR, TA
8. Market development				
9. Value chain governance				
10. External and policy				PI
Sum of scores				8

Total score on quality issues: 8. GE refers to the need to replant older trees; TA under production refers to the urgent need to solve the potato taste issue; CM refers to the need to improve application of agro-inputs; TA under harvesting refers to the need to harvest more uniform batches of berries; PR to the need for additional coffee washing stations; TA under processing to the need to improve coffee washing techniques so that the beans' quality (taste) is not affected; and PI to the need to implement more widely the better pricing schemes that reward producers for quality.

The VQA Matrix for Rwandan Coffee

Table 8.4 shows coffee's competitiveness score according to Track 2: a little worse than tea (more issues and more serious issues to be resolved), but better than some of the other products analyzed later. In the case of coffee, the only market of relevance is the export market for partially processed product, and the main issue to be resolved is the potato taste problem, which requires agricultural research efforts for the crop.

8.6.3 The Banana Family

The Field Findings

Banana is by far the dominant fruit crop in Rwanda in terms of value of production, and it is a basic crop for subsistence and domestic markets. Bananas are also export crops to a degree. Four members of the banana family are consumed or produced in Rwanda: cooking bananas, beer bananas (sometimes referred to as wine bananas), apple bananas, and plantains. Although some commentaries treat plantain and cooking bananas as interchangeable terms, they are genetically different and are prepared differently in households.

Apple bananas (also known as baby bananas, dessert bananas, and sweet bananas) have the highest unit value and are being exported successfully in organic form to Europe. The elements of the value chain are largely in place, although one link in particular needs to be improved (the management of the airport cold storage facility). In addition, there is a large export market for organic dried apple bananas. Representatives from the Belgian Technical Cooperation and RHODA attended the European trade fair Biofach 2009 and found that one importer was interested in 8 containers of organic dried apple banana per year from Rwanda, a volume that the country is currently in no position to supply.

Another factor in favor of apple bananas is that recently market tests have been carried out in Germany for a Rwandan liquor based on these bananas, and the results were very favorable. However, this crop is the most vulnerable of the banana family to the feared *Fusarium* wilt disease (*Fusarium oxysporum f.sp. cubense*), and this disease has been debilitating for its production. It is present in all producing areas of the country. To date researchers have not developed an effective approach for controlling this disease or significantly moderating its effects. Judicious irrigation and mulching to reduce drought stress on the plants helps defend them against *Fusarium* wilt to a degree, and accordingly techniques of integrated pest management (IPM) have been developed to mitigate somewhat the effects of this fungus on all bananas including apple bananas.

Recent findings of banana researchers include the following (Murekezi & Van Asten, 2008):

- Survey results from 5 sites in Rwanda show that 85% of farmers reported drought stress and 74% poor soil fertility as being major constraints to production, and researchers estimated drought stress losses that correspond to farmers' perceptions.
- *Fusarium* wilt is a major constraint, especially for apple banana and "exotic" beer banana as well.
- Although yield was positively correlated with amount of mulch applied, very few farmers applied external mulch and even fewer manure.

Provision of irrigation together with improved crop husbandry (proper mat spacing, composting, mulching, applying manure) would have far greater impact than irrigation alone, but evidently continuous on-farm technical assistance will be necessary to ensure it is practiced. There are examples in Kirehe of properly managed apple bananas in this sense.

Apple bananas definitely have good market prospects, both domestically and externally, and so they must be considered as a major option for selected zones in the watersheds. However, in light of this disease issue, the caveat for apple bananas is that intensive technical assistance must be provided on a continuous basis to growers, and cultivation practices must be monitored for a few years until farmers can clearly see the benefits of better crop husbandry. It is already a tradition to interplant apple bananas with cooking bananas and/or beer bananas, which are more resistant to *Fusarium*, and this tradition should be followed in the newly developed watershed areas.

With appropriate varietal trials, the FHIA 01 variety of bananas may be planted in place of Rwandan apple bananas if they would be destined for use in the production of banana chips. Tests of consumer preferences in Europe showed that chips produced from the FHIA 01 variety were preferred to all other options, including apple bananas (Van Asten, Florent and Apio, 2008). However, until a processor confirms interest in making chips from this banana for export, and the capacity to do so, this variety would not pass the value chain filter. It appears that for Rwandan consumers the taste of FHIA 01 does not match that of apple bananas.

Rwanda imports plantain, chiefly from the Democratic Republic of Congo (DRC) and produces a small amount, but in general, climatic conditions in Rwanda will not permit an acceptable import substitute to be grown locally, so it does not pass the production filter. On the other hand, for cooking bananas domestic supply falls considerably short of demand, and there are substantial imports that could be replaced. They come mainly from the DRC and secondarily from Uganda. If production were to increase sufficiently it would even be possible to engage in some exports since the prices of cooking bananas are higher in Burundi than in Rwanda, at both farm gate and retail levels. This type of banana is more resistant to *Fusarium* wilt, especially when it is irrigated and farmers practice mulching.

Cooking bananas are excellent for soil retention on slopes and terraces. Approximate calculations of the market indicate that about 1500 additional hectares of cooking bananas could be planted without affecting prices in the downward direction. Over the longer run the consumption budget share devoted to bananas declines as incomes rise, but banana consumption still will rise in absolute terms for quite a long time and current volumes of imports provide ample space for competitive import substitution by growth in domestic output. Cooking bananas benefit more from irrigation (suffer more from drought) than apple bananas do. An indication of the importance of cultivating cooking bananas under irrigation is the fact that their price rises much more in the dry season than in neighboring countries, which receive more rainfall throughout the year. Irrigation extends the length of the harvest and thus would give farmers opportunities to sell at prices that are higher than the ones they receive now. (Seasonal shortages are the main reason for imports of this product.)

In addition to monoculture plantings of cooking bananas, small areas in the watersheds could be devoted to evaluating intercrop systems with this kind of banana. Research in Uganda has shown that intercropping with coffee benefits both crops, and similarly intercropping with beans appears to be a productive combination provided the beans are managed in a no-till way. Also, as noted, intercropping can be implemented with apple bananas with irrigation and appropriate cultivation techniques. To illustrate the very substantial gains that may be had from intercropping bananas with coffee, in Arabica coffee—growing areas of Uganda, annual returns per hectare were US\$ 3421, US\$ 2092, and US\$ 1552 for intercropped banana-coffee, monocropped banana, and monocropped coffee, respectively (Van Asten, Mukasa and Uringi, 2008).

Exporters have also expressed interest in red bananas, but the issue that remains to be clarified for this variety is whether there would be a domestic market for the rejects—the fruit that does not meet export quality standards. As of this writing this issue is being investigated.

Finally, the importance of organic and fair trade certifications should not be overlooked. The apple bananas that Rwanda is currently exporting are certified as organic, and this has helped them secure a consistent market in Europe. In conclusion, the banana family is important for Rwanda and has been shown to be competitive both on domestic and export markets, but a number of issues hinder expansion of banana crops. None of the issues are insurmountable with the exception that *Fusarium* wilt can only be managed, not eliminated. The prospects for new banana export products are bright given Rwanda's strength in producing a number of types of bananas. Overall, issues faced by bananas are somewhat more daunting than those faced by coffee (Table 8.5), but the country has underlying competitive strengths in bananas.

Application of Value Chain Filters

- *Seeds and breeds*: no issues.
- *Production*: there is a need for more widespread use of mats, composting, and IPM; more irrigation; and more intercropping including of different banana varieties
- *Harvesting*: no issues.
- *Postharvest*: no issues.
- *Storage*: spoilage of bananas for export markets occurs in the airport storage facility.
- *Transportation*: no issues.
- *Processing*: need for processing capacity for banana liquor and expanded capacity for banana chips.
- *Market development*: (1) export markets can be developed for banana liquor, organic apple bananas, chips from FHIA 01 bananas, and fresh red bananas; (2) certifications required for exports (organic, fair trade).
- *Value chain governance*: value chain development needed, with reliable suppliers, for new export products.
- *External and policy*: no issues.

The Value Chain Quality Assessment Matrix for Rwandan Bananas

Drawing upon the filters applied to each of the value chain stages, the VQA matrix for bananas (Table 8.5) reflects the complexity of the markets and the range of varieties of these important crops. Obviously crop management for control of pathogens figures among the most important issues for these crops. The score of 12 is higher than the scores for tea and coffee, but per *market* the average score is only 4, which reflects the competitive strength of bananas. As the matrix shows, bananas are more competitive for the domestic market, especially cooking bananas, but apple bananas have a reasonable degree of competitiveness on export markets also.

TABLE 8.5 Rwanda: Value Chain Quality Assessment Matrix for Bananas

	Domestic Markets		Export Markets	
Value Chain Stage	Fresh	Processed	Fresh	Processed
1. Seeds and breeds				
2. Production	CM, PD		CM, PD	CM, PD
3. Harvesting				
4. Postharvest				
5. Storage			SP	
6. Transportation				
7. Processing				PR
8. Market development			NM	NM, CE
9. Value chain governance				VC
10. External and policy				
Sum of scores	2		4	6

Total score on quality issues: 12. In production, CM refers to the need for more widespread use of mats, composting, and IPM; more irrigation; and more intercropping, including of different banana varieties; PD to the prevalence of *Fusarium* wilt that can only be mitigated by these practices; NM to the need for more exporting of dried organic apple bananas, a banana liquor, and perhaps red bananas and FHIA 01 banana chips; PR for the need for processing capacity for banana liquor and expanded capacity for chips; CE to the importance of certifications for expanded exports; SP to the spoilage that occurs in the present airport storage facility; and VC to the need for new value chains for new variants of exported products.

8.6.4 Avocado

The Field Findings

Rwanda's ecosystems have inherent advantages in the production of avocado. Studies in Cameroon have shown for a range of varieties of avocado that quality characteristics (taste, percentage of oil) improve as altitude in which the crop is grown increases, which lowers temperatures (Ministère des Affaires Etrangères, France, 2002). In addition, the crop's water requirements are 1200—1600 mm of rainfall per year, well distributed through the seasons, and that characterizes the rainfall in most of Rwanda.

The local variety of avocado in Rwanda is superior in taste and consistency but the international market prefers the Hass variety, which also has good taste and in addition has a tougher rind and therefore is not as perishable and is easier to ship long distances without damaging the fruit. If someone were to make a sample shipment of Rwandan local avocadoes to Europe by air, carefully packed, and carry out market trials, demand for that avocado might well be created given its special qualities. Currently one Rwanda exporter is shipping to Germany very small quantities of organic apple bananas, passion fruit, and both local and Hass avocadoes (the latter growing near Butare in the southwest), but it is not yet clear whether a larger export market for local avocadoes exists. In the meantime any export of avocado would have to be of the Hass variety.

The domestic market for the local variety appears to be in balance regarding supply and demand. Therefore plans for expansion of the areas planted in avocado would have to concentrate on Hass avocadoes. It is a variety that originated in Guatemala and does best at higher elevations, at altitudes of 900—2400 m above sea level (Ploetz, Zentmyer, Nishijima, Rohrback and Ohr, 1994). Avocado is a crop whose prospects are closely tied to plans for improvements in the transportation network in East Africa. More than one entrepreneur has commented that avocado exports (Hass) would be a viable proposition via the proposed railway to Isaka, Tanzania, and from there by sea to Europe. In fact, there appears to be interest in planting Hass avocadoes in the next year or two given that the trees would begin to yield when the railway is scheduled for completion, in 2014.[4] This illustrates the importance for a landlocked country of international transportation links with neighboring countries.

Specifically, the Managing Director of East African Growers (EAG), which already has invested in Rwanda vegetables for export, came to Rwanda to explore the possibility of setting up production of Hass avocados (up to a total of 1000 ha), which they would hope to ship via the new railway link. Trees would come into production in 3—5 years' time. The Managing Director said that the company would even be willing to invest in installation of irrigation facilities, provided they were done correctly. He also insisted on improvements to management of the airport cold storage facilities, and improvement in air cargo services. If this project were to go ahead, EAG would set up an avocado conditioning and packing facility in Kigali, similar to the one they have in Nairobi, and would provide all the expertise to enable farmers to correctly produce and harvest the crop. The packing and conditioning facility would wax and pack the fruit so that they could be shipped in containers via the railway to Tanzania (estimated transit time of 3 days), and from there uploaded onto ships. Hass plantlets would be provided by EAG and distributed by MINAGRI to selected growers who are geographically clustered. Sanitary certifications would have to be obtained for export.

As an example of a functioning alliance between growers and exports, for its Rwandan French beans operation EAG provides all inputs on a credit basis (paid back by farmers when the crop is harvested) and has an excellent repayment rate; farmers benefit from having the assistance in procurement of these otherwise hard-to-access inputs, and the EAG has a memorandum of understanding (MOU) with RHODA whereby all contracts with farmers are witnessed and monitored by RHODA. This reinforces the farmers' respect for the contract with EAG. It also illustrates how governments can play a facilitating role for creating value chain alliances.

Avocado trees could be expected to start yielding about 0.9 tons/ha after three years and 21 tons/ha from year seven onward. As soon steps are taken to resolve the cold storage deficiencies and MINAGRI agrees to collaborate by providing extension services to farmers in coordination with the investor, the EAG is willing to initiate the project by investing in a nursery for avocado plants.

Application of Value Chain Filters

In the Rwandan case, avocado is a crop whose possibilities of future growth are closely tied to the development of the East African transportation network, specifically rail. Hence the transportation filter is by far the most important filter for this crop. The interest of commercial growers in exporting Hass avocadoes, and their previous experience in Rwanda and Kenya, suggests that they would take the lead in constructing an export value chain if the transport link were built.

- *Seeds and breeds*: no issues.
- *Production*: the prospects for Hass avocadoes depend on investments by companies like EAG that would supply plantlets, other inputs, and credit; the government would have to direct some of its extension resources to this crop in coordination with investor(s).
- *Harvesting*: no issues.
- *Postharvest*: no issues.
- *Storage*: the government needs to construct more cold storage facilities and put their management in private hands.
- *Transportation*: development of Hass avocadoes depends crucially on a railway link to seaports, and the preferred option appears to be one through Tanzania.
- *Processing*: no issues.
- *Market development*: an investor would develop the markets in Europe for Hass avocadoes but further export market explorations could be made for the local variety.
- *Value chain governance*: no issues; experienced investors would construct value chains from farmers to export markets.
- *External and policy*: the transportation policy framework and financing needs to improve in three countries of the region so that the mooted railway construction can move forward.

TABLE 8.6 Rwanda: Value Chain Quality Assessment Matrix for Avocado

Value Chain Stage	Domestic Markets		Export Markets	
	Fresh	Processed	Fresh	Processed
1. Seeds and breeds				
2. Production			CM	
3. Harvesting				
4. Postharvest				
5. Storage			SP	
6. Transportation			2SP	
7. Processing				
8. Market development			CE, NM	
9. Value chain governance				
10. External and policy			2PI	
Sum of scores			8	

Total score on quality issues: 8. In production, CM refers to the need for assignment of extension resources to this product; SP in storage refers to the spoilage that will occur without a better cold chain; SP in transportation refers to the spoilage that will occur with attempts at overland transportation to seaports that is not by rail; CE refers to sanitary certifications required for export; NM refers to efforts to develop export markets for existing avocado varieties in Rwanda; and PI refers the need for a more clear and effective policy framework in three countries for promoting investment in regional railways.

The Value Chain Quality Assessment Matrix for Rwandan Avocado

With the information in the filters applied to each of the value chain stages, the VQA matrix for avocado is as shown in Table 8.6. The overall score of 8 (for only one market) indicates an opportunity to realize the competitive advantage of avocadoes in Rwanda that is perceived by agribusinesses, but two constraints are severely binding: transportation infrastructure and the related policy frameworks.

8.6.5 Pineapple

The Field Findings

Pineapple is a significant horticultural crop in Rwanda, grown mainly in the warmer eastern and southern parts of the country. It is not as vulnerable to infestations and pathogens as crops such as passion fruit, tree tomato, and apple bananas, so from a viewpoint of the production filter it is a crop worth considering. On the surface it would appear that there are possibilities for increasing production to substitute for the rather substantial amount of pineapple imports that flow into the country, mainly from Uganda, DRC, and Burundi. According to an admittedly imprecise cross-border survey, in 2007, Rwanda imported more pineapple (about 43,000 tons) than it produced (about 31,000 tons). Some estimates of imports are higher.

There is a market for pineapple juice, pulp, and jam as well. Enterprise Urwibutso, among others, produces pineapple juice, and Shema Fruits is exporting 100 jerry cans (29 kg/jerry can) of fruit jam by surface to France per week. Pineapple jam, which is perceived by the importer to possess a unique taste characteristic, comprises about half these exports, and guava and mixed fruit jam comprise the rest.

In spite of these favorable circumstances, doubts have been expressed about the extent to which additional pineapple should be promoted. The doubts concern both markets and the value chain. The first doubt arises from the fact that extensive additional areas of pineapple have been planted in the past 2 years, and it is not yet known how much additional production will be brought onto the market from those areas. The second issue is that farm-to-market transportation of pineapples is still carried out in rudimentary fashion in most cases and, together with the lack of a cold chain, this results in considerable losses of fruit and degradation of its average quality when it reaches markets. At least one person involved in the trade in a managerial capacity said to the authors that pineapple supplies are insufficient because the production areas are far from Kigali and the fruit arrives in the capital in very poor condition.

Because of the state of rural access roads, in some cases pineapple can be obtained in better condition from Uganda, and also from DRC, for the areas of the country located close to those borders. Transportation conditions have created a market for pineapple that is semisegmented spatially. In addition, the circumstances generate a relatively large amount of fruit suitable only for processing as opposed to consumption in fresh form, so in relation to demand there can be a surplus of pineapple juice and jam on the local market, whereas fresh supplies are in balance with demand or in shortage.

From an international viewpoint, horticulture experts point out that owing to Rwanda's climate and the particular variety of pineapple that is produced in the country, it is at a comparative disadvantage with respect to major pineapple producers in countries such as Ghana, Ivory Coast, Cameroon, Southeast Asia, and even Costa Rica and Panama, where the MD2 variety, which is preferred in industrialized countries, is planted and exported by ship on a considerable scale. In addition to the varietal factor, Rwandan pineapple exports to countries outside the region currently have to go by air, and the relatively low value—weight ratio of pineapple (in relation to many other fruits and vegetables) means that the airfreight costs constitute another hurdle facing Rwandan competitiveness in the crop. Finally, it should be noted that because of climate the pineapples from Uganda and Burundi are considered to be of higher quality in regard to flavor.

However, it remains true that Rwanda exports small quantities of higher value—certified organic pineapple by air to Europe and that it possesses a pineapple processing industry. On the other side of the ledger, this exporter is not planning to increase the volumes of organic pineapple shipped by air. In a longer term perspective, a manager of a major fruit and vegetable operation in the region believes large potential markets exist in the Middle East and recommends looking for seasonal gaps in Asian production of tropical fruit including pineapple, as well as markets in the Middle East and European Union for certified organic dried tropical fruit.

There may be other sources of export demand. Enterprise Urwibutso has found importers interested in "pineapple concentrate" (which is the syrup with water content further reduced) in Switzerland and Singapore. To produce the concentrate the company needs first to invest in new equipment, which it is in the process of doing. The owner of the company said that to meet the future demands for pineapple concentrate, the area planted in pineapple would need to increase by 1700 ha. He would be interested in working with the hillside irrigation program to increase his pineapple supplies.

Among agriculturalists there is not a consensus about the supply—demand balance for pineapple, but the viewpoint of this study is that there can be room for additional expansion of pineapple plantings of the existing variety for the domestic fresh market and for processing, particularly if cold stores are built in the producing areas and road transport conditions are improved. Transportation and handling costs would absorb a good part of the price differential with respect to pineapples sourced from Uganda, so a modest increase in additional domestic supply to the capital city area should not affect prices appreciably, and the quality differences vis-à-vis fruit from neighboring countries, although they exist, are not great. It is recommended that any increase in pineapple acreage under the watershed projects be limited to about 500 ha (which should yield about 8000 tons of fruit under optimal production conditions and probably less) until the effects on the market of the recent plantings can be properly assessed. If processors begin to produce and export dried pineapple or pineapple concentrate, then the demand could rise significantly and the additional area feasible for pineapple might rise to 1000 ha or more.

Baby pineapples have been mentioned in some of the literature on Rwandan horticulture, but in recent years this product has been supplanted in international markets by the MD2 variety of pineapple and other full-sized varieties.

In brief, the main issues concerning pineapple are in the filters concerning transportation, cold storage, and market development. On the production side, pineapple is a profitable crop for producers and relatively easy to manage. Calculations have shown that the profit margin for producers is sufficient that modest declines in prices, if they were to occur, could be absorbed.

Application of Value Chain Filters

- *Seeds and breeds*: no issues.
- *Production*: no issues.
- *Harvesting*: no issues.
- *Postharvest*: no issues.
- *Storage*: more cold storage facilities are needed.
- *Transportation*: growth of the pineapple sector depends very much on improving the internal road network, especially from pineapple-producing areas to the capital city; also, vehicles for chilled transportation are needed.
- *Processing*: no issues; the needed new processing facilities are being built.
- *Market development*: (1) exploration is warranted of Middle East markets for fresh and processed pineapple and both Middle East and EU markets for certified organic dried pineapple; (2) specifically, international markets for pineapple concentrate should be investigated, starting with Switzerland and Singapore.
- *Value chain governance*: no issues.
- *External and policy*: the policy framework for domestic transportation and cold chains needs to be improved.

The Value Chain Quality Assessment Matrix for Rwandan Pineapple

The overall score for pineapple is a relatively modest 11 (Table 8.7), with most of the "demerits" relating to transportation and its policies. This signifies that if the internal transportation issues can be resolved there would be scope for expansion of production, both for fresh markets and processed markets, conditional on how much additional pineapple would be brought to market from the recent new plantings. However, per *market* its average score is only 3.75, as a reflection of its competitiveness in the domestic for fresh and processed fruit and export markets for processed fruit. That one dominant issue of transportation (road networks), however, could be expensive to resolve.

8.6.6 Passion Fruit

The Field Findings

Passion fruit is the fruit of choice for juices in the domestic market, as confirmed by processors such as Inyange Dairy and Urwibutso. Rwandan passion fruit also has a superior flavor that is well received in the European and regional markets. A manager of Inyange Dairy said that expressions of interest for pure (no sugar or added preservatives) pasteurized passion fruit juice have come from Oman and Dar es Salaam, but that unless they can guarantee the consistency of both the quantities and quality of the product they provide to importers, there is no point in trying to develop an export market. Their factory

TABLE 8.7 Rwanda: Value Chain Quality Assessment Matrix for Pineapple

	Domestic Markets		Export Markets	
Value Chain Stage	Fresh	Processed	Fresh	Processed
1. Seeds and breeds				
2. Production				
3. Harvesting				
4. Postharvest				
5. Storage	SP	SP	SP	SP
6. Transportation	2SP	2SP	2SP	2SP
7. Processing				
8. Market development			NM	
9. Value chain governance				
10. External and policy	2PI		PI	
Sum of scores	5	3	5	3

Total score on quality issues: 16. In storage, SP refers to the spoilage that will occur without a better cold chain; it affects fresh pineapple most but also causes some loss of pineapple destined for processing; SP in transportation refers to the considerable spoilage that occurs in domestic transportation because of deficient road networks and lack of vehicles with cooling capabilities; NM refers to the need for developing new markets, as mentioned earlier, in both Europe and the Middle East; and PI refers the need for a more clear and effective policy framework governing internal investments in transportation and the cold chain.

would have to be certified before attempting to sell on the international markets.

With its new ultra-high temperature (UHT) processing facility, Inyange Diary estimates that from July onward it will need 63 tons of passion fruit daily, an amount the sector will be hard pressed to provide. As another indication of the market's preference for Rwandan passion fruit juice, trucks arrive regularly from Uganda to purchase passion fruit grown in the northern part of Rwanda. Entrepreneurs claim to have found additional markets for Rwandan passion fruit juice in Canada, the United States, and elsewhere. In addition to its flavor, passion fruit juice has a longer shelf life than pineapple juice.

Although passion fruit passes the market filter easily, to date the volumes of juice exported to the extraregional market have been small. The limiting factors are production, which has been devastated in some areas by diseases, and postharvest handling. The main enemies of the plant are passion fruit woodiness virus (PWV) and the *Septoria* spot fungus. The plant is also vulnerable to the less serious diseases anthracnose (*Colletotrichum gloeosporioides*), brown spot (*Alternaria* spp.), and the cucumber mosaic virus.

These diseases represent very serious threats to the passion fruit sector in Rwanda, and because of their greater presence in northern parts of the country it is not recommended that the crop be planted in new watershed developments in those areas. Given the overarching importance of the disease issue for passion fruit, it is worthwhile to review how they affect the plant and what the appropriate responses could be for the hillside irrigation projects. Persons in the industry feel these disease issues may not yet receive the priority they warrant in MINAGRI. Research by one of the present team members reached the following conclusions regarding PWV.

This disease was first identified in Rulindo in 2002 and has since become widespread in Kigali-Ngali and has been observed from Ruhengeri to Cyangugu. The primary symptom of this deadly viral disease is a thickening and hardening of the fruit's skin, the smaller fruit size, and the greatly reduced juice content. The foliage is affected and plant growth is stunted. Depending on the age of the plant when it is infected, the onset of the viral infection may mean that no marketable fruit is produced. The virus is transmitted by insects and by mechanical means such as cultivation tools. As long as the plant is not stressed it can be infected and show no symptoms, but once the growing conditions become less than optimal owing to drought, insufficient plant nutrition, cool temperatures, and other conditions, the virus manifests its symptoms and there is no remedy save uprooting the plant and burning it or destroying it completely. It is important to note that PWV has seriously affected passion fruit plantations in Kenya, to the extent that many producers there have abandoned the crop. Researchers around the world have not found a PWV-resistant variety of passion fruit, so the only course of action is to prevent it.

Regarding the *Septoria* spot fungal disease, it is found throughout Rwanda's passion fruit growing areas. The pathogen attacks both leaves and the fruit, producing brown spots that remain small (c. 2 mm) on the leaves and stems but which can coalesce to cover large areas of the fruit. Affected fruit ripens unevenly and sometimes rots entirely, and the juice can become unacceptable for processing. Even a light infection can reduce the value of the crop sharply or render it useless. *Septoria*, like other fungi, is favored by high humidity and therefore attacks when the plant is not properly pruned and trellised. What makes this infection especially troublesome is that it can be transmitted by seeds. Given that the infection begins early in the plant's life through the seeds and spreads quickly throughout nurseries by water-borne dispersal (and failure of most producers to remove infected seeds), use of a fungicide, if a registered one existed for passion fruit in Rwanda, would not be effective. Again, the best approach is prevention: careful selection of the fruit from which seeds are to be extracted, and proper inspection and hygiene in nurseries and in the field.

Disease management techniques for passion fruit are centered on the following measures aimed at prevention:

1. *Selection of the growing environment to minimize stress*: Passion fruit should be grown at altitudes of 1200–1700 m. Higher elevations up to 1900 m can be tolerated only if soil fertility is high, drought stress can be avoided, and careful crop management is employed. Passion fruit requires well-drained soils, and therefore valley bottoms, which are humid most of the time, should be avoided.
2. *Seed selection and extraction*: Only seed of known origin should be used. Seeds should be selected from plants that show no symptoms of disease, grow vigorously, and produce purple fruit. It is preferable to let the fruit mature fully on the plant before using it for seed. If a producer has no access to passion fruit other than that found in the market, then only fruit with absolutely no disease symptoms should be used. Before extracting the seed, the fruit should be dipped in sodium chloride solution to kill surface pathogens. Technical assistance can guide the farmers in seed selection procedures.
3. *Nursery hygiene*: Nurseries should be located far from passion fruit fields, and their location should be rotated. Seeds should be planted at least 1 cm apart, and seedbeds should be separated by at least 10 cm. Closer spacing favors the propagation of diseases. The nursery should be inspected *daily*, and any plant showing signs of disease should be uprooted and destroyed.
4. *Hygiene at the field level*: Plant rotation and separation are essential. New plantings should be located at least 50 m and upwind from existing ones. Once plants cease to be productive, they should be uprooted and burned. Land on which passion fruit has been grown must be devoted to other crops for at least 3 years before being returned to passion fruit.
5. *Pruning and other practices*: One of the factors that predispose Rwanda's passion fruit to disease is that producers do not practice regular pruning. Without pruning, the plant develops a dense canopy that favors the emergence of fungal infections. It is helpful to keep the soil covered with straw or other organic material, weed regularly (and avoid injuring the plant during weeding), and ensure proper fertilization.

More detailed prescriptions of measures to reduce the incidence of passion fruit diseases can be provided to farmers by technical experts, but these comments are sufficient to show the nature of the requirements for a sustainable passion fruit sector in Rwanda that will capitalize on the country's comparative advantage in this valuable crop.

Inyange Dairy's experiences provide good illustrations of how to develop alliances between growers and the processing industry. The firm, which first provided technical and material aid to selected groups of passion fruit growers, has continued to provide seedlings and technical assistance to growers. Now it does it only in southern regions of Rwanda (Ruzizi, Nyamasheke, and Gikongoro) due to the high levels of passion fruit disease in northern Rwanda. The company believes it is important to build up trust among farmers to ensure that they will sell their product to Inyange. Ugandan traders do represent competition for procurement of passion fruit, but because the Ugandans do not adhere to an agreed upon price, instead dropping their offer as soon as other buyers have left the scene, the Rwandan farmers are beginning to appreciate the loyalty of (as well as technical assistance provided by) the likes of Inyange, and they are increasingly willing to sign long-term contracts with the company. The company negotiates a price range defined by the minimum and maximum prices they will pay for the fruit, according to seasonality of production. The importance of this crop is indicated by the fact that the company would be willing to invest up to RWF 100,000 million/year in technical and other assistance to passion fruit producers.

The experiences with passion fruit of another agroprocessing entrepreneur, Sina Gerard, illustrate well some of the issues surrounding this crop. Sina initiated his now-extensive business with chili sauce, and has exported small volumes of passion fruit via a joint venture with the Kenyan-based fresh produce company VegPro. This initiative was abandoned, however, when it became apparent that Sina could not procure the volumes of passion fruit of the required quality to meet the Belgian-based importer's (Star Fruit) demand. Sina expanded into production of strawberry-based processed fruit products when the passion fruit disease epidemic seriously reduced production of this crop in northern Rwanda, starting in 2003. Although passion fruit juice is preferred to all other fruit juices by Rwandans (and Ugandans), the impact of the disease has made it too difficult for Sina and other processors to procure sufficient fruit to meet consumer demand.

Sina now has a new factory that is capable of producing 1500 L of juice/hour; 60% of production is composed of passion fruit, 25% of strawberry, and 15% of pineapple juice/syrup. The factory has been ISO certified for the following products: processed passion fruit, oil of *Capsicum chinense*, banana, and strawberry wine. Sina hopes to achieve organic certification for all his juice products, banana wine, and chili oil with assistance from the local branch of Ceres, and financial support from RHODA. When asked, he asserted he had quality control measures and could meet traceability requirements at the producer level.

Sina currently exports small volumes of his passion fruit products, chili oil, and banana wine both regionally and to markets in Belgium, Germany, and the Netherlands (probably targeting the Rwandan, Burundian, and eastern Congolese communities in these countries). He also found importers interested in his products in Canada, the United States, as well as the East African Community. EU exports are sent via the airline SN Brussels, given the small volumes shipped.

In sum, passion fruit is an attractive product for markets and entrepreneurs in spite of its issues, but systematic and continuous efforts at disease management, good plant nursery management, and sound alliances between producers and processors are the keys to sustaining and developing this crop in Rwanda. The government may wish to consider a compensation scheme for producers in areas in which all passion fruit has to be uprooted, and not replanted for several years, because of pathogens. Otherwise, some producers may be reluctant to destroy their plants, and this will prejudice production in the entire zone.

Application of Value Chain Filters

- *Seeds and breeds*: availability of disease-free seeds is an issue; well-managed passion fruit nurseries that rotate their locations are needed.
- *Production*: mitigating the effects of multiple diseases is an important and difficult challenge.
- *Harvesting*: no issues.
- *Postharvest*: no issues.
- *Storage*: no issues; agroprocessors manage this part of the value chain.
- *Transportation*: no issues; agroprocessors manage this part of the value chain.
- *Processing*: no issues; the needed new processing facilities exist.
- *Market development*: with consistency of quantities and quality new markets can be developed.
- *Value chain governance*: no issues.
- *External and policy*: policies may be considered for assisting producers in areas where uprooting of all passion fruit plants is required.

The Value Chain Quality Assessment Matrix for Rwandan Passion Fruit

The overall score for passion fruit is 18, indicating overall lack of competitiveness or marginal competitiveness at best under the present conditions (Table 8.8). It signals the presence of serious issues blocking full realization of the crop's potential. The issues affect all four markets, domestic and export, fresh, and processed. Management of crop diseases is the most important issue because it can require occasional destruction of the plants in entire zones, without replanting for several years. It also means careful attention to seeds and good nursery management, both of which require large changes in the way producers now operate. However, again, looking at only one market, say the domestic market for fresh or processed passion fruit, the crop appears to have strong competitive potential if the diseases can be kept to a manageable level.

8.6.7 Tree Tomato

The Field Findings

Tree tomato or tamarillo[5] (*Cyphomadra betacea*) is grown mostly in the northwestern and western parts of the country. It is a member of the family *Solanaceae*, which also include potatoes, aubergine, and sweet peppers. Tree

TABLE 8.8 Rwanda: Value Chain Quality Assessment Matrix for Passion Fruit

Value Chain Stage	Domestic Markets		Export Markets	
	Fresh	Processed	Fresh	Processed
1. Seeds and breeds	GE	GE	GE	GE
2. Production	2PD	2PD	2PD	2PD
3. Harvesting				
4. Postharvest				
5. Storage				
6. Transportation				
7. Processing				
8. Market development			NM	NM
9. Value chain governance				
10. External and policy	PI	PI	PI	PI
Sum of scores	4	4	5	5

Total score on quality issues: 18. See the comments under the Application of Value Chain Filters. GE refers to both the need for planting disease-free seeds and the need for special nurseries for passion fruit plants whose locations are rotated frequently because crop diseases reside in soils.

tomato is a perennial bush with a woody trunk that grows to heights between 2 and 5 m in favorable conditions, and it requires a cool climate and optimal temperatures for this plant are in the range of 14–20°C. It is susceptible to wind damage because of its shallow root system but adapts well to a variety of soil types, preferring those of medium texture with good drainage and considerable organic material. It does not require irrigation, but availability of water can extend its productive period, which typically lasts up to 3 years for most of the plants, and improve yields and resistance to viruses.

Although it is a relatively new product on international markets, there is strong demand for tree tomato in fresh form, particularly in the United Kingdom, Germany, the Netherlands, and Spain, and especially if it is organic. Fair trade certification also helps open markets for it. After being imported, it is processed into juices, concentrates, jams, gelatins, and sweets. If processing facilities and adequate transport were available, it could be exported also in the form of fruit pulp or concentrate. Other principal exporters of tree tomato include Kenya, Zambia, and New Zealand.

Some of the actors in the value chain believe that MINAGRI is advising farmers to spray this crop as a routine production technique, which results in high pesticide residues. As alluded to earlier, these residues were found in a sample of tree tomatoes sent by EAG to the United Kingdom, and the shipment was declared unacceptable. For organic tree tomato to succeed, the government would have to delineate zones for organic production—which would be the case for other products in those zones as well. Research has found that tree tomato responds well to organic pesticides, so reduction of agrochemical use need not have deleterious effects on production levels.

The Rwandan horticulture strategy developed by the On the Frontier Group (OTF Group, 2006) emphasizes that success in exporting tree tomatoes will depend on installing a cold chain, developing professional relationships with buyers, becoming a reliable supplier, complying with export certification requirements as well as obtaining organic certification, implementing a traceability system, and giving attention to the quality of packaging.

Another challenge is imparting appropriate practices for postharvest handling. After being harvested, the fruit should be washed, disinfected, and waxed, but producers usually are not accustomed to doing that. For local transportation, the fruit should be placed in containers that are rigid with separations inside, and plastic boxes are the international standard for that purpose. These containers also should be washed and disinfected after every use. Then for international shipments, rigid cardboard boxes typically are used with a capacity of 2—2.5 kg each.

After the unfortunate experience with the trial shipment to the United Kingdom, EAG has sent additional samples to Dubai, and more to the United Kingdom, and is awaiting the response of the importers. EAG believes Rwandan tree tomatoes to be of superior quality and potentially capable of finding export markets. However, it should be noted that diseases (especially viral) usually destroy the crop after 2—3 years. If it were grown for export, a careful IPM program would have to be put in place, windbreaks installed, and a rotation program followed. However, none of these problems are insurmountable.

Application of Value Chain Filters

- *Seeds and breeds*: no issues.
- *Production*: spraying of agrochemicals is excessive; organic production zones and IPM programs are needed; irrigation is beneficial; windbreaks and crop rotations are needed.
- *Harvesting*: no issues.
- *Postharvest*: after harvest, the fruit should be washed, disinfected, and waxed, but producers usually are not accustomed to doing that.
- *Storage*: cold storage needed.
- *Transportation*: appropriate containers are needed and they have to be washed and disinfected after every use.

- *Processing*: processing into fruit pulp and concentrate would increase domestic value added in this crop, but capabilities for that do not yet exist.
- *Market development*: efforts to develop markets in Europe and the Gulf States should be reinforced and export certifications obtained.
- *Value chain governance*: professional relationships with buyers are so far lacking; Rwanda needs to become a reliable supplier of the crop and develop traceability systems for it.
- *External and policy*: no issues.

The Value Chain Quality Assessment Matrix for Rwandan Tree Tomato

This matrix was presented in Section 6.4 to illustrate the Track 2 methodology and is repeated here as Table 8.9 for convenience. The total score in the matrix is a very high 35, which indicates that this crop is not competitive on any significant scale in the present form, although it is closer to competitiveness on domestic markets than on export markets. The matrix shows the presence of many issues of importance for the crop. However, it is a high-value crop and appears to have potentially strong export markets, so it may be worthwhile to try to solve these issues on a pilot scale, working with producers in a given

TABLE 8.9 Rwanda: Value Chain Quality Assessment Matrix for Tree Tomato

Value Chain Stage	Domestic Markets		Export Markets	
	Fresh	Processed	Fresh	Processed
1. Seeds and breeds				
2. Production	CH	CH	2CH, PD	2CH, PD
3. Harvesting				
4. Postharvest	HY	HY	2HY	HY
5. Storage	SP	SP	SP	SP
6. Transportation	SP	SP	SP, 2 PK	SP, 2 PK
7. Processing		PR		2 PR
8. Market development			2CE, NM	
9. Value chain governance			2VG	2VG
10. External and policy				
Sum of scores	4	5	14	12

Total score on quality issues: 35. See comments under the Application of Value Chain Filters.

area of the country and a particular exporter to develop latent competitive strengths. Once the ability to surmount the difficulties were demonstrated, then the pilot experience could be scaled up.

8.6.8 Tomato

The Field Findings

Tomato is Rwanda's second largest vegetable in volume produced and area cultivated (after cabbage) and is the largest in value. It is sold on the domestic market fresh and in processed form. Rwanda is a net importer of tomatoes, mainly from the DRC and Burundi, although it also exports some to those countries and others in the region. There appears to be excess demand on the domestic market since some processors are anxious to obtain greater supplies of fresh tomatoes for conversion into tomato paste. Hence, tomatoes pass the market filter easily, with the domestic market constituting the main potential outlet for additional production.

Processors, specifically the company SORWATOM, still suffer from the competition with imported tomato paste subsidized by the Italian Government, but they feel that if they could produce sufficient volumes of processed tomatoes they would be competitive. It is believed that eastern DRC is a large potential market for Rwandan tomato paste since it is a common ingredient in Congolese cuisine. There is also some potential for exporting tomato paste to Burundi.

However, there are significant concerns about the quality of the tomatoes produced, at least as they are received at processing plants and supermarkets. This concern is attributable to both production practices (mainly insufficient disease control) and deficient postharvest handling, so there are issues for these filters.

At the field level, tomatoes suffer drought stress during the dry season, and pests and diseases pose a serious constraint to production. Insects destroy young seedlings on a large scale in some fields; *Phytophthora* (late blight) is particularly severe during the rainy season and when overhead irrigation is used. There are also viral diseases that can drastically reduce production. One appears to be tomato yellow leaf curl virus from a description of the symptoms. The national research agency Institut des Sciences Agronomiques du Rwanda (ISAR) has been carrying out tests on 40 varieties of tomatoes, so it is possible that some of these problems can be reduced through research efforts. Irrigation can ameliorate the impact of the dry season on production, but it should not be sprinkler irrigation because this method tends to propagate diseases more rapidly in tomato fields.

In regard to the value chain, a problem that has to be overcome for tomatoes is the lack of strong, functional alliances between producers on the one hand and processors and other buyers on the other hand. SORWATOM, for example, is concerned about side selling of the fruit by the cooperatives that are partners of company. SORWATOM provides inputs (seeds, fertilizers, pesticides) on credit since the producers cannot access credit themselves, in the expectation that producers will repay at the time of sale. However, many traders arrive at farmers' fields at harvest (which is spread out over 1 month) and offer a higher price than the company had agreed to in advance. Evidently, contract farming has not taken root firmly for this crop. However, to complete the picture it must be said that processors need to ensure timely purchases from farmers so that the crops do not get damaged sitting in the sun after the harvest. It would be worthwhile to explore the option of making some payments to farmers prior to harvest, so that they do not harvest prematurely for need of cash. The horticulture development authority RHODA can play a role in helping broker alliances and monitoring their performance to see where problems emerge and devise solutions.

SORWATOM has been working with a MINAGRI-sponsored project to produce tomatoes under sprinkler irrigation at the LUX-Development irrigation scheme in Gashora (Bugesera). However, the overhead irrigation proved detrimental to production during the last rainy season when it, combined with the rain, resulted in extremely high levels of late blight infection. The company would prefer gravity-fed irrigation and is hoping that such a system can be set up at the Gashora scheme.

In conclusion, Rwanda has potential to increase the production of tomatoes, but important issues at the farm level and in the value chain need to be resolved before that can happen. Also, increases in production volumes are likely to be accompanied by somewhat of a decline in prices, but improved productivity at the farm level and more stable selling arrangements could compensate for that development.

Another product that is worth exploring on a pilot scale is cherry tomatoes. They would have a secure market within the country, in supermarkets.

Application of Value Chain Filters

- *Seeds and breeds*: no issues.
- *Production*: optimal irrigation methods are not yet implemented; disease control is inadequate.
- *Harvesting*: no issues.
- *Postharvest*: cooling facilities are not available to take out the field heat; coordination is lacking between producers and buyers in regard to timing of collection of the harvest.
- *Storage*: no issues since the product is taken directly to processors and supermarkets.
- *Transportation*: no issues.
- *Processing*: processing facilities exist; higher volumes and better quality of production are needed to supply the processors.
- *Market development*: no issues.

- *Value chain governance*: value chain relationships are weak and both producers and buyers violate terms of agreements.
- *External and policy*: no issues.

The Value Chain Quality Assessment Matrix for Rwandan Tomato

The total score in the matrix is a high 20 (Table 8.10), which indicates that this crop is not close to resolving its barriers to competitiveness. The matrix shows the presence of many issues of importance for the crop, on both domestic and export markets. However, it is a high-value crop and appears to have potentially strong export markets, so it may be worthwhile to try to solve these issues on a pilot scale, working with producers in a given area of the country and a particular exporter to develop latent competitive strengths. Once the ability to surmount the difficulties were demonstrated, then the pilot experience could be scaled up.

TABLE 8.10 Rwanda: Value Chain Quality Assessment Matrix for Tomato

Value Chain Stage	Domestic Markets		Export Markets	
	Fresh	Processed	Fresh	Processed
1. Seeds and breeds				
2. Production	CM, PD	PD, CM		CM, PD
3. Harvesting				
4. Postharvest	HH, SP	HH, SP		HH, SP
5. Storage				
6. Transportation				
7. Processing				2 PR
8. Market development				
9. Value chain governance	2VG	2VG		2VG
10. External and policy				
Sum of scores	6	6		8

Total score on quality issues: 20. See comments in the Application of Value Chain Filters.

8.6.9 French Beans and Peas

The Field Findings

In the East African context, Kenya is a significant producer and exporter of French beans and peas. However, there is a seasonal window from January to

March when, due to its being drought prone, Kenyan production of vegetables often falls short of demand. Not long ago persons who trade in these products estimated that Rwanda could probably export the following products and volumes over this period: French beans 5 tons/week (delivered two times per week) and the same volumes and frequencies for sugar snap peas and mange tout peas. All crops would have to be GlobalGAP certified, would have to be put into and maintained in cooled facilities (i.e., a cold chain) right after harvest, and be delivered to Nairobi not more than 2 days after harvest. The ideal arrangement would be to harvest in the morning and ship to Nairobi in the afternoon, they said.

The role of airfreight costs has not yet been properly analyzed, but the buyers said if such an arrangement were found to be commercially profitable and feasible, at least one company would provide the bulk shipping containers to the Rwandan producers (at cost) and expect them to conduct an initial grading on farm to reduce transport costs.

This potential has now been more than fulfilled for the case of French beans, by another company. On May 8 2009 *The New Times* wrote:

East African Growers (EAG) Rwanda has exported about 50 tons of French beans to Belgium in a period of six months. The company, which was targeting five tons per week, has doubled the volume to 10 tons per week. East African Growers Rwanda's main customers are Star Fruit.

"With no doubt it's a success story because it was our target for the first six months and we are on track," Aimable Gakirage, Director General of EAG, said in an exclusive interview.

According to Gakirage, currently the firm manages to utilize only 10 ha of the total 50 ha it owns. He said they intend to expand their capacity but it depends on the cargo space, which is their biggest challenge.

"We are constrained by direct cargo space from Kigali otherwise we have a wide range of products that can be exported to the United Kingdom", Gakirage said.

Recently President Paul Kagame visited the Gasabo-based site and he hailed the initiative where he urged local leaders to learn from EAG and use cooperatives for developmental activities.

Among the challenges addressed to the President include expensive cargo space and cold room charges. The president urged the Ministry of Agriculture to look into the issue and ensure the problem is solved. Kagame suggested that the project should be expanded to other areas.

Gakirage said that the company pays $2.2 per kilogram [for transportation to Europe] yet these are not direct flights compared to Kenya where the same amount is paid for direct flights.

"We cannot compete with people from Kenya or other places with direct flights; it is expensive and inconvenient because it also involves reloading," Gakirage emphasized.

Gakirage is a Rwandan who worked in the horticulture sector in Kenya for many years before returning to Rwanda. He therefore had significant experience in the export horticulture industry; this, together with his Rwandan roots has been key to the success of this venture, which is an illuminating experience for the potential for other horticulture crops.

Initially EAG tried producing mange tout and sugar snap peas in addition to beans. However, heavy rainfall destroyed the pea crops. The varieties grown would have to be produced under tunnels to avoid damage from heavy rainfall, but then they could encounter problems of high temperatures and increased incidence of diseases. Those in charge of the experiment would like to have first conducted more varietal trials to determine those best suited to Rwanda, but they felt working with cooperatives that farmed land communally was an obstacle. Now that the EAG has succeeded with the bean exports, the Rwandan government is providing more support, and they may go ahead with more extensive varietal trials for sugar snap peas. The government is also very interested in promoting GlobalGAP certification of Rwandan horticultural producers.

The EAG exports all packaging materials and most inputs (seeds especially) to Rwanda from Kenya. The biggest constraints are lack of irrigation (resulting in seasonal gaps in production); lack of awareness on the part of farmers of market standards and food safety, which results in a very high percentage of rejects in the beans they produce; and a lack of a business orientation in some of the public sector support programs. EAG would be interested in renting a packhouse if one existed, which they believe (together with training of farmers in quality issues and standards) could enable them to expand their exports. They insist the packhouse would have to be under their own management rather than being a government-controlled enterprise.

A big advantage Rwanda has over Kenya for these crops is that there are no "briefcase brokers" for these products who manage to induce farmers to side sell by offering higher prices, or other incentives, which leaves the exporting company short in its supply along with the loss in its investment in provision of inputs (a big problem in Kenya).

The EAG does not think that the airfreight costs—direct from Kigali to Brussels—are too expensive. Gakirage cited a rate of $1.30/kg because for the moment outgoing costs are lower than incoming owing to empty space on the planes. An additional advantage for Rwandan exports is that there is little competition for cargo space on the Kigali to Brussels route and therefore little danger of fresh produce being offloaded, which frequently occurs with exports from Nairobi.

EAG has a client in Belgium interested in importing 30 t/week of French beans, provided they are GlobalGAP certified and meet market and food safety standards (e.g., HACCP certification of the packhouse). Their current exports from Rwanda (about 10 t/week) make it into the Belgian market only because the EAG has convinced the importers that the produce meets GlobalGAP and other requirements, given that they are following the same procedures used by their certified producers in Kenya.

One factor that has made the EAG's French bean export venture in Rwanda a success is that they are able to sell the nonexport grades on local markets at a profitable price, given that green beans are already a common food in the Rwandan diet. If podded pea production were to take off, its success would in part depend on whether or not a local market could be found for nonexport grade, perhaps in the hotel and restaurant industry.

This experience also underscores the importance of a strong alliance between the exporter and the producers, and that such an alliance can help deal with production issues as well. In short, Rwandan exports of French beans have proved themselves to be competitive in export markets, and tests are underway to see if the same is true for sugar snap peas. The question addressed in this assessment is to what extent can French bean exports be expanded, and what problems have to be overcome to allow that.

Application of Value Chain Filters

- *Seeds and breeds*: no issues.
- *Production*: more irrigation needed; farmers need more training in good agricultural practices to reduce the percentage of rejects.
- *Harvesting*: no issues.
- *Postharvest*: packhouse needed for production on a larger scale.
- *Storage*: no issues.
- *Transportation*: no issues.
- *Processing*: no issues.
- *Market development*: no issues.
- *Value chain governance*: value chain relationships are weak.
- *External and policy*: no issues.

The Value Chain Quality Assessment Matrix for Rwandan French Beans and Peas

The total score in Table 8.11 is a relatively low 10 (5 per market), which indicates that the products have relatively few hurdles to realizing their competitive potential. The assessment refers only to French beans, with the possibility that the assessment will be similar for sugar snap peas after pilots are concluded. The production issues refer to implementing more appropriate irrigation modalities and training farmers in good agricultural practices,

TABLE 8.11 Rwanda: Value Chain Quality Assessment Matrix for French Beans and Peas

	Domestic Markets		Export Markets	
Value Chain Stage	Fresh	Processed	Fresh	Processed
1. Seeds and breeds				
2. Production	CM, CH		CM, CH	
3. Harvesting				
4. Postharvest				
5. Storage	SP		SP	
6. Transportation				
7. Processing				
8. Market development				
9. Value chain governance	2VG		2VG	
10. External and policy				
Sum of scores	5		5	

Total score on quality issues: 8. See comments under Application of Value Chain Filters.

especially for application of pesticides. The spoilage that occurs can be corrected with availability of a packhouse. Above all, value chain relationships need to be strengthened by educating farmers on the multiple and long-term benefits of a contract farming regime.

8.6.10 Chili Peppers

The Field Findings

International buyers expressed interest to the members of this study team in purchasing chili peppers from Rwanda. There is good potential for certified organic chili peppers, fresh and dried, on international markets. In the dried form, for example, bird's eye chili (BEC) peppers, they would have the obvious advantage of going by surface freight. In fresh form, local demand exists for good-quality *C. chinense* (Habanero type), which is currently imported by the Nakumatt supermarket from Uganda.

Various types of chili peppers are commonly found growing throughout Rwanda, albeit cultivated only on a small scale for the most part. In many instances, isolated plants have sprouted spontaneously from seed dispersed by birds and other animals. Given the suitable climate for *Capsicum* production, the abundance of labor, and its relatively nonperishable nature, dried BEC chili (*Capsicum fructescens*) was identified as a potential export from Rwanda during the US Agency for International Development (USAID)−funded project called Agribusiness Development Activity for Rwanda (ADAR). This crop, if managed properly, can produce fruit over a 2-year period; moreover, harvesting the chili peppers requires a large pool of labor, which makes its production comparatively less competitive in countries such as South Africa and Zambia where labor costs are much higher than in Rwanda. The high levels of solar radiation in Rwanda (due to proximity to the equator) and the right temperature conditions are conducive to the production of a fruit with excellent color and high levels of capsaicin, the chemical that gives chilies their pungency.

Several entrepreneurs and two cooperatives received assistance from the USAID-funded ADAR project over the period 2004−06 for production and export of organically certified BEC. In the case of BEC, in Rwanda the low use of agrochemicals and reliance on organic fertilizers and cultural practices for soil fertility maintenance provide an advantage to producers seeking organic certification. The added value conferred by certification helps to compensate for Rwanda's high transport costs. The large supply of low-cost labor in Rwanda also favors production of crops such as BEC that require large numbers of workers who do not necessarily have many skills or much expertise. Extra efforts must be made at the managerial level, however, to ensure control of product quality.

Samples of the organic BEC produced by entrepreneurs and cooperatives assisted by the ADAR project were sent to importers in Germany and the United States in 2005. Their evaluation was that the crop was of high quality, and both importers expressed an interest in importing at least one container load of certified organic BEC every couple of months from Rwanda. Unfortunately, budget cuts resulted in early termination of the ADAR project, and with lack of adequate technical and financial assistance all producers eventually let their organic certification lapse and ceased exporting their product.

Given that high-quality organic BEC can be produced in Rwanda, and in view of the continued interest on the part of European and North American importers in procuring this product from Rwanda, this is a crop that appears appropriate for the hillside irrigation project. In addition to close, hands-on technical assistance, financial support for organic certification and purchase of solar tunnel driers (required for production of high-quality dried fruit that is aflatoxin free) would be required for successful production of an export quality crop.

Other types of chili peppers, e.g., *C. chinense* (referred to as Habanero and Scotch Bonnet in English and "piment lantern" in French) have also grown well in Rwanda's climatic conditions. In the case of *C. chinense*, demand exists for the fresh product both on the local market and for processing. The Nakumatt supermarket in Kigali currently imports prepackaged *C. chinense* from Uganda, although it could be produced locally, citing the higher quality and better presentation of the Ugandan product as reasons for importing the fruit. Yet with simple plastic greenhouses or "tunnels" the same product could readily be produced in Rwanda.[6] Fruit of lower quality would find a ready market in the processing sector. Sina Gerard is already producing an oil (sauce) from *C. chinense*, which he sells on both local and export markets (albeit in small quantities for the exported product). A bottled *C. chinense* sauce is made in Burundi and imported to Rwanda, and the same sauce could be produced locally, provided that volumes of production were increased and an interested entrepreneur could be identified.

Fresh Habanero peppers also have export potential that is worth exploring but are more demanding than the dried version in postharvest handling.

In sum, the main requirements for success with chili pepper exports are sustained hands-on technical assistance to farmers and attention to two particular issues in the value chain: installation of solar tunnel dryers at the farm level and obtaining and maintaining organic certification. Markets are not a limitation since exports of fresh chili peppers to Europe have been well received and a sauce also is being exported. Domestic demand exists as well, and current imports from Uganda and Burundi could be replaced with cost-competitive domestic production.

Application of Value Chain Filters

- *Seeds and breeds*: no issues.
- *Production*: intensive technical assistance needed for farmers.
- *Harvesting*: no issues.
- *Postharvest*: solar tunnel dryers needed at the farm level.
- *Storage*: no issues.
- *Transportation*: good packaging needed for *C. chinense*.
- *Processing*: no issues.
- *Market development*: organic certifications needed for export.
- *Value chain governance*: financing needed for certification costs and tunnel dryers.
- *External and policy*: no issues.

The Value Chain Quality Assessment Matrix for Rwandan Chili Peppers

The VQA matrix indicates few obstacles to competitiveness for Rwandan chili peppers, with an average score of less than 4 per market. The high quality of Rwandan chilies has been validated in European markets. They are a crop that requires careful management and assistance to farmers, plus postharvest drying. Finance is critical for this value chain, including for covering the cost of organic certifications. Once these conditions are met, the prospects are bright for significant expansion of chili pepper production (Table 8.12).

TABLE 8.12 Rwanda: Value Chain Quality Assessment Matrix for Chili Peppers

	Domestic Markets		Export Markets	
Value Chain Stage	Fresh	Processed	Fresh	Processed
1. Seeds and breeds				
2. Production	CM	CM	CM	CM
3. Harvesting				
4. Postharvest	HH	HH	HH	HH
5. Storage				
6. Transportation	PK			
7. Processing				
8. Market development			CE	CE
9. Value chain governance	VG	VG	VG	VG
10. External and policy				
Sum of scores	4	3	4	4

Total score on quality issues: 8. CM refers to the need for intensive, hands-on technical assistance and close supervision of production practices; HH refers to the humidity that needs to be removed from the crop at farm level with tunnel dryers; PK refers to the need for better packaging of the fresh product for local supermarkets; CE denotes the need to renew organic certifications; and VG refers to the need for value chain finance, for both the certification costs and the installation of tunnel dryers.

8.6.11 Macadamia

The Field Findings

Macadamia is a tree that originated in Australia. The German botanist Ferdinand von Müeller discovered it in 1848, and commercial exploitation of the tree began shortly after that. The first macadamia tree planted for cultivation is still yielding nuts because its productive period can exceed 200 years. Macadamia nuts are the highest valued nuts in the world.

They are produced commercially primarily in Australia, Hawaii, South Africa, Kenya, Malawi, Colombia, Costa Rica, Guatemala, and Brazil. To date the United States, Australia, and Japan are the principal consuming nations for macadamia nuts, but European and Asian demand for them is increasing. Taiwan consumes more macadamia nuts and oil per capita than any other country. The nut is not only appreciated for its taste and use in cooking but also is a star ingredient in antiaging creams produced by leading brands such as Lancôme. In addition, more than 40% of its composition is oleic acid, a substance similar to olive oil that helps reduce cholesterol. More than 80% of the fatty acids in macadamia nuts are monosaturated. All these factors contribute to the increasing international demand for the crop.

The possibility of planting macadamia in Rwanda has been mentioned by ISAR in its *Participatory Diagnostic Report for the Cyabayaga Watershed* (2006) and in the RHODA Business Plan, but in both cases without elaboration. Since the crop is not yet well known in Rwanda, a few comments are offered in this section about its cultivation.[7] Macadamia grows well with annual precipitation of 1500—3000 mm, well distributed throughout the year and with no more than 2 months of drought. Thus irrigation during the dry season can benefit this crop. Its preferred temperature range is 18—29°C. Normally these conditions are found at altitudes of 400—1000 m above sea level, but the crop can be cultivated up to 1200 m.

Establishing nurseries is important for macadamia production, and grafting rather than propagation by seeds is recommended. The trees begin producing in the fifth year, and yields continue to increase until the 50th year.

The nuts that have fallen should be collected at least once every 2 weeks, and then they are carefully dried until the humidity drops to 3.5%. The commercial product is toasted and vacuum-packed nuts. Macadamia trees benefit watersheds by helping conserve sources of water. They are also favored hosts for bees.

Macadamia has significant long-term potential for Rwanda. Among other considerations, its high unit value reduces the importance of international transport costs. However, commercial-scale production would require an investor in the processing industry and a program of training and supervision of producers. The government may wish to try to interest more investors in this crop, and in the meantime it is listed as one of the crops for which it would be worth carrying out trials.

Experimental cultivation of this crop was started at Rwamagana in 2007. Since the original draft of this case study a small but thriving processing and export industry for macadamia nuts has developed in Rwanda, confirming the crop's suitability for the country. NORLEGA Macadamia Rwanda Ltd., FRESHCO Rwanda Ltd., the Rwanda Nut Company Ltd., and Farm Gate East Africa Ltd. are companies producing roasted, well-packaged nuts. NORLEGA in particular has established strong value chain ties to producers. In 2015 a Kenyan company announced plans to plant 1 million macadamia trees and involve 20,000 farmers in its cultivation. It is estimated that in 10 years Rwanda can be earning $200 million per year from macadamia exports. The crop's strong competitiveness is now clearly established, and management of the nurseries, production, and processing segment of the value chain are well developed. Issues that may have arisen for the filters of the value chain have been resolved, or are being resolved, by the actors in the value chain, and therefore there are no issues of significance to report.

Macadamia is an outstanding example of a smallholder crop that can give an enormous boost to their standards of living while having secure links to world markets.

Application of Value Chain Filters

- *Seeds and breeds*: no issues.
- *Production*: no issues.
- *Harvesting*: no issues.
- *Postharvest*: no issues.
- *Storage*: no issues.
- *Transportation*: no issues.
- *Processing*: no issues.
- *Market development*: no issues.
- *Value chain governance*: no issues.
- *External and policy*: no issues.

The Value Chain Quality Assessment Matrix for Rwandan Macadamia

Obviously macadamia is an extremely competitive crop for Rwanda, with no obstacles to fulfilling its potential. The recently developed value chains for this crop are examples of the *direct relationship* type of value chain mentioned in Section 6.2. The farmer has more limited options to leave the relationship, and side selling is virtually nonexistent because of the high-value, specialized nature of the crop, but at the same time it can provide more economic benefits to the farmer than, say, cultivating tomatoes (Table 8.13).

TABLE 8.13 Rwanda: Value Chain Quality Assessment Matrix for Macadamia

	Domestic Markets		Export Markets	
Value Chain Stage	Fresh	Processed	Fresh	Processed
1. Seeds and breeds				
2. Production				
3. Harvesting				
4. Postharvest				
5. Storage				
6. Transportation				

Continued

TABLE 8.13 Rwanda: Value Chain Quality Assessment Matrix for Macadamia—cont'd

Value Chain Stage	Domestic Markets		Export Markets	
	Fresh	Processed	Fresh	Processed
7. Processing				
8. Market development				
9. Value chain governance				
10. External and policy				
Sum of scores	0	0	0	0
Total score on quality issues: 0.				

8.6.12 Irish Potato

The Field Findings

Rwanda is a major net exporter of Irish potato, mainly to the DRC and Burundi. The crop is widely grown throughout the country, and within Rwanda it is consumed entirely in fresh form. It is a major staple in the diet of Rwandans. Although the high levels of rainfall in Rwanda favor the development of potato disease (especially late blight, *Phytopthora infestans*), Rwanda's relatively cooler climate compared to the DRC and Burundi give it a regional advantage in the production of this crop. High temperatures are not conducive to good tuber formation due to excessive rates of respiration (thus reduction in assimilate available for tuber development) and inhibition of tuberization in some cultivars. Early growth is favored by temperatures around 22°C and later growth by temperatures in the region of 18°C. The best tuber production is achieved where day temperatures are warm (not hot) and night temperatures are cool.

The optimal soil temperature for tuberization is in the range of 15−20°C. Rainfall in the range of 500−700 mm evenly distributed over the growth period is optimal. Altitudes over 1000 m above sea level are normally required for successful growth. Economic yields are usually obtained only at these elevations, although cultivars have been developed for production at lower altitudes. High light intensity such as is found in Rwanda (given its altitude and proximity to the equator) favors tuber formation, and low light intensity has the reverse effect of inhibition of tuber production.

Thus, the prevailing conditions in the higher altitudes (cool temperatures, high light intensity, even distribution of rainfall over the rainy period) found in much of northern and western Rwanda correspond to a good production environment for Irish potato. Although the product is bulky, if harvested at the mature stage, after tuberization of the outer skin has taken place, the tubers are not highly perishable compared with those of vegetables like tomato and fruits like banana. Therefore they are suited for being stored for as long as several months provided they are not exposed to high temperatures or moist conditions.

Irish potato, although considered a staple crop, enjoys a higher status in consumer preferences than most other root and tuber crops (cassava, sweet potato), and consequently it is much more commonly featured on hotel and restaurant menus than the latter crops. Hence it can be assumed that increased production of potato in Rwanda would find a market, both domestically and regionally.

With respect to the production filter, following strict crop rotations with non-solanaceous crops is quite important. It is recommended to crop the same land with potatoes or other solanaceous crops only once every 3 years to avoid high levels of diseases such as late blight and nematodes (e.g., *Meloidogyne* species) or excessive pesticide spraying for these diseases, especially in areas such as Ruhengeri and Gisenyi, which have a tradition of growing the crop. Irish potato can be intercropped with plants such as beans, cereals, and cassava. Expansion of Irish potato production is therefore recommended but only on a moderate scale under the hillside irrigation project. Although Irish potato is a highly competitive crop in Rwanda, the need for rotation places limits on how much its production can be expanded (Table 8.14).

Application of Value Chain Filters

- *Seeds and breeds*: no issues.
- *Production*: Irish potato needs to be planted in 3-year rotations and/or intercropped with beans, cereals, or cassava. Care must be taken to utilize crop rotations rather than overusing pesticides, which is a common problem with potatoes throughout the world.
- *Harvesting*: no issues.
- *Postharvest*: no issues.
- *Storage*: no issues.
- *Transportation*: no issues.
- *Processing*: no issues.
- *Market development*: no issues.
- *Value chain governance*: no issues.
- *External and policy*: no issues.

The Value Chain Quality Assessment Matrix for Rwandan Irish Potato

The matrix shows that Irish potato has few obstacles to realizing its competitive potential in Rwanda, and that the potential appears strong, as evidenced by the scale of its production and its net exports. However, because of its vulnerability to diseases that remain in the soil it has to be planted in rotations or in intercropping systems, and this requirement may limit the rapidity with which its production can be expanded.

TABLE 8.14 Rwanda: Value Chain Quality Assessment Matrix for Irish Potato

	Domestic Markets		Export Markets	
Value Chain Stage	Fresh	Processed	Fresh	Processed
1. Seeds and breeds				
2. Production	CM, CH		CM, CH	
3. Harvesting				
4. Postharvest				
5. Storage				
6. Transportation				
7. Processing				
8. Market development				
9. Value chain governance				
10. External and policy				
Sum of scores	2		2	

Total score on quality issues: 4. CM refers to the need for crop rotations and/or intercropping. CH refers to the need for care in application of pesticides for nematodes and pathogens that attach potatoes.

8.7 RECOMMENDATIONS OF THE RWANDA STUDY

Taking into account market limitations, crop disease issues, and other factors, the original study indicated approximate acreages that would be appropriate for each crop in the newly developed hillside irrigation zones. It also highlighted a number of investments and complementary programs needed to ensure that the newly planted crops represent profitable and sustainable options for Rwandan farmers, drawing on the crop-by-crop analyses presented earlier. The conclusions were as follows:

- Quality is the single most pervasive concern for development of the recommended crops. It embraces production methods including approaches to disease control, postharvest handling procedures, the use of cold chain facilities, product and process certifications, and even the type of packaging. To improve quality as well as yields at the farm level, for many crops a complementary program of intensive, hands-on technical assistance will be needed for several years.

- Critical infrastructure needs to include coolers and dryers at the field level, cold chain facilities, greenhouses or tunnels, and improved rural access roads. In addition, it is vital to improve the management of the cold store at the international airport by putting its management in private hands.
- Financial and technical assistance for obtaining organic and other certifications is needed for groups of small farmers. In many cases, financial and technical assistance is required for market exploration including for sample shipments.
- In general, alliances between buyers (or processors) and producers need to be strengthened and made into multifaceted relationships. Positive examples are found in the operating modes of EAG, SORWATHÉ, and Inyange Dairy, but the approaches of these enterprises need to be replicated more widely. Market linkages are very important for small farmers.
- A continuous program of trials of new products and new varieties should be made into an integral part of the hillside irrigation projects. Those trials should be carried out in a participatory manner with farmers. A number of products could become significant lines of export for Rwanda, but trials are needed to determine their feasibility in Rwandan conditions and the best ways of managing those products.

The study emphasized that keys to success for Rwandan agriculture will include promoting crops that are of high quality and the producers becoming known as a reliable, steady supplier of those crops. The study pointed out that modern agriculture, especially for high-value crops, is a knowledge-intensive business, and Rwandan farmers will need links not only with markets but also with the requisite kinds of technical information and expertise on a continuing basis, including from the private sector. The hillside projects and related activities provide opportunities to make that kind of expertise available to farmers.

8.8 BROADER CONCLUSIONS FROM THE QUALITY ASSESSMENT MATRICES

The VQA matrices have served as indicators of competitiveness on the quality axis as well as organizing frameworks for issues all along the value chain that affect product quality and inhibit fulfillment of competitive potential. It is worthwhile to review briefly the overall patterns of those issues. The VQA matrix scores by crop are summarized in Table 8.15, first in total and then in average by the markets relevant to each crop and finally in each crop's best market (i.e., where it is most competitive).

The total score for each crop is an indicator of the number and severity of issues that need to be resolved before the crop can become fully competitive in the quality dimension, and with regard to producers' ability to reliably supply required quantities as well. These total scores appear to classify the crops into four groups, ranked from the lowest number of issues to the highest.

TABLE 8.15 Rwanda: Summary of Value Chain Quality Assessment Matrix Scores by Crop

Crop	Total Score	Average score by market	Score in best market
Tea	5	5	5
Coffee	8	8	8
Bananas	12	4	2
Avocado	8	8	8
Pineapple	16	4	3
Passion fruit	18	4.50	4
Tree tomato	35	8.75	4
Tomato	20	6.67	6
French beans	10	5	5
Chili	15	3.75	3
Macadamia	0	0	0
Irish potato	2	2	2

Crop groupings by the total VQA score is as follows:

1. Macadamia, Irish potato, tea
2. Avocado, coffee
3. French beans, bananas, chili peppers, pineapple
4. Passion fruit, tomato, tree tomato

Opinions of informed observers of Rwandan agriculture coincide with this ranking in that the crops with greatest competitive potential appear to include macadamia, Irish potato, and tea, followed by avocado and coffee. However, this ranking reflects only one dimension of competitiveness. Other relevant considerations include cost competitiveness, absorptive capacity of the relevant markets, and the cost of providing solutions to the issues identified in the matrices. The questions raised by this ranking, and the answers provided for them, are as illuminating as the ranking itself. For example, the prospects for avocado hinge largely on the construction of a railway to a port in Tanzania, and obviously this is a major investment decision that goes far beyond the ambit of a single crop, and even well beyond a single sector in the economy. For Irish potato, Rwanda undoubtedly has an advantage over other countries in the region owing to its climate. However, prospects for sustainably expanding the production of this crop depend on finding suitable crop rotations in each locality where it is cultivated. (If solutions to a crop's issues could be costed, those numbers would correspond to the vertical distances in the graph of Fig. 4.4.) In the case of macadamia, undoubtedly issues once existed about sourcing good genetic material, training farmers in cultivation techniques, purchasing processing machinery, developing appropriate packaging, and so forth, but entrepreneurs have already found solutions to them, probably owing to perceptions of macadamia's strong competitive potential and high value. This is a reminder that the VQA matrices are dynamic constructs: they will change over time as solutions are found.

Given that the analysis covers from 1 to 4 markets for each crop, the overall VQA score may not be as informative as the average score per market. That score is shown in the second numerical column of Table 8.15, and it provides a somewhat different crop ranking.

Crop groupings by VQA average scores over markets is as follows:

1. Macadamia, Irish potato, chili peppers
2. Pineapple, bananas
3. Passion fruit, tea, French beans
4. Tomato, avocado, tree tomato, coffee

In this second "ranking," macadamia and Irish potato maintain their high standing, but surprisingly avocado and coffee drop to the lowest rung. In part this is because each of the latter crops has basically only one market, and all its issues weigh on prospects for that market rather than being spread over multiple markets. Chili peppers rise to a higher rung, in accordance with investors' perceptions of the crop. Bananas rise also, mainly because of the competitive strength of Rwandan cooking bananas for domestic and regional markets, with relatively few issues for that banana variety.

A third possible ranking classifies crops by their VQA scores in their "best" markets, that is, for each crop the market with the fewest issues.

Crop groupings by VQA scores in the best markets is as follows:

1. Macadamia, Irish potato, bananas
2. Chili peppers, pineapple, passion fruit, tree tomato
3. Tea, French beans
4. Coffee, tomato, avocado

The biggest surprises here are the rise in ranking of pineapple, passion fruit, and tree tomato. Pineapple rises because, in spite of its lower flavor rating than pineapple from some neighboring countries, the high international transportation costs in relation to product value mean that Rwandan pineapple has more opportunities for the domestic fresh market (and processing market as well) if rural farm-to-market roads can be improved so that damage to and losses of the fruit are reduced. Rwanda is perceived in international markets as a source of high-quality passion fruit and tree tomato, so in spite of struggles with pathogens it is worthwhile to seek solutions to the barriers facing these

TABLE 8.16 Rwanda: Frequency of Quality Issues by Type

Value Chain Stage	No. of Issues
1. Seeds and breeds	5
2. Production	39
3. Harvesting	1
4. Postharvest	14
5. Storage	12
6. Transportation	13
7. Processing	6
8. Market development	13
9. Value chain governance	12
10. External and policy factors	9

crops. For tree tomato, the best outlet is the domestic fresh market, and for passion fruit it is both the fresh and processed domestic markets.

The numbers of issues in the matrices also can be summed across crops to gain an idea of the kinds of issues that are most pervasive (Table 8.16). Not surprisingly, the most frequent issues occur in the production stage of the value chains: disease management, provision of irrigation, crop rotations (Irish potatoes) or intercropping (bananas), and other crop management issues. This finding reinforces the need for effective extension services and technical assistance providers who have access to the specialized kinds of knowledge needed for proper management of high-value crops. Following production issues, five other value chain stages have comparable numbers of issue occurrences: postharvest management, storage, transportation, market development, and value chain governance. These latter issues are not typically prominent in national agricultural strategies or in the priorities of national research and extension agencies, so these findings constitute a call for giving them more importance to promote the growth of high-value smallholder crops in Rwanda, and in other countries as well.

ENDNOTES

1. These and other data on demand parameters were compiled for an ongoing model-based analysis of the role of agriculture in Rwandan development by Xinshen Diao of the International Food Policy Research Institute (IFPRI).

2. It is estimated that about 500,000 farmers are at least partially engaged in coffee production. See: Republic of Rwanda, Ministry of Agriculture and Livestock Resources (2008b, p. 30).
3. CTC refers to the process of producing made tea: cut, tear, and curl.
4. As this book went to press, this railway was still in the planning stage, but in 2015 Rwanda, Burundi, and Tanzania formally invited expressions of interest to design and build it.
5. This crop is also referred to as Japanese plum, and internationally it is known as tamarillo.
6. Production under plastic may not be necessary; this is the method employed in Uganda and trials would need to be conducted to determine whether or not such infrastructure is required to achieve the same quality for *C. chinense* production in Rwanda.
7. The material in this section is based in large part on personal communications from Hector Rodriguez in Panama.

Chapter 9

El Salvador: Crop Competitiveness and Factor Intensities

9.1 INTRODUCTION

The case study for El Salvador is adapted from Norton and Angel (2004). It was carried out only a few years after the end of that country's civil war. After being the motor of the economy in previous decades, the agricultural sector was virtually stagnant from the early 1990s onward. Rural poverty, although declining, was still extreme for many households. Influential voices were questioning whether agriculture had a future in the country, or what kind of future would it have beyond continuing to provide a subsistence living for very poor rural families. For this reason, one of the principal motives for this study was to evaluate whether the sector had competitive potential, and if so, in what lines of production.

Other reasons for including the study in this volume include: (1) its analysis of intersectoral price movements and the consequent intersectoral resource transfers; (2) its illustrations of the application of the long-run competitiveness (LRC) methodology accompanied by analysis of how sensitive the products' LRCs are to changes in prices, yields, and wage costs; and (3) its development of a practical procedure for quantifying the knowledge content of production costs, which allowed a more complete analysis of factor intensities of production. This procedure in effect developed a proxy for the use of skilled labor in addition to unskilled field labor.

As in the case of the studies for Colombia and Rwanda, the Salvadoran economic landscape has changed since this case study was carried out, so a number of facts mentioned may no longer be true and issues that were once important may have been solved—and others may have arisen.

Table 9.1 shows estimates of the percentage of the Salvadoran population living in poverty in rural and urban areas, according to national criteria for the poverty line. In some parts of the country, especially those most impacted by the civil war, poverty was much more widespread and severe. In the four departments of Chalatenango, La Unión, Cabañas, and Morazán, the human

The Competitiveness of Tropical Agriculture. http://dx.doi.org/10.1016/B978-0-12-805312-6.00009-X

TABLE 9.1 El Salvador: Incidence of Poverty, 1991, 1995, and 2002

Poverty Category	National	Urban	Rural
Total poverty			
1991	66.0%	60.0%	71.0%
1995	54.0%	46.0%	64.0%
2002	43.0%	34.0%	56.0%
Extreme poverty			
1991	33.0%	32.0%	34.0%
1995	22.0%	15.0%	30.0%
2002	19.0%	12.0%	29.0%

Dirección General de Estadísticas y Censos, El Salvador.

development index was lower than that of Papua New Guinea, and in three more departments it was lower than that of Equatorial Guinea. For the entire rural population, per capita income levels were approximately one-third of the levels in urban areas. All these poverty estimates include a valuation of crops retained for home consumption.

Rural poverty was made worse by declines in the purchasing power of agricultural products, which meant declines in real incomes of the rural population. Also, it is likely that this decline in real agricultural prices played an important role in the weak performance of agricultural production. Fig. 9.1 shows that nominal agricultural prices increased at a much slower rate than those of the manufacturing and service sectors. A similar pattern emerges when the prices of basic grains are divided by the consumer price index, giving a ratio that shows the purchasing power of grain harvests (Fig. 9.2). Although these trends may have been caused in part by the declining international prices of agricultural products, they were undoubtedly reinforced by the country's policy of maintaining a fixed exchange rate in the 1990s, which led to an appreciation of the real exchange rate. The strong nature of the exchange rate was then cemented in place by the dollarization of the currency in 2001 at the same fixed rate of 8.75 colones per dollar.

The 14 crops analyzed in this study are Booth avocado, sugar apple (or custard apple, *Annona squamosa*), coconuts, Persian limes, loroco (*Fernaldia pandurata*), tomato, rice, coffee, cashews, sesame, cotton, indigo, beans, and corn. In addition, an LRC for dairy production was calculated for both a pasturage system and a stabled system. For rice and corn, different technologies of production were analyzed, and coffee was analyzed for plantings at three different altitudes.

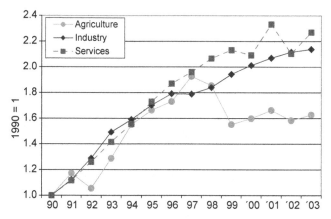

FIGURE 9.1 El Salvador: gross domestic product deflators by sector, base 1990. *Central Reserve bank.*

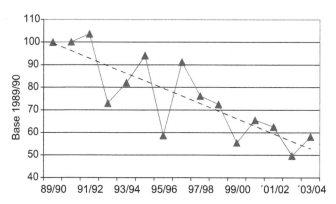

FIGURE 9.2 El Salvador: index of real producer prices for basic grains.

9.2 NATIONAL FACTORS AFFECTING COMPETITIVENESS

In the national context, the principal positive factors for competitiveness in El Salvador are:

1. The rapid expansion of the coverage of land titling, which now is greater than in many developing countries and provides farmers incentives to invest in their land.

2. The country's geographical closeness to the United States market, which is the largest in the world. However, this advantage is somewhat negated by the high port costs in El Salvador and the typically small size of shipments to other countries, which increases their unit transportation costs.

3. A relatively well-educated workforce, although there still is much to be desired in that regard, and a workforce known for being hard working. At the time of the original study the average schooling of the agricultural labor force was only 3.1 years, but it was sure to increase since 80% of children between 7 and 15 years were enrolled in school.
4. The dollarization of the currency, which reduces the cost of finance for investments and production; however, its effects on the real exchange rate, which drive down real agricultural prices, may be a disadvantage that is stronger than the advantages of a fixed currency.
5. The progress toward a Central American customs union, which is widening the potential market for Salvadoran exports.
6. Relatively extensive rural infrastructure. According to the original study, 71% of the rural population had access to electricity and 84% to sanitary facilities.

On the negative side, the main factors are, in addition to the disadvantages mentioned earlier:

1. The small size of the average farm, although size is not much of a disadvantage for fruit and vegetable production because of the availability of family labor (high labor to land ratio).
2. A legal structure of many agrarian reform cooperatives that is not entrepreneurial in character; among other things this had led to significant amounts of idle land in those cooperatives. Because of this factor, along with the falling trend of real agricultural prices, it is estimated that 15–30% of the agricultural land is idle in spite of the heavy population pressure on the land.
3. The country's vulnerability to earthquakes and floods, which can deter some investors and cause serious setbacks to farmers. According to the 2014 World Risk Report of the United Nations University, El Salvador is the eighth most risky country in the world, out of 171 countries ranked.
4. The high crime rate, which also discourages investments.
5. The fact that less than half of the rural population had access to potable water (at the time of the original study), which translates into a high incidence of health issues.
6. Existing irrigation schemes cover only 9–22% of the potentially irrigable area, depending on the source of the estimation, and this makes Salvadoran farmers less resilient in the face of droughts, perhaps exacerbated by climate change.
7. The vast majority of smallholders do not have access to crop loans, and medium-term finance for investments is practically nonexistent.
8. An adverse macroeconomic climate. The aforementioned appreciation of the real exchange rate has drastically lowered the purchasing power of agricultural harvests. In all countries, agriculture is the sector that is most sensitive to the real exchange rate because it is the most tradable sector.

Fig. 9.1 shows how agricultural prices have lagged movements in other prices. This is equivalent to taking purchasing power, or real income, away from farmers. Fig. 9.3 shows clearly how this relative price trend, driven by the fixed exchange rate policy, has taken resources away from agriculture cumulatively, in favor of the industrial and service sectors. The magnitude of the effect has been very large.

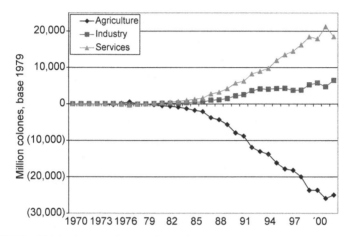

FIGURE 9.3   El Salvador: annual economic transfers among sectors.

9.3 TRACK 1 ASSESSMENTS OF CROP COMPETITIVENESS

9.3.1 Sources of Information

Information on yields and costs of production has been drawn from various data banks. The principal sources have been: the Ministry of Agriculture and Livestock, above all its manual of costs of production, and field data collected by the Multisectoral Investment Bank, the Salvadoran Foundation for Coffee Research, and the Inter-American Institute for Cooperation in Agriculture (IICA). This last source was particularly useful for information on perennial crops.

The data from these sources have been collected and reviewed by experts in their respective fields, to provide assurance of internal consistency of the data and generate best estimates of farm budgets for each crop. However, in the future, efforts should be made to run independent tests of the sensitivity of the results to original data from different sources.

In all cases, the original farm-level data were complemented with other kinds of information to ensure that the list of inputs was complete, including for example, imputed costs of land, farm administration, and unforeseen expenses.

9.3.2 Long-Run Competitiveness Calculations for El Salvador

The LRC values calculated in this study, by application of Eq. (5.8), show that definitively El Salvador has lines of production that are competitive over the long run, some of them highly competitive. However, the results are characterized by a wide dispersion of LRC values. Some products are highly competitive, whereas others apparently have no chance of being competitive. In contrast, few products are located on the verge of competitiveness. In this regard, the perspectives for Salvadoran agriculture differ markedly from those of Colombia, which, as shown in Chapter 7, does not have such a bipolar dispersion of degrees of competitiveness.

By the type of product, the message is clear: generally basic grains lack competitiveness, with the exception of rice cultivated under the most advanced technology, and on the other hand, fruit and vegetables and other niche products such as coconut and loroco display a high degree of competitiveness (Table 9.2). This appears to be a widespread pattern of competitiveness for tropical agriculture.

TABLE 9.2 El Salvador: Long-Run Competitiveness Coefficients by Crop

Strong Competitiveness		Marginal Competitiveness		Lack of Competitiveness	
Booth avocado	0.59	Rice advanced technology	0.93	Sesame	2.92
Sugar apple	0.66	High-altitude coffee	0.95	Cotton	1.31
Coconut	0.69	Cashews	1.03	Indigo	1.60
Persian lime	0.44			Rice, semiadvanced	1.33
Loroco	0.71			Rice, traditional	2.38
Tomatoes	0.08			Medium-Altitude coffee	1.20
				Low-altitude coffee	1.52
				Winter beans	1.81
				Corn, advanced technology	1.99
				Traditional corn	3.95

In the case of rice, the results underscore the importance of cropping technology. Adoption of appropriate and advanced technologies of production improves the crop's competitiveness substantially. In the Salvadoran context, rice under an advanced technology has irrigation and is transplanted. That is the only technology of production that shows a degree of competitiveness, although it is slight and could be undercut quickly by fluctuations in international prices.

Recently (as of the time of this original case study) international grain prices have experienced an upward trend and some observers feel the upward pressures will continue. This trend could work in favor of irrigated rice in El Salvador, since the analysis here was carried out with prices for the year 2003.

However, the case of corn is more difficult. Even in a scenario in which international subsidies were removed from the corn market, El Salvador would not have LRC in the crop, as shown in Table 9.3. The first column in this table presents hypothetical options in regard to the international price of corn, expressed in multipliers of the actual base period border price. The first row, with a multiplier of 1.0, represents the LRC with no price change, and each row after that shows the LRC with a 20% additional increase in the price.

TABLE 9.3 El Salvador: Effects of Price and Yield Changes on the Competitiveness of Corn

Multiplier	LRC, Advanced Technology	LRC, Traditional Technology
1.0	1.99	3.95
1.2	1.49	2.75
1.4	1.20	2.11
1.6	1.00	1.71
1.8	0.86	1.44
2.0	0.75	1.24

LRC, long-run competitiveness.

The hypothetical price changes could result from reductions in international subsidies, from new trends in world markets, or from changes in El Salvador's tariff policies. However, it has to be pointed out that corn grown with the traditional farming technology would not be competitive even with a tariff of 100%.

The multipliers in Table 9.3 can also be interpreted as hypothetical yield increases, ranging from 20% to 100%. [Changes in the border price and yield have the same numerical effect in Eq. (5.8)]. It can be seen that very large yield increases would be required to make corn competitive, 80% for cultivation with advanced technologies and more than 100% for the traditional technology.

These results lead to clear implications for policies oriented toward low-income farmers: the route of tariff protection for grains does not represent a way out because the protection would have to be extremely high and would have negative repercussions for livestock and agroindustries and on consumer well-being. Another implication is that poor producers need training and support to learn how to manage other crops and training for rural nonagricultural employment, along with straight-out social support. In addition, it is evident that fostering the growth of labor-intensive crops also will offer more employment opportunities for the rural poor off of their own farms.

The case of indigo merits further comment. The crop was cultivated by Mayans centuries before the Spanish conquest, and the Spaniards set up indigo plantations to supply European textile industries with the dye. In the 19th century El Salvador was a major exporter of indigo, and labor conditions were akin to slavery on the plantations. After the civil war of 1970–82 there was renewed interest in the crop, and as of the publication of this book, El Salvador is exporting small quantities of very high-quality indigo. It has developed a niche market based on its quality and being hand processed by Mayan women. The LRC calculations shown earlier refer to indigo of an average quality and therefore do not accurately reflect the specialized export potential of this crop. As commented in Section 5.6, for all crops there always will be producers who are less and more competitive than the ones whose farm budgets are used for these calculations, and indigo is a case in point. Hence it is incumbent on the analyst to be alert for discontinuities in crop technology spaces, especially for options that are proven in the field and that are more competitive.

9.3.3 Additional Sensitivity Analyses

Tests with the LRC calculations revealed that in general the coefficients are quite sensitive to changes in prices and yields and less sensitive to changes in the cost of labor. This confirms the importance of production technologies and markets. A policy that aimed at reducing labor costs would not rescue the noncompetitive nature of the less competitive crops and, in addition, would result in rural income levels that were even lower than they are today.

The sensitivity of the coefficients to changes in prices represents a warning about the need for continuous improvements in productivity since in the long run the international trends in real agricultural prices have been mostly downward, until very recent years. This negative trend is most pronounced for fruits, vegetables, and every kind of niche product. Offsetting this somewhat is the fact that the markets for these specialized products have been increasing more rapidly than those of traditional products.

Table 9.4 shows the responsiveness of the LRC coefficient to changes in prices, yields, and labor costs, for all other crops. The first two columns provide insights regarding the crops for which productivity improvements could be most important, in the sense of moving them into competitive

TABLE 9.4 El Salvador: Sensitivity Analysis of the Long-Run
Competitiveness Coefficients

Crop	Initial Value	Variation in Price or Yield		Variation in Labor Cost	
		+20	−20%	+20%	−20%
Rice, advanced technology	0.93[a]	0.73[b]	1.29	0.97[a]	0.89[b]
Rice, semiadvanced	1.33	0.98[a]	2.05	1.40	1.26
Traditional rice	2.38	1.67	3.61	2.50	2.07
Corn, advanced technology	1.99	1.49	2.96	2.17	1.80
Traditional corn	3.95	2.75	6.99	4.39	3.51
Winter beans	1.81	1.44	2.42	2.03	1.58
Sesame	2.92	2.20	4.34	3.29	2.54
Cotton	1.31	0.97	2.05	1.36	1.27
Indigo	1.60	1.31	2.06	1.69	1.52
High-altitude coffee	0.98[a]	0.80[b]	1.25	1.12	0.83[b]
Medium-altitude coffee	1.20	0.98[a]	1.52	1.34	1.05
Low-altitude coffee	1.52	1.26	1.93	1.70	1.34
Coconuts (mayapan variety)	0.69[b]	0.56[b]	0.91[a]	0.76[b]	0.63[b]
Cashews	1.03	0.83[b]	1.35	1.12	0.93[a]
Booth avocadoes	0.59[b]	0.48[b]	0.78[b]	0.64[b]	0.54[b]
Sugar apple	0.66[b]	0.54[b]	0.85[b]	0.70[b]	0.61[b]
Persian lime	0.44[b]	0.36[b]	0.56[b]	0.48[b]	0.40[b]
Loroco	0.71[b]	0.55[b]	0.98[a]	0.75[b]	0.67[b]
Tomatoes	0.08[b]	0.07[b]	0.11[b]	0.09[b]	0.08[b]

[a]*Narrowly competitive.*
[b]*Strongly competitive.*

territory or strengthening marginal competitiveness. In this sense, the three crops that would benefit most from productivity improvements are rice, coffee, and cashews. A 20% drop in the border price would erase the competitiveness of advanced technology rice, coffees, and cashews, and it would weaken the competitiveness of loroco and coconuts. In contrast, upward and downward swings in the cost of labor do not have much effect on competitiveness. A significantly lower cost of labor moves cashews from slightly noncompetitive to slightly competitive and improves the competitiveness somewhat in the case of high-altitude coffee and rice under the advanced production technology.

The results in Table 9.4 suggest that the sensitivity of competitiveness to changes in prices, yields, and labor costs vary substantially among crops. A more convenient way to see the variations over crops is in the form of elasticities of the LRCs with respect to changes in those factors, which are shown in Table 9.5. The first impression that can be obtained from Table 9.5 is that competitiveness is proportionally more sensitive to changes in border prices and yields than to the cost of labor, as mentioned earlier. The second impression is that the elasticities vary significantly among crops. This implies that the effects on competitiveness of improvements in crop productivity, as

TABLE 9.5 El Salvador: Elasticities of Competitiveness with Respect to Border Prices, Yields. and Labor Costs

Crop	Versus Border Prices, Yields	Versus Labor Cost
Rice, advanced technology	1.08	0.22
Rice, semiadvanced	1.32	0.26
Traditional rice	1.49	0.25
Corn, advanced technology	1.26	0.45
Traditional corn	1.52	0.56
Winter beans	1.02	0.61
Sesame	1.23	0.63
Cotton	1.30	0.19
Indigo	0.91	0.28
High-altitude coffee	0.92	0.71
Medium-altitude coffee	0.92	0.58
Low-altitude coffee	0.86	0.59
Coconut (mayapan variety)	0.94	0.51
Cashews	0.97	0.44
Booth avocadoes	0.93	0.43
Sugar apple	0.91	0.30
Persian lime	0.91	0.45
Loroco	1.13	0.28
Tomatoes	0.63	0.63

well as the effects of changes in international prices, are markedly different among crops.

The coffees, followed by sesame, tomatoes, corn, and beans, are the crops whose competitiveness is most sensitive to changes in the wage rate. Cotton, rice, indigo, and loroco are the least sensitive. Productivity improvements, or changes in international prices, can do the most for competitiveness in the cases of corn, rice, sesame, and cotton. These results may suggest that two crops that are borderline competitive (advanced technology rice) or not far from competitiveness (cotton) would benefit most from research on productivity improvements.

The circumstances of coffee are unique given the extreme volatility of international coffee prices. Accordingly, a separate analysis was carried out for the sensitivity of the competitiveness of high-altitude coffee, with respect to variations in both prices and costs. Table 9.6 shows scenarios of coffee prices ranging from 250 to 350 colones per hundredweight for green coffee berries (depulped coffee), and costing first all inputs and second only variable costs.

TABLE 9.6 El Salvador: Analysis of the Sensitivity of Coffee's Competitiveness to International Price Changes (Long-Run Competitiveness Values for High-Altitude Coffee; Prices in Colones per Hundredweight, Green Coffee)

Coverage of Costs	Price = 250	Price = 300	Price = 350
Full costing[a]	1.20	0.98	0.83
Without some fixed costs[b]	1.00	0.81	0.69

[a]Land at 50 colones/manzana/yr.; unforeseen @ 3% of total cost; interest on half of direct costs @7.5%/yr. (For other crops the annualized cost of land is higher and the unforeseen outlays are at 5%.)
[b]All these items listed are at zero. One manzana is 0.7 ha.

In the year 2002 the farmgate price of green coffee berries was about 200 colones per hundredweight for the better coffees. What Table 9.6 shows is that coffee producers barely survived economically with prices that low, and only by ignoring the opportunity cost of land (which already was lower than that for some other crops, because of the sloping nature of most coffee land) and also some interest costs (or ignoring the opportunity cost of their own funds) and any unforeseen costs. The situation looks much better when the coffee price rises to 300 colones, which shows how the fortunes of coffee producers are at the mercy of fluctuations in world prices.

Irrigation is another vital input for some crops. The foregoing analyses have incorporated a long-term price of irrigation water at 50% higher than the prevailing price, in light of the scarcity of water and the fact that water fees are well below its opportunity cost, and even below the cost recovery level needed for maintaining irrigation systems. The case of Persian lime was used to test the sensitivity of competitiveness in the face of hypothetical changes in the cost of water. This crop's LRC coefficient is 0.44 when water is priced at US$300 per land unit of five manzanas. When the water tariff is doubled to $600 the LRC is affected only slightly, rising to 0.47. When the water tariff is tripled to $900, the LRC becomes 0.49. These results imply that access to water is much more important than its cost to producers, and also that producers could afford to pay higher water fees more commensurate with the cost recovery levels needed for maintaining irrigation systems.

9.3.4 Factor Intensities of Crop Production

These calculations of the LRC coefficients demonstrate that the most competitive crops are the most intensive in the use of labor, in yet another of the numerous illustrations of Ricardo's thesis about comparative advantage. This also means that facilitating the development of these crops will have the advantage of generating more rural employment. However, the competitive crops generally are also intensive in the use of fixed capital assets. And it turns out they are the most intensive in the use of technical and administrative knowledge. To examine the role of knowledge inputs, an index has been developed that indicates the intensity of requirements of this kind of input, covering scientific, technical, administrative, and marketing knowledge.

The components of the index are the following:

- The degree to which the product is perishable, which demands more continuous attention and more knowledge of postharvest management.
- The use of irrigation, which demands complex and specialized skills.
- Whether the crop is exported, which requires knowledge of how international markets function and the strict quality requirements of those markets, as well as awareness of the entire value chain.
- Whether the product is easily damaged and requires special packaging or other treatment to protect it during transportation and marketing processes (as opposed to bulk commodities).
- Whether the producer of the product typically *purchases* technical assistance, which would represent full recognition of the need to have access to specialized technical information.
- Whether the crop requires transplanting in the field, which would imply more delicate and specialized tasks.

For each of these factors, a crop has been assigned a value of 2 (intensive) or 1 (basic). In a few special cases an intermediate value of 1.5 has been assigned. The overall index of the degree of knowledge inputs, then, is the sum of the values of these six components. The values of the index are shown for the crops of this study in Table 9.7. This table shows that generally there is an association between the intensity of knowledge inputs and LRC, although the association is not perfect.

Table 9.8 presents quantifications by crop of the intensity of use of all the main factors of production: labor, investments in fixed capital, and knowledge inputs. The striking aspect of this table is that the degree of competitiveness tends to be associated with higher intensity of use of *all three* basic factors of production. Again, the association is not perfect but it is strong. In short, competitive agriculture in El Salvador is the most modern part of the sector, the most sophisticated in requirements of capital and specialized skills in addition to unskilled field labor. In a world in which competitiveness is based in good measure on quality, a surplus of unskilled rural labor may be necessary for an international comparative advantage in agriculture, but it is no longer sufficient. These results suggest that the way forward for the sector does not lie in growing basic crops with rustic production technologies but rather in modernization of the sector, following the path opened up by the agricultural sectors of Brazil and Chile.

In regard to capital as a factor of production, tests showed that variations in the interest rate did not affect very much the LRC values. It appears that the availability of capital for fixed investments is much more important than its cost in terms of interest rates. This finding also underscores the importance of security of land tenure, without which many investments will not take place, and also the security of contractual relationships and the need to develop instruments of medium-term finance in developing-country agriculture.

9.4 THE COMPETITIVENESS OF DAIRYING

9.4.1 The Role of Dairy Production in El Salvador

The case of milk production is of special interest because within the range of livestock activities it represents a relatively intensive use of land, and land is a scarce resource in El Salvador when compared with neighboring countries. Another important reason for analyzing dairy production is its economic value to the country. In 2001, milk was the leading product in the agricultural sector, with a value of US$153 million. It was followed in importance by corn, with $140 million in production, and then by sugarcane and coffee, with values of production significantly lower. From 1983 to 2001, national milk production grew rapidly, increasing from 150 million liters to approximately 390 million liters.

In spite of its volume of dairy production El Salvador is a net importer of dairy products. According to FAO data, imports satisfy about one-third of the national demand. Exports of dairy products are minimal, consisting largely of small quantities of powdered milk and yogurt. From 1980 to 2000, milk

TABLE 9.7 El Salvador: Index of the Intensity of Knowledge Inputs Into Production

Crop	Perishable	Irrigated	Exported	Packaged	Technical Assistance	Transplantation	Total
Rice, advanced technology	1	2	1	1	1	2	8
Rice, semiadvanced	1	1	1	1	1	1	6
Traditional rice	1	1	1	1	1	1	6
Corn, advanced technology	1	1	1	1	1	1	6
Traditional corn	1	1	1	1	1	1	6
Winter beans	1	1	2	1	1	1	6
Sesame	1	1	2	1	1	1	7
Cotton	1	1	1	1	1	1	6
Indigo	1	1	2	2	1	1	8
High-altitude coffee	1.5	1	2	2	1	1	8.5
Medium-altitude coffee	1.5	1	2	1.5	1	1	8
Low-altitude coffee	1.5	1	2	1.5	1	1	8
Coconuts	1	2	2	1.5	1	1	8.5
Cashews	1.5	2	2	1.5	1	1	9
Booth avocadoes	2	2	2	2	1	1	10
Sugar apple	2	2	2	2	1	1	10
Persian limes	2	2	2	2	1	1	10
Loroco	2	2	2	2	2	1	11
Tomatoes	2	2	2	2	1	1	10

TABLE 9.8 El Salvador: Intensity of Factor Use by Crop

Crop	Labor, Person-Days per Manzana	Fixed Capital Investment, US$ per Manzana	Index of Knowledge Inputs
Rice, advanced technology[a]	72.0	n.a.	8
Rice, semiadvanced	33.1	n.a.	6
Traditional rice	74.5	n.a.	6
Corn, advanced technology	51.9	n.a.	6
Traditional corn	40.2	n.a.	6
Winter beans	55.1	n.a.	6
Sesame	40.0	n.a.	7
Cotton	46.6	n.a.	6
Indigo	52.6	874.39	8
High-altitude coffee [a]	12.8	n.a.	8.5
Medium-altitude coffee	28.3	n.a.	8
Low-altitude coffee	12.8	n.a.	8
Coconuts[b]	106.0	1691.71	8.5
Cashews	125.3	1500.34	9
Booth avocadoes[b]	110.4	1505.07	10
Sugar apple[b]	149.9	2442.64	10
Persian limes[b]	69.3	1562.66	10
Loroco[b]	276.4	639.89	11
Tomatoes[b]	353.3	n.a.	10

n.a., Not available.
[a]Narrowly competitive.
[b]Strongly competitive.

consumption per capita increased at a rate of 3.4% per annum, above the rates of the other Central American countries, Mexico, Brazil, Argentina, Australia, New Zealand, the European Union, and several other countries. Nevertheless, the level of milk consumption per capita in El Salvador is still below the levels of Mexico, Costa Rica, Argentina, and other more developed countries, which shows the potential for further increases in the domestic market for dairy products.

The fundamental question is, what role will national production have in satisfying the increasing demand for milk? The answer will depend on the competitiveness of national production vis-a-vis imports. At the present time fluid milk enjoys tariff protection of 40%, and for powdered milk it is 15−20% and for cheeses 15−40%. It will be important to know if these levels of protection are necessary to sustain the competitiveness of national dairy production. Of course dairy products are among the most highly subsidized products in developed countries, and it appears that these subsidies will not disappear in the near future. Therefore it has to be asked whether it is worthwhile to continue imposing a tax on Salvadoran consumers via tariff protection for dairy products. Does this protection improve the viability of the country's dairy sector?

Approximately 43% of the country's milk production comes from stabled herds, another 43% from mixed operations with stables and pastures, and the remaining 14% from purely pasture operations. Yields vary substantially according to the production system, the breed of cows, and the quality of farm administration. At one extreme are subsistence operations with one milking a day and a yield of less than 3 L/day. At the other extreme, the majority of the more technologically advanced operations produce between 15 and 22 L/day, milking each cow three times a day.

Official sources of information on costs of production do not cover the field of milk production. However, in 2002, the international nongovernmental organization Technoserve carried out a detailed study on the structure of production and the markets for dairy products. The costs of production and the majority of other data cited here come from that study.

9.4.2 Production Parameters by Dairy System

Two milk production systems have been analyzed: pasture-based and stabling systems, each with a herd of 100 cows to facilitate comparisons. Costs are decomposed into fixed investments, other fixed costs, and variable costs. Tables 9.9 and 9.10 present the costs in US dollars for each system and each type of input.

The treatment of investments in this study differs from the approach used in the Technoserve study. Here it is assumed that 80% of fixed investments is financed though bank credits, at an interest rate of 6.5%, and that the same interest payment is made each year during the 10-year useful lifetime of the

TABLE 9.9 El Salvador: Fixed Investments by Milk Production System

Item	Pasture-Based System	Stabling System
Cows in milking herd	100	100
Price per cow	1500	1500
Value of the cows	150,000	150,000
Infrastructure	22,000	34,000
Cows per hectare	5	10
Land price per hectare	6500	6500
Value of the land	130,000	65,000
Mechanical milker	–	16,571
Chopper	–	2629
Cooling tank	–	5143
Pounding mill	–	2286
Mixer	–	2857
Caloric stress	–	5486
Total investments	302,000	283,971

investments. This approach represents the "consol" model of financing, which does not require the costing of depreciation. In addition, it is assumed that 50% of the variable costs and other fixed costs is financed with credit at an interest rate of 8.5% (because of the shorter term of those loans). These financing costs are added to the other costs of production before applying a percentage profit margin.

Another key aspect of this analysis is the use of a discount rate for incomes and future costs. All are discounted uniformly at a 6% annual rate, which is the same as the percentage profit margin. Similar discounting was applied earlier in the assessment of perennial crops.

The price of milk is higher for stabling systems because of their advantage in milk quality when taken to market.

9.4.3 Results of the Analysis

The coefficient of LRC has been calculated for each production system following the same procedure that was used for crops. A sensitivity analysis of the results was also carried out under hypothetical variations in some key parameters. Table 9.11 presents these results. The broad conclusion is that the era of expansion of milk production in El Salvador appears to be coming to an

TABLE 9.10 El Salvador: Other Costs and Yields by Milk Production System

Item	Pasture-Based System	Stabling System
Variable costs		
Labor	16,457	28,800
Feed	44,983	100,000
Animal health	2857	5714
Fuel	2505	3507
Artificial insemination	—	1500
Unforeseen outlays	6% costs	6% costs
Expected profits	6% of total cost	6% of total cost
Other fixed costs		
Salary of owner	12,000	18,000
Electricity	1920	8400
Advisory services	—	2760
Maintenance, repairs	2052	4692
Income		
Production/cow/day (L)	12	22
Annual sales (L)[a]	346,602	650,870
Price per liter	0.30	0.35
Total income	103,981	227,805

[a]It is assumed that the cattle consume little milk: 5.3% of the total in the pasture-based system and 3% in the stabling system.

end, although yield improvements could change that picture. Regarding the alternative production systems, it can be seen that the stabling system had a slight advantage over the pasture-based system, although increases in the price of petroleum can erode this advantage.

The principal conclusion that emerges from these calculations is that milk production in the most common systems does not have LRC in El Salvador, although it may not be far from attaining it. The potential competitiveness of this sector depends very largely on increases in the productivity of the operations. For example, increasing the yield of stabled cows (in terms of milk sold) from 17.8 to 25 L/day would have the effect of reducing the LRC coefficient from 1.09 to 0.74. (Like all the scenarios, these increases in productivity are illustrative and do not include increases

TABLE 9.11 Analysis of the Competitiveness of Milk Production (LRC Values)

Scenario	Stabling System	Pasture-Based System
a. Base LRC value[a]	1.09	1.16
b.1. Milk price = $0.50/L	0.73	
b.2. Milk price = $0.45/L		0.77
c.1. Milk price = $0.25/L	1.62	
c.2. Milk price = $0.20/L		1.74
d. Land price = $1000/ha	1.07	1.10
e. Combination (b) + (d)	0.72	0.73
f. Short-run interest rate = 5%	1.07	0.83
g. Long-run interest rate = 4%	1.06	0.81
h.1. Milk sales at 25 L/day	0.74	
h.2. Milk sales at 13 L/day		0.85
i.1. Milk sales at 30 L/día	0.61	
i.2. Milk sales at 16 L/día		0.69
j. Feed cost reduced by 25%	1.07	

LRC, long-run competitiveness.
[a]The base scenario includes a milk price of $0.35/L for stabled cows and $0.30/L for cows at pasture; a land price of $6500/ha; a yield (in milk sales) of 17.8 L/day for the stabling system and of 9.5 L for the pasture-based system; a short-run interest rate of 8.5%; and a long-run interest rate of 6.5%.

in the costs of feed and other inputs that might be necessary to facilitate higher yields.) Similarly, increasing the yield in the pasture-based system from 9.5 to 13 L/day would reduce the LRC from 1.16 to 0.85.

A more modest increase in the productivity of stabled cows, from 17.8 to 20.0 L/day, would place these operations in the competitive space with an LRC of 0.95. In the case of pastured cows, raising the daily yields from 9.5 to 11.5 L would also make those operations competitive by reducing the LRC from 1.16 to 0.96.

By the same token, the sector's competitiveness is quite sensitive to variations in the price of milk. Raising the price by means of an increase in tariff protection to the 40–50% range (scenarios b.1 and b.2 in Table 9.12) would make the country's milk production firmly competitive. On the other hand, eliminating the current tariff protection would make it decidedly uncompetitive.

In regard to the investments, the cost of land does not affect the outcomes very much. In Table 9.11 it can be seen that under an assumption of much less costly land the LRC changes only slightly. For that reason, varying the number of grazing animals per hectare also does not have much effect on the LRC.[1]

9.4.4 Toward the Future of Dairying in El Salvador

With the aforementioned results in mind, perhaps some perspectives on the future of dairying in El Salvador can be ventured:

1. Milk production is struggling to be competitive. Above all, it needs increases in the productivity per cow.
2. Under current conditions and without improvements in productivity, only the most efficient producers (especially those with better control over costs) can have encouraging future prospects.
3. It will be important to find ways to make milk processing plants more efficient so that they can increase the prices paid to dairy farmers.
4. It would be helpful to seek out niche markets for processed products, such as cheeses and yogurts with special characteristics and milk-based drinks, that El Salvador could export, thus increasing producer profits.
5. As long as international milk markets remain so highly subsidized, with the result of world prices well below the free market level, import tariffs are critical for the survival of the dairy sector in El Salvador. This should include raising the tariffs on powdered milk to the same level as those on fresh milk. However, all tariff measures are transitional, because the sector needs transformations to survive in the long run.

9.5 CONCLUDING COMMENTS

The analysis presented in this chapter represents a mere beginning of a process of investigating the competitiveness of Salvadoran agriculture. It illustrates the application of a useful analytic tool and brings to the table some of the considerations that are relevant to the sector's competitiveness. The ideal process would be to continue applying the LRC tool to other farm budgets, in close communication with producers, to be able to evaluate more products, more variants of production technologies, and the effects of different panoramas regarding prices.

As useful as additional analysis would be, the unavoidable and central question is, what are feasible paths toward greater competitiveness in the sector, roads to restore its dynamism of past eras? Although coffee and sugar will continue being very important products, it can no longer be expected that they will represent strong forces for growth. Their comparative advantage is not sufficient to raise hopes very much. (Indeed, as of the writing of this book

TABLE 9.12 El Salvador: Principal Agricultural Exports, 2003

Customs Code	Item Description	Value (US$ 000)	Volume (mt)
9011130	Coffee berries (depulped)	105,094	80,142
17011100	Cane sugar	46,602	266,293
21069090	Other food preparations	27,298	22,344
19041090	Processed cereal products	19,185	14,035
22021000	Water, including mineral or sweetened waters	17,535	38,397
19049090	Cereal flakes	15,068	14,370
22029090	Other sparkling beverages	12,515	16,871
17049000	Nonchocolate candies, except chewing gum	9913	8443
19053190	Cookies	9850	7208
20098090	Other fruit or vegetable juices	9,603	16,377
19041010	Rice flour pellets	9449	4997
22071010	Ethyl alcohol	7,868	23,857
17031000	Sugarcane molasses	7,397	141,607
11010000	Wheat flour	7,055	23,933
3034900	Other frozen tuna products	6960	6458
11022000	Corn flour	5,870	19,677
19,059000	Other bakery products	5840	7865
3061319	Frozen wild shrimp	5154	598
22030000	Beer	5,134	10,770
10011000	Wheat[a]	5,086	24,670
4090000	Natural honey	4858	1859
15171000	Margarine	4763	6304
3061390	Frozen lobsters	4290	521
16041410	Cooked and frozen tuna steaks	3663	1204
21039000	Other sauces	3606	1933
4070090	Eggs in their shells	3391	4267
1051100	Hens and cocks	3242	519

Continued

TABLE 9.12 El Salvador: Principal Agricultural Exports, 2003—cont'd

Customs Code	Item Description	Value (US$ 000)	Volume (mt)
7133390	Other beans	3232	2163
3062390	Fresh shrimp	3046	133
3055900	Dried fish	2874	108

aReexport.
Central Reserve Bank.

Salvador's once proud coffee sector has fallen far behind its counterparts in other Central American countries, and climate change is putting more pressure on it.) Rather, the study concluded that strategy for these crops should be to try to consolidate their position by increasing value added per hectare, instead of trying to promote a significant increase in production volumes. However, on the other hand there do not appear to be other products that have the potential to attain the scale of coffee and sugar production in the foreseeable future. The shrimp subsector raised expectations, but it faces a very competitive export market that is characterized by a downward trend in prices in addition to having to constantly struggle against the threat of shrimp diseases.

Processed agricultural products figure prominently among the most promising lines of production, as demonstrated by the data in Table 9.12. They show that, in rankings with respect to export values, agroindustrial products occupy 14 of the 15 places following coffee and sugar. In all there are 19 categories of processed products that export more than US$5 million each (excluding processed wheat, which is reexported) and 28 categories (without wheat) that export more than US$3 million. This suggests that the path toward sparking growth and greater competitiveness in the sector could be diversification plus policies that foster agroprocessing in general.

The fact that the most competitive products generally are the most intensive in knowledge as well as capital and labor suggests that the future of El Salvador's agriculture lies in the direction of specialized products of the gourmet type and for niches in consumer preferences. Each of these kinds of products typically has relatively small export volumes—although there is the possibility some of them could truly experience a takeoff (Peru is exporting hundreds of millions of dollars of asparagus alone)—but taken together they could represent substantial export earnings.

Another benefit of a strategy of this nature is that it favors the development of a more skilled labor force. Of necessity those who work with specialized products, especially those destined for export, have to get engaged in a continuous and swift process of learning. They have to learn how to learn, to continue producing competitive products. In the Central American context, it always has been recognized that El Salvador has advantages in industrial activities and in activities that require more skilled labor. This study suggests that now is the moment to extend this strategy to the agricultural sector.

A goal could be established to promote greater exports of a wide range of products, say coffee +40, in which the aim would be that at least 40 products generate export earnings of at least $5 million each, with the knowledge that some of them will exceed this threshold by a wide margin. Peru has followed a strategy somewhat like this with spices, vegetables, and fruits.

Achieving this kind of goal, in the context of the small- and medium-scale producers in El Salvador, would require considerable efforts to organize the producers, provide training, and give incentives for the formation of clusters that require similar types of knowledge. A successful element of Peru's strategy has been the use of competitive funds not only for agricultural research but also for marketing efforts for new crops and varieties. As can be seen in the sensitivity analyses reported earlier, attempts to regain competitiveness through reduction of real labor costs are not likely to succeed. More likely they would deepen rural poverty and complicate efforts to lift the skill levels of the labor force through training programs.

It also is worth bearing in mind that achieving increases in the output of export products requires accompanying research on product varieties and quality management, markets, packaging and processing. It goes without saying that a sophisticated research infrastructure cannot be established for each small-volume product, but the agriculture research system can be more oriented to products that have comparative advantage. Also, small but valuable efforts can be undertaken to widen the understanding of international markets and systematize that information in forms that are accessible to producers. In the final analysis, technological advances of various kinds are essential for economic development, in agriculture as well as in other sectors.

ENDNOTE

1. Other scenarios that do not affect the LRC value appreciably include substantial variations in the cost of labor, the owner's salary, the percentage of costs in the line item for unforeseen outlays, and the percentage profit margin. Even setting profits at zero barely affects the results.

Chapter 10

Colombia: Crop Competitiveness by Region Evaluated via Tracks 1 and 2

10.1 PURPOSES OF THE STUDY

The main purpose of this study, adapted from Norton and Argüello (2007) and Norton, Argüello, Samacá, and Martínez (2008),[1] was to contribute to the identification of the products with the greatest competitiveness and profitability, attributes that normally translate into greater potential for future growth. A related principal purpose was to identify the locations in which each product's competitiveness stands out, along with the technologies of production that are most competitive for each product. Track 1 methodology was the first tool used for these objectives. Then, for the products with outstanding competitive potential, the study carried out a qualitative review of the barriers along the value chains that have to be overcome to fully realize that potential. Equally the study aimed to identify the products or production localities that lack competitiveness. The study has been adapted for this chapter with its qualitative information reinterpreted in the Value Chain Quality Assessment (VQA) matrix framework of Track 2 assessments of competitiveness.

The findings of the study were designed for translation into guidance for projects of the US Agency for International Development and other international projects regarding which crops should receive the most support in each locality, especially because these kinds of project resources are always scarce when compared with the needs. The findings were also intended as inputs for national strategies for improving the value chains that are essential for producers to be able to receive returns on their production efforts. The study includes an unusually detailed analysis of the value chains for high-value chains based on insights provided by marketing agents involved in those chains. On the cost-price side, since the analysis was carried out at a very disaggregated spatial level, by county, a large number of Track 1 competitiveness calculations were carried out. One of the unusual features of this case study is the manner in which the results were presented so that they could provide the best guidance for local project decisions (see Section 10.4).

The Competitiveness of Tropical Agriculture. http://dx.doi.org/10.1016/B978-0-12-805312-6.00010-6

The quantitative part of the analysis was carried out on the basis of field information on farm budgets, both those that were recommended and proven and in other cases those that were most widely used in the locality. As mentioned in the introductory parts of this book, crop and livestock management practices vary widely among producers even in the same area, so a result that finds a product to be competitive in that area does not mean that all producers of it are, and vice-versa for products found to be not competitive. Nevertheless, the strategic gamble for each zone has to be based on technologies that are viable for most of its farmers.

These considerations point to another use of this kind of assessment: the evaluation of new proposals for products, crop varieties, or technologies of production or new locales for producing the same product. If field experts feel they have found a set of farm management practices that can make competitive a product that was not competitive previously, Track 1 analysis can be used to confirm or reject those proposals.

The cost-price analysis employed in this study has included detailed attention to the valuation of factors of production. Among the costs of production included are: the opportunity cost of land, labor, and financial resources; the cost implications of risk; the cost of farm administration (although it may be implicit in a farmer's operations); and a minimum required return on investment without which no investor would be interested in the product. In this sense, the assessments of competitiveness are conservative. An effort has been made to include all possible types of costs and if, in spite of this, the product still turns out to be competitive in the calculations, it is more likely that it truly is.

Another purpose of this study is to help understand why, in some cases, products that are apparently not competitive in given localities keep being produced there. The analysis suggests that often this is because farmers do not value their endowment of land when they are owners of the farm, they do not fully value family labor, or they do not assign an opportunity cost to their own financial resources. However, in the long run the economic sustainability of production patterns will depend on an appropriate valuation of these resources.

The persistence of cropping patterns or livestock management practices that are not promising for the long run also may occur because their profitability arises from paying salaries for farm labor below the legal minimum for rural salaries. In some cases the analysis of this study showed that raising the salary to the legal minimum takes competitiveness away from a product in a given location. Hence the value of this analysis of long-run competitiveness (LRC). Of course, it is important to take into account criteria derived from different approaches. When field experts or project specialists disagree with the competitiveness assessments the reasons for the differences should be explored, and perhaps new assessments should be carried out on the basis of new information to reach a consensus.

10.2 PREPARATORY STAGES FOR CROP ASSESSMENTS

As part of the process of evaluating agricultural competitiveness in Colombia, and before Track 1 assessments could be made, candidate products had to be winnowed or passed though screenings relating to data availability and quality. Only when the farm budget data were deemed acceptable could Eq. (5.8) be applied to determine competitiveness of a product on cost-price grounds. Subsequently a number of the most competitive products were subjected to Track 2 assessments to reach judgments of total competitiveness and identify barriers to fulfillment of competitive potential.

The steps in the assessment were the following:

• First screening: A list of candidate high-value products was compiled from knowledgeable persons in the sector. *This is the screening for relevance of the products.*

• Second screening: It was determined whether cost of production data existed for each of the candidate products, at least in one location in the country. *This is the screening for information availability.*

• Third screening: For products that passed the second screening, the reliability of farm budget information was evaluated with consistency criteria and on the basis of field experience of team members. *This is the screening for quality of information.* For example, if one product differed sharply from similar products in some input requirements, or those requirements differed markedly for the same product in different locations, questions were asked. In the absence of clear explanations of the discrepancies, those farm budgets were discarded as outliers.

• Fourth screening: Application of the long-run competitiveness (LRC) criterion for competitiveness, for those crops that passed the third screening. *This is the screening for cost-price competitiveness, the Track 1 assessment.* Those crops or livestock products that had an LRC value of less than 1.0 in at least one location were selected as competitive on cost-price grounds. (As shown later, the complete analysis included a range of uncertainty for crops that could not be declared either competitive or uncompetitive).

• Fifth screening: Application of Track 2 assessments. *This is the screening of competitiveness on quality grounds.* Some of the crops that passed the fourth screening and that required only reasonable efforts to overcome value chain obstacles could be considered fully competitive, at least in some locations.

Table 10.1 shows the complete list of candidate products and the results of applying the first through fourth screenings. The process reduced the number of products from 61 to 48 on the basis of availability of data, from 48 to 37 on the basis of the quality of information, and from 37 to 29 on the basis of the Track 1 competitiveness evaluation.

TABLE 10.1 Colombia: Screening of Crops for Competitiveness Assessment

First Screening: Relevant Crops		Second Screening: Availability of Information		Third Screening: Quality of Information		Fourth Screening: LRC Acceptable in Atleast One Location	
1	Avocado	1	Avocado	1	Avocado	1	Avocado
2	Chili peppers	2	Chili peppers	2	Chili peppers	2	Chili peppers
3	Broccoli	3	Broccoli	3	Broccoli	3	Broccoli
4	Cacao	4	Cacao	4	Cacao	4	Cacao
5	Lulo	5	Lulo	5	Lulo	5	Lulo
6	Mango	6	Mango	6	Mango	6	Mango
7	Passion fruit (maracuyá)	7	Passion fruit (maracuyá)	7	Passion fruit (maracuyá)	7	Passion fruit (maracuyá)
8	Blackberries	8	Blackberries	8	Blackberries	8	Blackberries
9	Plantain	9	Plantain	9	Plantain	9	Plantain
10	Gooseberries	10	Gooseberries	10	Gooseberries	10	Gooseberries
11	Honey	11	Honey	11	Honey	11	Honey
12	Baby banana	12	Baby banana	12	Baby banana	12	Baby banana
13	Coffee	13	Coffee	13	Coffee	13	Coffee
14	Rubber	14	Rubber	14	Rubber	14	Rubber
15	Dragon fruit	15	Dragon fruit	15	Dragon fruit	15	Dragon fruit

Continued

16	Strawberries	16	Strawberries	16	Strawberries	16	Strawberries
17	Guava	17	Guava	17	Guava	17	Guava
18	Heliconias	18	Heliconias	18	Heliconias	18	Heliconias
19	Banana passion fruit (granadilla)	19	Banana passion fruit (granadilla)	19	Banana passion fruit (granadilla)	19	Banana passion fruit (granadilla)
20	Cantaloupe	20	Cantaloupe	20	Cantaloupe	20	Cantaloupe
21	Cassava	21	Cassava	21	Cassava	21	Cassava
22	African palm	22	African palm	22	African palm	22	African palm
23	Papaya	23	Papaya	23	Papaya	23	Papaya
24	Pineapple	24	Pineapple	24	Pineapple	24	Pineapple
25	Tomato	25	Tomato	25	Tomato	25	Tomato
26	Watermelon	26	Watermelon	26	Watermelon	26	Watermelon
27	Onion	27	Onion	27	Onion	27	Onion
28	Milk	28	Milk	28	Milk	28	Milk
29	Tree tomato	29	Tree tomato	29	Tree tomato	29	Tree tomato
30	Eggs	30	Eggs	30	Eggs		
31	Lettuce	31	Lettuce	31	Lettuce		
32	Coconut	32	Coconut	32	Coconut		
33	Tilapia	33	Tilapia	33	Tilapia		

TABLE 10.1 Colombia: Screening of Crops for Competitiveness Assessment—cont'd

First Screening: Relevant Crops		Second Screening: Availability of Information		Third Screening: Quality of Information		Fourth Screening: LRC Acceptable in Atleast One Location	
34	Cheese	34	Cheese	34	Cheese		
35	Silk	35	Silk	35	Silk		
36	Trout	36	Trout	36	Trout		
37	Bream (mojarra)	37	Bream (mojarra)	37	Bream (mojarra)		
38	Garlic	38	Garlic				
39	Cauliflower	39	Cauliflower				
40	Asparagus	40	Asparagus				
41	Jeijoa fruit (feijoa)[a]	41	Jeijoa fruit (feijoa)[a]				
42	Soursop	42	Soursop				
43	Gulupa (*Passiflora pinnatistipula*)	43	Gulupa (*P. pinnatistipula*)				
44	Tahitian (Persian) lime	44	Tahitian (Persian) lime				
45	Macadamia	45	Macadamia				

46	Cashew	46	Cashew
47	Aromatic plants	47	Aromatic plants
48	Carrots	48	Carrots
49	Essential oils		
50	Annatto (achiote)		
51	Artichoke		
52	Cranberry		
53	Palm oil biodiesel		
54	Sugarcane ethanol		
55	Raspberries		
56	Chinese peas		
57	Small French beans (habichuela)		
58	Bay leaf		
59	Oregano		
60	Ornamental plants		
61	Blackberries		

The numbering of products does not indicate degrees of competitiveness.
[a]Also known as pineapple guava.

A striking aspect of Table 10.1 is the lack of reliable cost information for many potentially promising products, and in many cases the lack of any information at all. The mere fact that a product's potential has been commented, or even that it is included in investment programs, does not necessarily mean that adequate farm budget data have been compiled for that product. Of the original 61 products identified for analysis, information could not be found for 13, and reliable information could not be found for another 11. Thus 24 potentially worthwhile and competitive products were eliminated from the assessments on the basis of information availability and quality. This result points to the need for greater investments of effort in collecting reliable farm budget data.

Of the 37 high-value products evaluated with the Track 1 methodology, 29 turned out to be competitive and only 8 were not competitive. This indicates Colombia's general strength in high-value agricultural products. However, the quantitative assessments differed by region, in some cases substantially as noted in the following section.

10.3 COST-PRICE COMPETITIVENESS BY REGION IN COLOMBIA, TRACK 1

A major purpose of the Track 1 competitiveness calculations in this study was to ensure that the diversity of regions and technologies of production—cultivation practices—is as well represented as possible in light of the data availability (Table 10.2). For this reason, LRC values were calculated for all the farm budgets that appeared to have consistent data. As can be seen in Tables 10.1 and 10.3, a total of 309 LRC coefficients were calculated for the 37 products, for cases that had an acceptable quality of data on production practices. These calculations covered 26 regions, most of them representing a single county or department and some of them covering multiple departments. (Some of the data locations were specified as broad regions that covered multiple departments for which there were farm budgets from other sources.) For four cases, the location was not specified in the data. The products with the largest number of cases (location × technology) were tomato with 25, cacao with 22, lulo[2] with 20, cassava with 19, and blackberries with 18. For seven of the products there was only one case each, and the remaining products had between 2 and 16 cases each.

Incorporation of the spatial dimension of cost competitiveness assessments changes the overall picture (Table 10.4). In that table, allowance was made for errors in data by establishing a range of marginal competitiveness or uncertainty, defined by an LRC value between 0.9 and 1.1. A further distinction was made between highly competitive products (LRC less than 0.6) and competitive products (LRC between 0.9 and 0.6). It was observed that 29 of the 37 products assessed were competitive in at least one location (78%), whereas of the 309 cases (product × location) assessed, 193 or 62% turned out to be

TABLE 10.2 Colombia: Number of LRC Coefficients Calculated by Department and Product, Part 1

	Antioquia	Atlántico	Bolívar	Boyacá	Caldas - Risaralda	Caquetá	Casanare	Cauca	Cesar	Chocó	Córdoba	Cundinamarca	Huila	Llanos	Magdalena	Meta	Nariño	Putumayo	Quindío	The Santanders	Sucre	Tolima	Valle	Coffee zone	Central zone	Western zone	Unspecified location	Cesar	Total
Avocado	2				2			2												2			3						11
Chili peppers	2	2																					8						16
Honey	1		1																				2						4
Baby banana															1							1							3
Broccoli	1																												3
Cacao			2		1			2	1			1	1		2					6	2						3		22
Coffee																1	1					1				1			4
Rubber	2			3		2			2		2		2									1	5						9
Onions	1											1								2		1							15
Coconut																													1
Strawberries	2							1									1						1						4
Granadilla	2				1			1															3						7
Guava												1	2										3						5
Heliconias																													1
Eggs																				1									1
Milk																				5									5

TABLE 10.3 Colombia: Number of LRC Coefficients Calculated by Department and Product, Part 2

Product	Antioquia	Atlántico	Bolívar	Boyacá	Caldas - Risaralda	Caquetá	Casanare	Cauca	Cesar	Chocó	Córdoba	Cundinamarca	Huila	Llanos	Magdalena	Meta	Nariño	Putumayo	Quindío	The Santanders	Sucre	Tolima	Valle	Coffee zone	Central zone	Western zone	Unspecified location	Cesar	Total
Lettuce																												1	1
Lulo	2			1	5				1			1	3		1		1		1	1			4						20
Mango	2							1	2				1								2		3						9
Maracuyá	1	1			2			1	2				2			1				1	1		2						14
Cantaloupe													1				1			3									7
Bream					5																								1
Blackberries	1			1	3										1				1				4						18
Oil palm					3														1				3		1		1		11
Papaya	2							4	2		1	2	2	1	2	2				2			3						13
Pineapple	1								1						1					2			2						10
Dragon fruit	1												1																4
Plantain	2				3	1	1		1		1		1				1		1	1			1	2					16
Cheese																													1
Watermelon	1										2									1									7
Silk																				1									4
Tilapia																													2
Tomato	2				1			4	1			3	1							2			6	3		1			25
Tree tomato	2				1			1	1			1	5										3			1			10
Trout	1																												1
Gooseberry	1				1			2								1				1					1	1			5
Cassava	2	2							1		1		1		2					3	2	4	1						19
Total	33	6	4	4	31	2	1	21	13	1	8	15	26	1	9	4	4	1	6	35	8	4	56	5	2	3	4	1	309

TABLE 10.4 Colombia: Frequency Distribution of LRC Values by Product

Product	Highly Competitive ≤0.60	Competitive >0.60 and ≤0.90	Uncertain >0.90 and ≤1.10	Not Competitive >1.10	Total
Avocado	9	1	1		11
Chili peppers	2	10	4		16
Honey	1		3		4
Baby banana		1		2	3
Broccoli	3				3
Cacao	5	10	4	3	22
Coffee		1	1	2	4
Rubber	5	4			9
Onions	3	7	3	2	15
Coconut			1		1
Strawberries	2	1		1	4
Granadilla		4	2	1	7
Guava	4		1		5
Heliconias		1			1
Eggs				1	1
Milk		1	2	2	5
Lettuce				1	1
Lulo	8	6	4	2	20
Mango	1	6	1	1	9
Maracuyá	2	4	3	5	14
Cantaloupe	2	5			7
Bream				1	1
Blackberries	1	7	5	5	18
Oil palm		6	5		11
Papaya	8	2	2	1	13

Continued

TABLE 10.4 Colombia: Frequency Distribution of LRC Values by Product—cont'd

Product	Highly Competitive ≤0.60	Competitive >0.60 and ≤0.90	Uncertain >0.90 and ≤1.10	Not Competitive >1.10	Total
Pineapple	4		3		10
Dragon fruit	2	2			4
Plantain	1	6	9		16
Cheese				1	1
Watermelon	2	5			7
Silk			3	1	4
Tilapia				2	2
Tomato	1	15	6	3	25
Tree tomato	1	3	5	1	10
Trout				1	1
Gooseberry		1	1	3	5
Cassava	3	11	4	1	19
Total	71	122	73	42	309

highly competitive or competitive (Table 10.4). Hence competitiveness is strongly conditional on the location of production. These results also indicate which products are most sensitive to location and production methods in regard to their competitiveness. For example, avocado, broccoli, rubber, guava, cantaloupe, and watermelon are cost competitive in virtually all cases, and at worst uncertain in very few. Avocado and guava stand out for being highly competitive in almost all cases. In contrast, among the products with competitiveness in at least one case, baby bananas, coffee,[3] milk, and gooseberries are products that are uncertain or not competitive in most cases, so for those products, location and technology of production are particularly important.

Overall, location and technology turned out to be important for the majority of the crops. Cacao is highly competitive or competitive in 15 cases and not competitive in 3. For onions, this ratio is 10 and 2, and for lulo it is 14 and 2. Some other crops fell proportionately more in the columns of competitive (but not highly competitive) and uncertain, so in those cases attention to costs of production and yields is particularly important.

10.4 COMPETITIVENESS ASSESSMENTS FOR PROJECT DECISIONS

These kinds of spatial results by product can be organized in different ways and reviewed through different lenses. As a further step in this Track 1 analysis for Colombian agriculture the relatively large number of departments were grouped into geographical corridors that are more uniform ecologically. Then, to better guide project investments, the results were disaggregated by counties (*municipios*) within those corridors. The more aggregate corridors provide a picture of how Colombian ecological zones relate to competitiveness by crop, but both the corridors and departments are too large to provide the best information for project decisions on crops, which are inherently local.

The corridors were defined as the following groupings of municipios of the following departments or parts of departments[4]:

1. *Southwest Colombia*:
   Departments of Nariño, Cauca, and Valle del Cauca.
2. *Macizo/Putumayo*:
   The north of Putumayo and the entire Departments of Huila and Tolima.
3. *Magdalena Medio/The Santanders/Boyacá*:
   Portions of Santander, the south of Bolívar, the south of Cesar, and the entire Department of Norte de Santander.
4. *Urabá/Northwest of Antioquia/Chocó*:
   The subregion of lower Cauca, the largest part of Urabá, Córdoba, parts of Chocó, and the northwest of Antioquia.
5. *The Coffee Axis*:
   The south of Antioquia plus the Departments of Risaralda, Quindío, and Caldas.
6. *Atlantic Coast/Sierra Nevada*:
   The north of Bolívar, all or the largest part of Atlántico and Magdalena, and parts of the northwest of Cesar and La Guajira.

The corridor with the highest proportion of highly competitive cases was Urabá—Antioquia—Chocó, with 14 of 41 cases at that degree of competitiveness. However, in all the corridors more than half of the cases turned out to be either competitive or highly competitive, and thus each corridor has an important potential to be developed. The original report contained five tables displaying the results for each of the corridors, and in this chapter one of them is reproduced as Table 10.5 for the sake of illustration.

For project decisions, a possibly more useful arrangement of the results is shown in Tables 10.6—10.8 (which are three of the eight tables of this type in

TABLE 10.5 Colombia: LRC Values by Product for the Urabá–Antioquia–Chocó Corridor

Product	Highly Competitive ≤ 0.60	Competitive >0.60 and ≤ 0.90	Uncertain >0.90 and ≤ 1.10	Not Competitive >1.10	Total
Avocado	2				2
Chili peppers			2		2
Honey			1		1
Broccoli	1				1
Cacao		1			1
Rubber	2	2			4
Onions	1				1
Strawberries	1			1	2
Granadilla			1	1	2
Lulo	1	1			2
Mango		1		1	2
Maracuyá				1	1
Blackberries				1	1
Papaya	2		1		3
Pineapple	1				1
Dragon fruit	1				1
Plantain		1	2		3
Watermelon		2			2
Tomato	1		1		2
Tree tomato		1	1		2
Trout				1	1
Gooseberry				1	1
Cassava	1	2			3
Total	14	11	9	7	41

TABLE 10.6 Colombia: Products and Municipios With Cost-Price Competitiveness, Part 1

Department	Corridor	Municipios	LRC	Data Year
Avocado, Highly Competitive				
Cauca	Southwest Colombia	Popayán, Timbío, El Tambo, Cajibío, Morales and Piendamó	0.25[a]	2006
Cauca	Southwest Colombia	Popayán, Timbío, El Tambo, Cajibío, Morales and Piendamó	0.27[b]	2006
Valle	Southwest Colombia	Caicedonia	0.43	2005
Valle	Southwest Colombia	Caicedonia	0.45	2002
Caldas	Coffee Axis	Anzerma	0.46	2006
Risaralda	Coffee Axis	Pereira	0.51	2006
Antioquia	Urabá-Antioquia-Chocó	La Ceja, el Retiro and Guarné	0.52	2004
Santander	Magda. Medio-Santanders-Boyacá	Bucaramanga, San Gíl, Cúcuta and Ocaña	0.57	2004
Antioquia	Urabá-Antioquia-Chocó	Montebello	0.59	2005
Avocado, Competitive				
Santander	Magda. Medio-Santanders-Boyacá	San Vicente de Chucuri	0.68	2005
Chili Peppers, Highly Competitive				
Valle and Cauca	Southwest Colombia	Various (see below)[c]	0.57	2005
Valle	Southwest Colombia	Yumbo	0.58	2005
Chili Peppers, Competitive				
Atlántico	Atlantic Coast-Sierra Nevada	n.r.	0.64	2006
Sucre	Outside of the corridors	Montes de María and Sabanas de Sucre (Ovejas, Los Palmitos and Sincé)	0.75	2003
Valle and Cauca	Southwest Colombia	Various (see below)[c]	0.77	2005
Cesar	Outside of the corridors	Valledupar	0.80	2006
Valle	Southwest Colombia	n.r.	0.81	2006
Sucre	Outside of the corridors	Ovejas, Galerasel Roble, Sincé and Los Palmitos	0.81	2005
Valle	Southwest Colombia	Jumbo (3 cases)	0.81, 0.83, 0.89	2006
Magdalena	Atlantic Coast-Sierra Nevada	Sitio Nuevo and Remolino	0.85	2006

The abbreviation n.r. means "not reported."
[a]Data from MIDAS.
[b]Data from ADAM project (originally from the Colombia Corporation International, CCI).
[c]Dagua, Víjes, Yumbo, Yotoco, Darién, Jamundí, Versalles in the Dept. of Valle; Caloto, Puerto Tejada, Santander de Quilichao, Villarrica in the Dept of Cauca.

the original report), listing the municipios in which products are competitive or highly competitive. This provides local leaders as well as project managers with guidance as to where development efforts will have the greatest potential. The highly competitive cases are shaded in the tables, and in some cases notes to the table describe data sources and the cropping system used. These three tables and the other five in the original report are structured to identify significant competitive potentials both by location (department and municipio) and by crop or livestock product. For example, the tables show that products that are highly competitive in the Department of Cauca include avocado and chili peppers.

TABLE 10.7 Colombia: Products and Municipios With Cost-Price Competitiveness, Part 2

Department	Corridor	Municipios	LRC	Data Year
Honey, Highly Competitive				
Sucre	Outside of the corridors	Sabanas, Montes de María and Golfo de Morrosquillo	0.50	2005
Baby Bananas, Competitive				
Valle	Southwest Colombia	Buenaventura	0.88	2007
Broccoli, Highly Competitive				
Santander	Magda. Medio-Santanders-Boyacá	Irrigation District Chichamocha	0.19	2000
Cundinamarca	Outside of the corridors	Bogotá	0.41	2006
Antioquia	Urabá-Antioquia-Chocó	Marinilla	0.57	2005
Cacao, Highly Competitive				
Santander	Magda. Medio-Santanders-Boyacá	Bucaramanga	0.40	2006
Cundinamarca	Outside of thecorridors	El Peñón	0.54	2006
Norte Santander	Magda. Medio-Santanders-Boyacá	El Tarra	0.58	2005
		San Jacinto and Carmen de	0.58	2005
Bolívar	Magda. Medio-Santanders-Boyacá	Bolívar		
Caldas	Coffee Axis	Manizales	0.60	2006
Cacao, Competitive				
Magdalena	Atlantic Coast-Sierra Nevada	n.r.	0.66	2006
Santander	Magda. Medio-Santanders-Boyacá	Rionegro	0.68	2004
(Corridor)[a]	Northwest Colombia	n.r.	0.68	2006
Norte Santander	Magda. Medio-Santanders-Boyacá	San José de Cúcuta (Corregimientos: Palmarito, Banco de Arenaand Buena)	0.69	2004
Bolívar	Magda. Medio-Santanders-Boyacá	Santa Rosa del Sur	0.70	2005
Chocó	Urabá-Antioquia-Chocó	Unguía	0.75	2006
Magdalena	Atlantic Coast-Sierra Nevada	Santa Marta, Veredas: Quebrada el Sol, Linderos, Guacoche, Paz del Caribe, Alto don Diego, Mendihuaca and La Unión	0.79	2006
Meta	Outside of the corridors	Lejanías, Cubarral and el Dorado	0.81	2005
Córdoba	Outside of the corridors	Tierralta and Valencia	0.84	2006
Huila	Macizo-Putumayo	Paicol, Tesalia, Agrado and Garzón	0.87	2004
Tolima	Macizo-Putumayo	Ríoblanco and Ataco	0.89	2005

[a]*Technology representing all of the corridor; Ministry of Agriculture was the source of data for this technology.*

For Antioquia they are avocado, broccoli, onions, strawberries, lulo, papaya, dragon fruit, and tomato. For Caldas they include avocado, cacao, guava, lulo, papaya, and pineapple. Some other products are at least competitive, even if not always highly competitive, in all these departments (Tables 10.9–10.12b).

These tables also provide information organized by product, to identify the locations (municipios) with the best results for a given product. This way of organizing competitiveness information may be the one that is most helpful to project designers. In addition, the extremely competitive cases can be sought out, say those with an LRC value in the range of 0.2–0.4, to establish priority

TABLE 10.8 Colombia: Products and Municipios With Cost-Price Competitiveness, Part 3

Department	Corridor	Municipios	LRC	Data Year
Heliconias,Competitive				
Cundinamarca	Outside of the corridors	Pacho	0.77	2006
Milk, Competitive				
Norte Santander	Magda. Medio-Santanders-Boyacá	Toledo and Labateca	0.70	2006
Lulo, Highly Competitive				
Caldas	Coffee Axis	Manizales	0.33	2000
Valle	Southwest Colombia	Trujillo and Dagua	0.37	2000
Antioquia	Urabá-Antioquia-Chocó	El Jardín and San Andrès	0.39	2000
Quindío	Coffee Axis	Salento and Pijao	0.39	2000
Risaralda	Coffee Axis	Belén de Umbría and Mistrató	0.42	2000
Cesar	Outside of the corridors	Valledupar	0.50	2003
Cundinamarca	Outside of the corridors	Pandi and Arbeláez	0.55	2003
Risaralda	Coffee Axis	Dosquebradas, Santa Rosa de Cabal	0.58	2005
Lulo, Competitive				
Huila	Macizo – Putumayo	Nátaga, Garzón and Saladoblanco	0.61	2002
Huila	Macizo – Putumayo	Nátaga, Garzón and Saladoblanco	0.70	2006
Risaralda	Coffee Axis	Belén de Umbría and Mistrató	0.73	2006
Valle	Southwest Colombia	Trujillo and Dagua	0.77	2005
Antioquia	Urabá-Antioquia-Chocó	El Jardín and San Andrès	0.82	2005
Santander	Magda. Medio-Santanders-Boyacá	n.r.	0.90	2003
Mango, Highly Competitive				
Magdalena	Atlantic Coast-Sierra Nevada	Ciénaga and Santa Marta	0.57	2003
Mango Competitive				
Huila	Macizo – Putumayo	Tello and Campo	0.62	2006
Antioquia	Urabá-Antioquia-Chocó	Santa Bárbara	0.66	2005
Cauca	Southwest Colombia	Patía	0.76	2006
Sucre	Outside of the corridors	Ovejas, Galeras, el Roble, Sincé and Los Palmitos	0.80	2005
Caldas	Coffee Axis	La Dorada	0.88	2006
Sucre	Outside of the corridors	Ovejas, Galeras, el Roble, Sincé and Los Palmitos	0.90	2005
Maracuyá, Highly Competitive				
Quindío	Coffee Axis	Montenegro	0.51	2000
Valle	Southwest Colombia	Roldanillo and Toro[b]	0.60	2000
Maracuyá, Competitive				
Caldas	Coffee Axis	Viterbo	0.61	2000
Valle	Southwest Colombia	Roldanillo and Toro[c]	0.72	2005
Cauca	Southwest Colombia	Patía	0.76	2006
Valle	Southwest Colombia	Roldanillo and Toro[d]	0.85	2006

[a]Planted with chili peppers.
[b]Technology with yield of 20 mt/ha and unit costs of 242 pesos/kg.
[c]With yield of 22.67 mt/ha and cost of 405 pesos/kg.
[d]Yield of 21.3 mt/ha, costs of 332 pesos/kg.

products for inclusion in development efforts. In this study of Colombia, these very strongly competitive cases turned out to include (among others):

• Avocado in the municipios of Popayán, Timbío, El Tambo, Cajibío, Morales, and Piendamó in the Department of Cauca.

- Broccoli in the irrigation district of Chicamocha in the Department of Santander.
- Lulo in the municipio of Manizales (Caldas) and the municipios of Trujillo and Dagua (Valle).
- Guava in the municipios of Manizales (Caldas) and Toro (Valle).
- Watermelon in the municipio of Villavieja (Huila).
- Papaya in La Unión (Valle).
- Dragon fruit in localities in Antioquia.

Results like this indicate logical starting points in terms of crops to promote in each of these localities, or the initial crops for which value chain reviews should be undertaken. For Colombia, these and other examples that were in the original report make it evident that fruits of various kinds are the products that most often are extremely competitive, broccoli being the only exception.

10.5 MULTIVARIATE SENSITIVITY ANALYSIS OF COMPETITIVENESS

Sensitivity analysis of competitiveness parameters is worthwhile for a number of reasons. First, *robustness*: factor prices, exchange rates, crop yields, and other variables can change over time and it is important to know how robust the LRC estimates are. Second, *data errors*: there always is a degree of uncertainty surrounding the raw data and hence the calculations of LRC, so for this reason also it is helpful to know how robust the estimates are. Third, *policy*: policymakers may wish to implement changes in prices via tariffs or the exchange rate, in wage rates, in capital costs via agricultural finance programs, and in other variables. The sensitivity analysis provides an understanding of how such changes may affect the country's agricultural competitiveness, in either direction. And fourth, *development projects* at the field level may aim to increase productivity, and so the approximate degree to which such efforts affect competitiveness can be seen through the sensitivity analysis.

This section presents four of 12 tables developed in the original version of this Colombia study to illustrate the kinds of explorations that may be made through sensitivity analysis. The four tables cover blackberries, passion fruit, mango, and cacao, in each case for all the producing locations for which sufficient data were available to support competitiveness calculations. Scenarios in which LRC values indicate competitiveness are shaded to facilitate interpretation of the tables. In the base case for all crops, the intertemporal discount rate was assumed to be 6% and the cost of capital (borrowing) was assumed to be 12%. The effects of variations in these assumptions are shown in the tables. Wage rates varied considerably by locality, as did the annualized cost (rental rate) of land.

TABLE 10.9 Colombia: Sensitivity Analysis of the Competitiveness of Blackberries

Location of Production by Municipios (Department)	Initial LRC Value	Initial Price/kg.	LRC With Wage at 20,400	LRC With Land rental 2,000,000	LRC With Land Rental 508,800	LRC With Discount Rate 10%	LRC With Discount Rate 15%	LRC With Cost of Capital 6%	LRC With Cost of Capital 18%	LRC With Price/kg 1274	LRC With Devalu. of 10%	LRC With Revalu. of 10%
Pasto (Nariño)	0.59	1039	0.98	0.79	0.62	0.62	0.65	0.56	0.63	0.48	0.54	0.66
Quinchiá (Risaralda)	0.70	1420	1.16	0.86	0.70	0.71	0.74	0.65	0.74	0.78	0.63	0.78
Guática. Belén de Umbría.Quinchía (Risaralda)	0.78	1190	0.90	1.01	0.83	0.80	0.82	0.73	0.83	0.72	0.70	0.88
Aranzazu. Aguadas (Caldas)	0.78	1229	0.88	1.03	0.83	0.80	0.83	0.73	0.83	0.75	0.70	0.88
San Vicente de Chucurí (Santander)	0.81	1440	1.02	1.06	0.84	0.84	0.88	0.77	0.86	0.75	0.74	0.91
Calarcá. Córdoba. Pijáo.Génova (Quindío)	0.88	1181	1.05	1.13	0.93	0.90	0.93	0.82	0.93	0.92	0.79	0.99
Isnos (Huila)	0.89	1340	1.15	1.12	0.92	0.92	0.96	0.83	0.94	0.89	0.80	0.99
Popayán (Cauca)	0.94	1250	1.10	1.17	0.97	0.97	1.01	0.88	1.00	0.97	0.90	0.98
Santa Rosa de Cabal (Risaralda)	0.95	1420	1.19	0.95	0.84	0.96	0.98	0.89	1.01	1.07	0.85	1.07
No reporta Municipio (Caldas-Risaralda)	0.97	1350	1.21	1.22	1.02	0.99	1.02	0.92	1.03	1.03	0.88	1.09
Floridablanca (Santander)	1.01	1375	1.19	1.17	1.03	1.04	1.08	0.95	1.06	1.09	0.91	1.13
Totoro (Cauca)	1.01	1200	1.31	1.22	1.04	1.04	1.07	0.95	1.08	0.96	0.93	1.12
Guarné. San Vicente.El Retiro (Eastern Antioquia)	1.14	1250	1.62	1.34	1.14	1.17	1.21	1.07	1.21	1.11	1.02	1.28
Popayán (Cauca)	1.15	1070	1.41	1.44	1.19	1.19	1.23	1.09	1.22	1.04	1.09	1.22
Caldono (Cauca)	1.19	1200	1.32	1.44	1.22	1.22	1.25	1.12	1.27	1.13	1.08	1.31
San Bernardo.Silvania (Cundinamarca)	1.24	1420	1.60	1.44	1.25	1.28	1.33	1.17	1.32	1.41	1.12	1.41
Silvania (Cundinamarca)	1.28	1420	1.71	1.41	1.28	1.30	1.33	1.20	1.36	1.46	1.14	1.46
Saboyá (Boyacá)	1.28	1210	1.64	1.47	1.28	1.32	1.36	1.21	1.35	1.21	1.16	1.42

Prices and values are in Colombia pesos in the year of the analysis. *Devalu.*, devaluation of the peso; *revalu.*, revaluation of the peso. The observed daily wage level, used in calculating the initial LRC, differed considerably over these locations but always was well below the legal minimum of 20,400 pesos. The subsequent tables show the prevailing ranges for actual wage levels, and the higher legal minimum wage was used in the scenarios. Land rental of 2 million pesos/ha. is the highest observed in these municipios, in Quinchiá. The rental of 508,000 pesos is the second highest observed in this group of municipios, in Quinchiá, and also in Risaralda.

TABLE 10.10 Colombia: Sensitivity Analysis of the Competitiveness of Passion Fruit

Location of Production by Municipios (Department)	Cropping System	Initial LRC Value	Initial Price/kg.	Actual Wage Rate by Location	Legal Wage Rate by Location	LRC with Legal Wage	LRC With 10% Rise in Actual Wage	LRC With 10% Rise in Land Rental	LRC With Discount Rate 10%	LRC With Discount Rate 15%	LRC With Cost of Capital 6%	LRC With Cost of Capital 18%	LRC With 10% Rise in Price	LRC With Devalu. of 10%	LRC With Revalu. of 10%
Montenegro (Quindío)	Maracuyá	0.51	380	8,400	13,005	0.63	0.53	0.51	0.52	0.53	0.48	0.54	0.46	0.46	0.56
Roldanillo I (Valle)	Maracuyá	0.60	383	8,000	13,005	0.74	0.63	0.61	0.62	0.63	0.57	0.64	0.54	0.55	0.67
Viterbo II (Caldas)	Maracuyá	0.61	330	8,500	13,005	0.75	0.63	0.61	0.62	0.63	0.57	0.64	0.55	0.55	0.67
Roldanillo III (Valle)	Maracuyá	0.72	600	15,000	19,075	0.82	0.76	0.73	0.72	0.72	0.68	0.76	0.65	0.65	0.80
Patía (Cauca)	Maracuyá	0.76	530	17,408	20,400	0.81	0.79	0.77	0.78	0.81	0.71	0.81	0.68	0.69	0.85
Roldanillo II (Valle)	Maracuyá	0.85	385	10,000	15,450	1.08	0.89	0.85	0.85	0.85	0.80	0.90	0.77	0.77	0.95
Viterbo I (Caldas)	Maracuyá	0.92	655	16,000	20,400	1.05	0.97	0.93	0.94	0.96	0.86	0.98	0.82	0.84	1.02
Suaza II (Huila)	Maracuyá	0.95	597	15,300	20,400	1.08	0.99	0.96	0.96	1.02	0.89	1.02	0.84	0.87	1.08
Suaza I (Huila)	Maracuyá	0.97	400	12,000	15,450	1.09	1.01	0.97	0.97	0.97	0.91	1.03	0.87	0.88	1.06
Valledupar (Cesar)	Maracuyá	1.30	400	12,000	15,450	1.53	1.38	1.31	1.30	1.30	1.22	1.38	1.17	1.19	1.45
Lebrija (Santander)	Maracuyá	1.32	350	12,000	15,450	1.52	1.39	1.32	1.32	1.32	1.23	1.41	1.16	1.20	1.46
Ovejas (Sucre) w/o irrigation	Maracuyá, chili	1.34	235	13,000	16,600	1.62	1.44	1.35	1.37	1.41	1.26	1.41	1.28	1.22	1.49
Ovejas (Sucre) w/o irrigation	Maracuyá	1.39	235	13,000	16,600	1.65	1.48	1.41	1.44	1.50	1.31	1.47	1.26	1.27	1.55
Dabeida (Antioquia)	Maracuyá	1.81	835	14,000	20,400	2.15	1.88	1.82	1.86	1.93	1.69	1.92	1.59	1.64	2.01
Girón (Santander)	Maracuyá	1.83	380	18,071	20,400	1.96	1.93	1.85	1.90	1.98	1.72	1.94	1.62	1.66	2.03
Ovejas (Sucre) w/o irrigation	Maracuyá, chili	2.01	235	13,000	16,600	2.34	2.13	2.01	2.14	2.32	1.86	2.15	1.89	1.83	2.23

TABLE 10.11 Colombia: Sensitivity Analysis of the Competitiveness of Mango

Location of Production by Municipios (Department)	Cropping System	Initial LRC Value	Actual Wage rate by Location	Legal wage Rate by location	LRC With Legal Wage	LRC With 10% Rise in Actual Wage	LRC With 10% Rise in Land Rental	LRC With Discount Rate 10%	LRC With Discount Rate 15%	LRC With Cost of Capital 6%	LRC With Cost of Capital 18%	LRC With 10% Rise in Price	LRC With Devalu. of 10%	LRC With Revalu. of 10%
Ciénaga (Magdalena)[a]	Mango	0.57	12,500	16,600	0.68	0.60	0.57	0.61	0.68	0.53	0.60	0.51	0.52	0.63
Tello and Campo (Huila)[a]	Mango	0.59	15,300	20,400	0.67	0.61	0.59	0.64	0.73	0.55	0.63	0.53	0.53	0.65
Santa Bárbara (Antioquia)	Mango	0.66	15,000	19,075	0.78	0.70	0.67	0.60	0.72	0.62	0.69	0.60	0.60	0.73
Patía (Cauca)	Mango	0.76	17,408	20,400	0.82	0.79	0.76	0.85	0.99	0.70	0.81	0.67	0.69	0.84
Ovejas (Sucre)	Mango, chili	0.80	12,700	19,075	1.08	0.86	0.80	0.82	0.85	0.76	0.85	0.78	0.73	0.89
La Dorada (Caldas)[a]	Mango	0.88	16,000	20,400	1.01	0.92	0.88	0.97	1.12	0.82	0.94	0.78	0.80	0.97
Ovejas (Sucre)	Mango	0.90	12,700	19,075	1.13	0.95	0.91	1.00	1.17	0.84	0.96	0.81	0.82	1.00
Anapioma (Cundinamarca)[a]	Mango	0.99	15,000	16,600	1.07	1.06	1.00	1.11	1.28	0.94	1.05	0.90	0.90	1.10
Sopetran (Antioquia)[a]	Mango	1.20	14,000	19,075	1.47	1.27	1.21	1.32	1.51	1.12	1.27	1.06	1.09	1.33

[a] The cost data did not report a municipio within the Department, so the cited municipio refers to the locations with the greatest amount of production.

TABLE 10.12A Colombia: Sensitivity Analysis of the Competitiveness of Cacao, Part 1

Location of Production by Municipios (Department)	Initial LRC Value	LRC With Legal Wage	LRC With 10% Rise in Actual Wage	LRC With 10% Rise in Land Rental	LRC with Discount Rate 10%	LRC with Discount Rate 15%	LRC With Cost of Capital 6%	LRC With Cost of Capital 18%	LRC With Higher Yield by 10%	LRC With Price Lower by 30%	LRC With Price Lower by 10%	LRC With Devalu. of 10%	LRC With Devalu. of 10%
Bucaramanga (Santander) II La suiza (CORPOICA)	0.40	0.48	0.42	0.40	0.43	0.48	0.37	0.42	0.38	0.45	0.41	0.36	0.44
El peñón (Cundinamarca)	0.54	0.63	0.57	0.55	0.61	0.72	0.5	0.57	0.50	0.71	0.61	0.49	0.60
San Jacintoy Carmen de Bolívar (Bolívar)	0.56	0.71	0.59	0.57	0.64	0.75	0.53	0.60	0.53	0.72	0.61	0.51	0.63
El Tarra (Norte de Santander)	0.58	0.68	0.62	0.59	0.68	0.83	0.55	0.62	0.54	0.75	0.63	0.53	0.65
Manizales (Caldas) Granja Lucker	0.60	0.57	0.64	0.60	0.63	0.69	0.57	0.63	0.56	0.79	0.65	0.55	0.67
Caso Magdalena Medio	0.66	0.69	0.68	0.67	0.71	0.77	0.62	0.70	0.60	0.93	0.73	0.60	0.74
Caso Min. Agricultura	0.68	0.77	0.71	0.69	0.73	0.81	0.64	0.72	0.62	0.94	0.75	0.62	0.75
Rionegro (Santander)	0.68	0.76	0.72	0.69	0.75	0.85	0.64	0.72	0.62	0.93	0.75	0.62	0.75
San José de Cúcuta (Norte de Santander)	0.69	0.78	0.74	0.70	0.76	0.86	0.65	0.73	0.63	0.97	0.77	0.63	0.77
Santa Rosa del Sur (Bolívar)	0.70	0.81	0.74	0.71	0.78	0.90	0.66	0.74	0.64	1.01	0.78	0.64	0.78
Unguía (Choco)	0.75	0.96	0.79	0.78	0.85	0.99	0.71	0.80	0.69	1.07	0.84	0.68	0.84

TABLE 10.12B Colombia: Sensitivity Analysis of the Competitiveness of Cacao, Part 2

Location of Production by Municipios (Department)	Initial LRC Value	LRC With Legal Wage	LRC With 10% Rise in Actual Wage	LRC With 10% Rise in Land Rental	LRC With Discount Rate 10%	LRC With Discount Rate 15%	LRC With Cost of Capital 6%	LRC With Cost of Capital 18%	LRC With Higher Yield by 10%	LRC With Price Lower by 30%	LRC With Price Lower by 10%	LRC With Devalu. of 10%	LRC With Revalu. of 10%
Santa Marta (Magdalena)	0.79	0.98	0.83	0.80	0.83	0.92	0.75	0.83	0.72	1.10	0.87	0.72	0.88
Lejanias, Cubarral and El Dorado (Meta)	0.81	0.82	0.81	0.82	0.93	1.14	0.76	0.86	0.74	1.18	0.91	0.74	0.90
Tierralta and Valencia (Córdoba)	0.84	1.00	0.90	0.85	0.92	1.03	0.79	0.89	0.77	1.20	0.93	0.77	0.94
Paicol, Tesalia, Agrado and Garzón (Huila)	0.87	0.97	0.91	0.88	0.92	0.99	0.82	0.92	0.81	1.16	0.95	0.79	0.97
Rio Blanco and Ataco (Tolima)	0.90	1.04	0.95	0.90	1.07	1.35	0.84	0.94	0.83	1.14	0.96	0.81	1.00
Tumaco (Nariño)	0.94	0.95	1.00	0.94	1.01	1.12	0.88	0.99	0.85	1.36	1.04	0.85	1.04
Consenso in the Secretaría Técnica de Cacao	0.97	1.12	1.02	0.98	1.04	1.13	0.91	1.03	0.88	1.38	1.07	0.88	1.08
Santander I	1.05	1.20	1.11	1.06	1.15	1.29	0.98	1.11	0.95	1.53	1.17	0.95	1.16
Huila I	1.06	1.29	1.13	1.07	1.25	1.57	0.99	1.13	0.96	1.61	1.20	0.97	1.18
Caso Fedecacao	1.12	1.21	1.19	1.13	1.20	1.31	1.05	1.19	1.01	1.63	1.25	1.02	1.24
Cimitarra and Landázuri (Santander)	1.49	1.64	1.56	1.50	1.63	1.81	1.39	1.58	1.35	2.15	1.66	1.35	1.65

A pattern that is evident in all the tables is that competitiveness is quite sensitive to wage rates. For blackberries, of the 10 competitive cases in the base scenario, only 3 remain competitive if the official minimum wage is paid instead of the (lower) prevailing wage rates. For passion fruit, 5 of 9 initially competitive scenarios remain competitive when the official wage is paid, and for mango it is 4 of 8. This may not be surprising because high-value crops are intensive in labor. For example, on average field labor costs for blackberries amount to 48% of total costs, and all labor costs together, including management and administration, account for 67% of total costs. Cacao is somewhat different, because on average all labor costs represent 56% of total costs (49% of the labor in cacao is used for harvesting and postharvest treatments). Partly for this reason, shifting to the official wage rate in cacao production removed competitiveness from only 3 of 18 cases. It is also relevant that cacao has a number of cases that are extremely competitive in the base scenario.

Competitiveness is not as sensitive to land costs as it is to wages. Although competitiveness was severely undermined in an extreme scenario constructed for blackberry with a more than four-fold increase in land costs for all but one location, for the other crops it was found that a 10% increase in land costs scarcely affected competitiveness. This result raises questions about the effectiveness of the policy of legislating a rural wage rate in conditions in which it is impossible to enforce it and much of the labor is provided by family members. A truly enforced legal minimum wage would cause production to drop for many crops, and presumably employment would fall as well. After labor, product prices have the biggest effect on competitiveness, and consequently the exchange rate has a powerful effect. A revaluation of the exchange rate (move in the direction of overvaluation or more overvaluation) undermines competitiveness in quite a few case for all crops.

Raising the cost of capital to 18% also has a strongly negative effect on competitiveness. On the contrary, lowering it from 12% to 6% has a significant effect in the opposite direction only for cacao, and a negligible or smaller effect for other crops. Cacao is highly sensitive to increases in the intertemporal discount rate, mainly because it is a long cycle crop and those increases mean that future income streams have a lower present value. This could have an implication for smallholders because it is widely assumed that their subjective discount rate is high owing to their urgent need for income in the short term. This is a subject that may be worth doing more research on, especially since long-cycle crops have some of the highest returns, some with very high returns, and they are well suited for smallholder cultivation.

Finally, a conclusion that emerges from these scenarios is that it is wise to use ranges of values when evaluating the cost-price competitiveness of products, all the more so because data are not available to give confidence intervals for the estimates of parameter values in farm budgets. From this perspective, LRCs in the range of 0.90–1.00 do not appear to indicate robust competitiveness under data uncertainty and possible changes in the external economic environment.

10.6 COLOMBIAN VALUE CHAINS: VIEWS OF MARKETING INTERMEDIARIES

To reinforce the understanding of value chains for high-value products, structured interviews were carried out with 14 marketing intermediaries and processors of different types, and they were supplemented by information from the extensive field experience of team members and previous studies of the value chains. An interview guide was developed to ensure that all the relevant topics were covered. With the marketing intermediaries the interviews not only concentrated on familiar issues but also explored the degree of vertical integration and coordination of the value chains, the nature of the final markets, and the marketing mechanisms. These conversations also shed light on appropriate strategies for value chains in the context of development projects. Often the projects concentrate resources and efforts at the level of producers without taking into account the need for development of entire value chains, as well as the need to foster sustainable relationships among the links of a value chain. The conclusions about the place of value chains in development projects are summarized in Section 10.6.5. The interviews provided insights that are not often found regarding the business of export marketing and other issues.

The next subsections summarize the findings from these interviews with marketing experts. The findings are organized in the following sequence: value chain structures, export marketing and logistics, situation of input supplies, storage logistics, and modalities of vertical coordination within the value chains.

10.6.1 Value Chain Structures: an Overview

The first and most evident characteristic of the value chains for the products of this study is that production and marketing activities aimed at the domestic market tend to be highly decentralized whereas the packing and processing stages, especially for export markets, tend to be relatively centralized. The domestic market has two parallel types of chains: large auction markets or wholesale markets on the one hand, and direct marketing links between producers and retail chains (especially supermarkets) or institutions on the other. The two competing modalities help create a dynamic domestic market that responds quickly to changes in demand.

Oligopolistic structures predominate in packaging and processing activities; usually between 3 and 10 firms control the largest share of the market by far. This reflects in part the economies of scale of those activities. Equally, a small number of firms have superior command of relevant technologies, greater financial leverage, or better connections with commercial actors in destination markets, which enables them to adapt to changes in the markets and to exert leadership over the upstream segments of the value chain. Inadvertently or not, government intervention also has been instrumental in promoting these oligopolistic structures. Centralization of commercial activities has given government accessible channels of influence over the export-oriented value chains and of provision of government services to encourage their development.

Second, vertical coordination between the links of a value chain can take different forms. In most cases, the actors are knitted together by the workings of a free market. Notwithstanding this, frequently the functioning of the open market is supplemented by other mechanisms such as seasonal or long-term contracts, integration through ownership of different links, coordination through cooperatives, and coordination by the government. Coordination via contractual mechanisms plays a predominant role for products like fresh and processed fruit and vegetables and for shrimp. This mechanism is characterized by supply of credit or inputs to the producers, setting prices in advance via an established formula, and specifications for product quality and the quantity and dates of delivery. Vertical coordination tends to be much tighter for export products. However, a tendency has been observed for the domestic market to incorporate more of these characteristics over time, which lays the basis for eventual increases in the exportable supply, either seasonally or for the longer run.

Third, for contracts to be fulfilled the coordination between exporters and the marketing agents in destination markets always involves a component of complementary operations on open markets, whether on spot markets or through consignments. However, the dominant mechanism continues to be contractual in its varied forms. Trade within units of the same enterprise occurs in a number of cases, whether within domestic firms or multinationals, but only in special circumstances is it the dominant form of value chain coordination. Long-run contractual ties have been proved to be fundamental for maintaining access to markets, for penetration of new and rapidly expanding marketing channels, for obtaining up-to-date information on market conditions and consumer preferences, for reducing uncertainty about receiving payments, and for keeping products in the desired markets (which bears directly on the possibilities for developing brands).

Fourth, the role of foreign firms has been important in the development and the continued presence of many of these export value chains, particularly in regard to provision of capital, technologies, training, and administrative skills, and especially in the takeoff phase of the value chains. In spite of this, domestic firms tend to play the greater role in the business of exporting. In contrast, the support of bilateral or multilateral development organizations does not appear to have played a relevant role in these processes. The exporting chains' initial dependence on technology imported from private sources (genetic material, tractors, processing plants, and equipment in general) has been a frequent and decisive factor in their development.

These interviews also brought out the important role of the domestic market in the development of export value chains. The domestic market can act as a stimulus for production increases and as a regulator and buffer for product supply, providing a permanent base of demand and absorbing the supply that cannot be exported for reasons of seasonality or product quality.

10.6.2 Export Marketing and Logistics

Marketing agents go mainly after markets with high incomes, relatively elastic product demand functions, and well-defined entry requirements. This last item refers not only to product quality standards but also to other regulations and trade practices. Export markets are varied, but the majority of those interviewed work with European markets, which they consider less difficult to penetrate that US markets. They also feel dealing with European markets on a continuing basis is an easier administrative task. In the opinion of some interviewees, the main issue with the U. S. market is what they call a lack of clarity and reliable information about the admissibility of products. The efforts of groups like the Center for Phytosanitary Excellence can help to reduce this informational gap, at least as far as food safety issues are concerned.

The main markets for exports of the products of this study are England, France, Holland (which re-exports a lot of products), Germany, Spain, Belgium, Italy, the United States (Miami), some Caribbean islands (especially the French-speaking ones, which represent a bridge to France), Canada, and the Middle East (for chili sauces). Most of those interviewed indicated the importance of maintaining a permanent business relationship with brokers, distributors, and other buyers of Colombian export products in those countries. In some cases, to ensure the quality of products upon arrival in the importing countries Colombian marketing entities have found it advantageous to have their own agents in the importing countries that receive and inspect the shipments before sending them on to destinations in those countries. Distributors in importing countries can manage the movement of the products to various regions in their countries, something that outside marketing entities cannot do. Some suppliers to supermarkets in those countries establish seasonal windows for purchases from each exporting country, thus taking advantage of production calendars to supply their own markets over the entire year.

Colombian marketing agents make use of both consignment shipments and shipments based on prior, signed contractual agreements. On the surface, consignment shipments are considered to be more risky because the price is not agreed upon beforehand and the sale agreement is not necessarily binding. However, the experience of some of the interviewees indicates that distributors in importing countries can place products at favorable prices, which benefits Colombian providers of goods in consignment. These same interviewees feel that it is not necessarily true that sales through contracts transfer the risk to distributors in importing countries. They say that if these distributors succeed in selling at higher-than-expected prices in the destination country the additional benefits remain only with them, but if they end up selling at lower prices they tend to discount the prices paid to the exporter by applying more rigorously or in an unfair way the quality standards. In any event, under both systems of export sales there are cases of shipments that were not paid for, apparently at least in some cases without justification.

An important role of distributors and other buyers of exports is the transmission of information about market conditions broadly. This covers not only the perceived situation of competing countries that supply the same products or near substitutes but also changes in markets and emerging market trends. For Colombian exporters the most valuable sources of information are of these kinds—that are directly linked to markets.

The process of opening new markets takes time and is costly. The interviewees generally mentioned periods of around 2 years to establish a new marketing channel, from the moment when contacts are made and the shipment of samples is initiated to the date when a program of regular shipments is agreed upon. These periods include the time necessary to comply with all the applicable regulations in each market: accessing information about the processes, obtaining licenses, assigning bar codes and other retail-level markers and packaging for the products, and so forth. Against this background when difficulties subsequently emerge in a marketing channel, or the channel is lost, the exporter and other marketing agents of the exporting country can incur significant losses and the reputation of the country as a reliable supplier of quality goods can suffer.

For high-value products, the volumes moved by exporters vary, typically between one 40-foot container every fortnight and seven or more containers per week. The agents that ship fewer containers appear to be less vertically coordinated with other links upstream in their value chain. They also tend to operate on slimmer margins and are more vulnerable to price fluctuations.

Export logistics have a number of important aspects that affect competitiveness. The first important issue for export logistics is domestic transportation to ports. The majority of those interviewed feel the availability of this transportation is adequate and when needed it can be refrigerated. However, it is necessary to monitor how temperatures are controlled since it is a relatively common practice for the truck drivers to shut off the flow of cold air to save on fuel or have more power for the vehicle. Some marketing operators make extensive use of thermography and temperature controls, but others have problems with their functioning because the devices get lost and it is not always possible to assign clear responsibility for their care since multiple agents can play a role in a logistics chain. A practice that appears to give good results is the inclusion in the transportation contract of clauses that cover the topic of temperatures and assign clear responsibility for it. Given the special requirements of each product, it is important that the transport conditions are clearly specified and are under the control of the exporter. The majority of marketing agents say refrigerated transport deficiencies are a major issue.

Breaking the cold chain can have serious consequences for the products. The rise in temperature shortens their shelf life and it can have disastrous consequences for the packaging as well as the products. The humidity generated can weaken packaging structurally and promote the growth of fungus on the products. It can also weaken the pallets to the point where the weight of the produce can collapse them, leading to further deterioration of the quality of the items being shipped.

In spite of the availability of transport, a significant number of marketing agents maintain that it is too expensive. From many agricultural areas of the country a journey in a tractor-trailer to Atlantic ports with 18 tons of net product costs around 2.7 million pesos, which is higher than the cost of shipping a 40-foot container from Cartagena, Colombia, to Spain (which costs around US$1100). The duration of the road trips also is a problem. The trip from the interior to an Atlantic port takes 2 days. This can be a critical period of time for products that need refrigeration but in practice are transported at ambient temperatures on the way to a port.

Often a cold chain also is broken in the ports, which causes serious product losses. An additional port problem for the shippers is the time required for customs and antinarcotic inspections. On occasions these inspections take days. Moreover, the antinarcotics inspections tend to be carried out at ambient temperatures, which is very damaging for products that have been maintained at 5°C or less. Sometimes exporters have recourse to specialized agencies that can move products through ports more rapidly, but this is too expensive for some exporters and for some agricultural products.

Another issue reported by the marketing agents is that shipments of agricultural products can be handled in ports as if they were dry goods, and this tendency introduces another element of risk in the logistics chain that can weaken competitiveness. In general, the agents interviewed felt the port of Santa Marta functions more or less well; that the port of Cartagena operates above capacity; which causes delays; and that the port of Buenaventura has many problems. This consensus provides clear priorities for public investments and strategies for port management.

Logistics tend to work better in airports than in seaports. Refrigerated containers move rapidly through airports, and the antinarcotics inspections appear faster in the airports. Of course air shipment is economically viable only for products with a high value in relation to volume or weight, such as fresh products (shipped more in the off-seasons in importing countries). When compared with neighboring countries Colombian airfreight rates appear competitive.

In spite of the advantages in speed of shipping by plane, for some products the transition from using airfreight to ocean freight represents an important change in market development. Shipping by sea allows exporters to reduce prices they offer in destination markets, which favors expansion of those markets. This appears to have been the case for gooseberries (*uchuvas*) whose market expanded greatly after it became possible to send them by boat. In addition, the relatively short times required for ocean shipment from Colombia to the US east coast and to Europe give the country an important advantage. Some commercial operators say investors in other Latin American countries want to develop business in Colombia precisely to benefit from this advantage. According to information supplied by several interviewees, a ship takes 26 days to go from Peru to northern Europe, whereas it takes only 6–9 days from Colombia.

Nevertheless, the lack of reliability of ocean transport has been a problem at times. Occasionally the frequency of ship arrivals has dropped, which has presented serious issues for perishable products. According to one interviewee, after the signing of the trade agreement between Chile and the United States many ships sailed north with full loads, with the consequence that for a period Colombian agricultural produce was stuck in ports up to 25 days. In addition to spoilage of produce these delays caused additional storage costs for exporters and led to failure to fulfill agreements on delivery dates.

10.6.3 Vertical Coordination in Colombian Value Chains

In recent years, exporters and marketing agents who do business with them, have drastically altered their approaches in light of the strict international standards on food safety and residues plus traceability. Now they work just with farmers who can guarantee use only of allowable agrochemicals and in permitted amounts. Because of this concern some marketing agents fear that products like dragon fruit will have to be taken off the market eventually, for failure to comply with these norms. However, it is not only regulations about food safety but also the need to ensure requisite volumes to meet market demands at the right moments that drive marketing agents into forging closer ties with producers. These forces are bringing about closer coordination between the agents and farmers regarding planting decisions, in regard to timing and areas planted.

Marketing agents that coordinate closely with farmers provide seeds and recommend agrochemicals or help farmers obtain them, with the cost discounted from payments for harvests. They offer farmers technical advice on field techniques, harvesting, and postharvest management to obtain products that comply with food safety requirements and are delivered in agreed volumes and at agreed times. However, for producers that are more widely dispersed in small plots, the control over cultivation techniques is looser and the marketing channels tend to be the traditional ones of sale at spot prices. The common variety of mango, whose production takes place in scattered locations, is marketed mostly in this manner. In such cases, coordination is limited to that occurring between the main marketing agent and lower intermediaries, and the latter organize collection points in production areas and transmit market demands and requirements to even lower level intermediaries or producers. On numerous occasions the main marketing agents[5] have found that producers do not adhere to the agreements or contracts, and the costs of inputs are not reimbursed to the marketers. This occurs most often for products with a relatively large domestic market, which are more competitive in the sense that producers have many options for channels through which their harvest may be sold. In these cases, the main marketing agents may turn to contracts with local intermediaries that operate in each zone, and the latter absorb the risk that producers may not provide the agreed quantities or quality of product. For passion fruit, for example, it has been practically impossible to implement contract farming schemes, in spite of the fact that the contracts offer prices 25% above costs, because of the marked fluctuations in prices and strong competition among producers and intermediaries. In the opposite case, with a small, easily saturated domestic market (e.g., for broccoli and asparagus), coordination between marketing agents and farmers is more effective.

When the domestic market for a product is relatively large the flows of products follow one of two basic options. In one case, commercial-scale producers or main marketing agents stay active in the domestic market and use it as an outlet when supplies exceed what is needed for export at that moment. This option requires that prices respond to product quality, which means that gaps between export and domestic prices tends to be closed, effectively integrating the domestic and export markets. The second option, which appears widespread at the time of this study, is that prices on the domestic market are higher than export prices at least at some times of the year. This situation works against the development of export markets. Of course whether domestic prices are higher than export prices depends in part on the exchange rate. Recently the Colombian peso appears to have been strengthened, to the detriment of agriculture, by partial influence of the "Dutch disease" syndrome brought on by exports of oil, minerals, and cocaine, which earn large amounts of foreign exchange when taken together.

It appears that creative efforts are needed to develop stronger mechanisms of contractual coordination under which producers and the different types of marketing agents all have incentives to strengthen the value chain. An example is the use of good agricultural practices, a prerequisite for obtaining the certifications that are increasingly important in markets and helping to guarantee that products move to their desired destinations. To the degree that producers perceive benefits from certifications, the implementation of food safety measures at the farm level is more easily assured.

A finding from the interviews that is of great relevance for development projects is the stated and firm preference of main marketing agents for working with smallholders. This preference prevails in spite of the need for groupings of farmers to obtain desired scales of production and to be able to provide technical assistance, and also in spite of the more costly administrative process of obtaining certification for small farmers and coordinating the operations of large numbers of producers.

The reasons mentioned for preferring to work with smallholders varied. First, they tend to be more careful in their cultivation practices in light of the opportunity they have, which represents an important part of their family income. Also, the supervision of the work on the farm is more effective because the work is carried out almost entirely by family members. Second, when the marketing agent obtains a commitment from small farmers, they are more likely to make good use of the advisory services provided to them. Finally, in the opinion of marketing agents, working with large numbers of small producers permits diversification of risks. The main risks arise from climatic conditions, crop infestations, and diseases, and also from the breaking of contracts. If any of these events affect a few of the producers in an area it does not necessarily put at risk the entire operation of a marketing agent. Associations of producers have begun to take shape in the past 6 years, and now they play an important role in some areas. Contracting with them appears to be yielding good results and facilitates communication.

In products like passion fruit and common varieties of mango traditional marketing relationships continue to be the norm. Buyers in formal value chains have adapted to these systems by coopting second-level intermediaries, who in turn coordinate with local intermediaries and who have adopted operations acceptable in more formal markets. In these cases prices are set initially through a prior evaluation of the harvests and then through the interaction of the offers of producers.

On the other side of the balance, marketing agents are of the opinion that medium- and large-scale producers comply more with requirements on quality and quantity when they are relatively specialized in the high-value products. However, on occasions it can be challenging to create a consensus between the differing agronomic criteria of the large farmers' agricultural advisors and those of the marketing agents. The trend appears to be clearly toward larger farmers developing their own marketing activities, including exporting, rather than going through intermediaries.

The conditions of transportation for bringing products to market depend on the degree of vertical integration or coordination between marketing agents and farmers. In cases of close coordination, as in the case of export of fresh asparagus, the transportation scheme follows planned routes for collecting harvests. This guarantees not only sufficient trucking capacity for the crop but also its collection in the appropriate moments. When the coordination is looser, the marketing agents place crop collection centers in the production zones, so that initial transportation of the crop is shorter and the grading and selection of the crops can be carried out early in the postharvest phase. For products to be exported in fresh form, the intermediaries try to ensure that the transportation on secondary roads is short. This allows use of trucks of smaller capacity, and it reduces the damage incurred by the products en route.

However, some intermediaries report that transportation of this type increases the risk of contamination of the products and they try to limit these runs to less than 30 km.

In some cases and especially for products destined to be processed, the transportation is handled by local intermediaries with a limited radius of action who buy the harvests and sell them to second-level intermediaries. In these instances, there appears to be a fleet of independent trucks that moves through the zones where harvests are being carried out, basically vehicles with capacities of 5, 10, and 15 tons. Usually the crops are placed in plastic boxes provided by the marketing agents or obtained with their help. In some cases sacks are used, especially for crops for processing that withstand the jolting during transport, as in passion fruit for juice.

10.6.4 Quality, Availability, and Cost of Inputs

The issues for this topic are different for products sold fresh and for processed products. For the former the most critical requirements concern product size, appearance, taste and smell, and related characteristics. In contrast, for processed products, the main issues concern product packaging and handling (including labeling) and physical and chemical characteristics of the products.

Beginning with genetic material, the interviewees were of the opinion that there often are difficulties in obtaining the right kinds of seeds for producing outputs destined for processing. For tropical fruit the situation is more complex. The agents reported serious deficiencies regarding research, propagation, and selection of genetic material. Some of the marketing intermediaries try to support farmers in seed selection, and others select the seeds themselves and then offer them to farmers. The majority feels that systematic research on these crops is lacking, and this lack represents an important limitation for their development and competitiveness. In the case of gooseberries, for example, farmers do not know in advance what the yields will be and the yields obtained can vary by up to 100% over farms. This situation can lead to over- or undersupply of contracted amounts, causing problems for the marketing intermediaries who buy the crop to fulfill commitments of their own.

Seeds of course can make a big difference in product quality and uniformity of the harvests as well as in yields. Seeds of good quality can cost 2 to 2.5 times seeds of poor quality, but this cost difference can be reduced significantly if a farm has his or her own seed nursery plot. Citrus seeds are known in Colombia for varying in quality and may be the cause of some of those crops' issues with disease and product quality. Agents in the value chain report that importing high-quality genetic material for citrus is very expensive.

Seed problems and cultivation issues lead to high percentages of harvests that are unsuitable for export and have to be sold on the domestic market. In citrus (Valencia oranges, Tahitian limes) typically only 20% of the harvest is suitable for export. For tree tomato, around 50% of the harvest can be exported; and for gooseberries the figure is higher at around 75—80%. For fruits in general, the exportable share of the harvest averages about 70%. In contrast, in Peru 95—98% of the mango harvest is exportable.

According to the interviewees, some other inputs also are of poor quality and are pricey. For example, wax used on some fresh products costs 2 million pesos per 200 liter can, whereas the imported version costs slightly less at US$900, including tariff and value-added tax, and is of better quality, but it is not imported regularly in sufficient quantities. In the fruit processing industry, most inputs are readily obtainable but glass bottles and tin cans sourced domestically are considerably more expensive than in competing countries. These issues along with others mentioned throughout this section illustrate the need for comprehensive development programs for entire value chains. Marketing intermediaries also feel that technical assistance to farmers is inadequate for both crop and postharvest management.

10.6.5 Value Chain Roles in Projects

This section presents reflections and interpretations based on the material gathered from the interviews with marketing agents, with focus on projects that work with smallholders who are often located in areas of difficult access; who depend on family labor, although they may also use hired labor; whose land tenure may not be secure; and who may market through cooperatives or looser groupings of farmers.

It is widely agreed that projects should be designed from markets backward to production, and not vice-versa. This approach requires detailed identification of existing opportunities regarding types of products; prices, quantities, and seasons or months when the products are demanded; standards for quality; admissibility; packaging; and other conditions relevant to gaining market access. In this context, a practical mechanism for establishing a market-based approach is to link the project with marketing agents who have experience and knowledge of how the markets work and who can guarantee sufficient coordination between themselves and the producers. This applies to supporting production for both the domestic and export markets.

Thus having marketing agents play a role in projects can be a requirement for their success, and the kinds of mechanisms of coordination between marketing channels and producers are a critical element. However, the interviewees in Colombia generally felt that the cost of working with development projects is too high in relation to the benefits they might receive from them, so evidently more work is needed in that area.

In light of the marketers' opinion regarding development projects, there is a tendency that the marketing agents transfer the transaction cost of relations with the project to third parties, specifically to the projects' own technical experts in the field. This in itself is not necessarily a problem and is a practice that follows from project design, but it must be asked if it does not weaken the relationship between marketing agents and producers, which is necessary for sustainability of the value chain in the long run and can be multifaceted. This default option may place too much responsibility in the hands of project experts in the field.

It is equally important to not create relationship structures in which producers that participate in the projects end up being dependent on governmental entities or activities for vital inputs, such as the case for improved seeds for cacao. This dependence always slows the rate of progress in the field because it inhibits the development of private input supply chains, which are the only sustainable mechanisms for inputs.

It is worth noting that significant numbers of the interviewees feel that priority should not be given to developing new markets or new products. On the contrary, they think that producers' incipient participation in established markets should be strengthened.

Of course, there are nuances to this view depending on the kinds of marketing chains the agents are involved in and their roles in them. Some agents in the sector of fresh tropical fruit believe that the best option is to fortify the position of products in niches already established, including products that have emerged relatively recently in the country, like gulupa (purple passion fruit; Passiflora edulis Sims). Others see the best option in consolidating the country's position in markets that, although once considered niches, today are evolving into volume markets—such as the case of fresh mango of the varieties Kent, Palmer, and Ataulfo. Those agents who are linked to the processed foods sector feel the way forward lies in products that generate more value added, such as frozen fruit and vegetables, aseptic fruit pulp that does not require chilling, and juice concentrates.

The interviewees widely agreed that the following two basic facts are very important for the satisfactory development of the projects: (1) all the links in a value chain including exporting have to gain something from the chain's development and (2) all the links must be competitive in the performance of their economic roles. Although it is clear that fulfilling these two conditions may at times be beyond the capacity of a project or its partners in the private sector, this statement underscores the importance of giving parallel support to the development of capacities to provide the services that are indispensable for the sound functioning of a value chain. A few examples include rental services for machinery, equipment, and tools; maintenance of these items; specialized transport services required by delicate products; provision of seeds of high quality propagated through tissue culture; provision of small-scale cooling devices that have an acceptable rate of return on their own; and providing packing materials of the quality and type required for high-value products. Equally, it affirms the importance of developing competitive markets for these complementary activities and profitable suppliers of them. This may depend in part on the possibility of generating and coordinating agglomerations of projects for similar crops.

With few exceptions the interviewees remarked on the lack of sufficient research in various essential fields. These fields range from more precise knowledge about the optimal agroecological zones for developing particular crops and the solution to problems of chemical residues that threaten export sustainability, to work on increasing the often woeful crop yields while respecting market-mandated restrictions on the use of agrochemicals.

Taken together, the foregoing observations point toward the value of having a program or project that concentrates on relatively few products and seeks to develop clusters. This would allow marketing agents or intermediaries to take advantage of their areas of specialization; it would facilitate closer coordination between higher level intermediaries, local marketing agents, and farmers; it would help channel together the requests for agricultural research in the different areas where it is needed; it would facilitate the logistics of postharvest management, handling of products, and marketing; and it would favor emergence and operation of enterprises that provide services to the value chains.

These recommendations are not new but rarely are seen all together, and the consensus around them among experienced marketing agents is striking. Taken together they suggest the need for reassessments of some basic aspects of the design of development projects.

Regarding the performance of the private sector downstream along the value chain from farms, several interviewees feel that a major obstacle to overcoming some of the problems mentioned is the small size of the average commercial operator and their limited numbers in light of the volumes of production to be managed. Development projects could improve this situation by helping equip the marketing agents to extend their range of operations and to foster their groupings or associations.

Undoubtedly there are important roles for the government in some areas of the projects. Some of the most important are, among others, developing a system for guaranteeing food safety for export products, and for food products sold on the domestic market as well; implementing agricultural traceability systems; building rural infrastructure; and encouraging and developing research efforts. In the Colombian case, two urgent issues that require government response are (1) improving the antinarcotics inspection system so that it does not interrupt the cold chain for fresh fruits and vegetables and (2) putting into effect a fumigation program for fresh products, which ensures that they arrive at their destination without phytosanitary problems picked up anywhere along the logistical chain including during international transport to the country of destination. Externally funded projects can help on these aspects by facilitating coordination among the relevant government agencies.

One of the valuable long-run functions of a development project for smallholders is to help develop social capital and inculcate in farmers a productive vision for the long run. Such a vision should include a full understanding of the value of dealing with markets characterized by relatively stable prices and larger volumes, and this model gives better results over time than responding to fluctuating spot prices, which results in fluctuating production and incomes also.

Passion fruit and blackberries appear to be examples of crops in which farmers in Colombia usually do not look beyond spot markets. This has generated a pronounced cyclical pattern in the prices and production levels of these fruits. For this reason, processed passion fruit appears to be the preferred outlet for production of that crop. In the case of blackberries, contract farming through groupings of producers appears to have a bright future if appropriate farm-level technologies are employed. In addition, the technological packages for cultivation of blackberries appear to be more developed, more capable of managing crop diseases among other things, than those for passion fruit. This is an illustration of the value of coordination along the value chain.

The interviewees also said that developing a solid farmer-to-market chain for high-value, niche products like blackberries can take 6–8 years. This reality represents a challenge for development projects. It indicates that obtaining tangible, sustainable results is not possible with shorter time horizons.

Finally, the experience of this study in Colombia suggests that development projects should devote part of their resources to generating, compiling, and analyzing data on costs of production, contributing to a growing data base in that area. Such an effort would facilitate very substantially the work of future efforts as well as the current project.

10.7 COLOMBIA: COMPETITIVENESS BY QUALITY CRITERIA, TRACK 2

This section provides a Track 2 analysis of six Colombian products selected from the 12 products that were most frequently competitive in various locations by the Track 1 criterion and for which quality analyses had been carried out in the original study. The Track 2 narratives in the original study are compressed and edited, with additional material in some cases, for this chapter. The coverage of the Track 2 analysis ranges from genetic material all the way to the markets. It is based largely on the field experiences of team members and interviews with agricultural banking experts, extension agents, an additional group of marketing agents, project field experts, and others. Information from these sources was supplemented by data from a census of producer practices and marketing issues for promising fruits (Departamento Administrativo Nacional de Estadística, Colombia, 2004a). This kind of census is all too rare in the tropics but provides extremely helpful insights about issues that need to be resolved to improve competitiveness. The crop analyses are adapted from the original case study (Norton & Argüello, 2007) and have been updated where appropriate.

10.7.1 Cacao

The Field Findings

The Crop and Its Markets

Cacao (*Theobroma cacao*) belongs to the family of the *esterculáneas*, whose principal characteristic is producing flowers and fruit on the trunk and on main branches. It is a plant that originated in the upper Amazon watersheds[6] and that grows in a tropical band of geography extending some 20 degrees of latitude above and below the Equator. Its range of altitudes is from sea level to 1200 m above sea level.

Cacao is cultivated in bushes 2−3 m high that have to be in shade, particularly in its early years of development. Cacao plants are found beneath larger trees such as cedar, teak, melina, avocado, rubber, mango, coconut, mamey, and plantain. Plantain works well because the amount of shade it provides can diminish as the cacao tree grows. Cacao trees will live to between 25 and 30 years, although with declining yields. The cacao fruit has a great diversity of forms, from long to almost round, and its color at maturity can vary between yellow, orange, and dark red. A mature fruit can weigh up to 1 kg, although the average weight is 400 g. The quantity of the fruit needed to obtain 1 kg of dry cacao varies from 9 to 33.

Within the fruit are found cacao beans covered by a white mucilaginous pulp. Each bean is 15−30 mm in length, 8−20 mm in width, and 5−15 mm in thickness. It is the basic raw material for the production of chocolate and its derivatives. The main varieties of cacao are forastero (or amazonic), criollo, trinitario (a hybrid), and national cacao of Ecuador. Other special varieties of cacao are found in specific regions. Criollo is considered the finest in flavor, but forastero has a higher fat content that industry sometimes prefers.

Cacao is a competitive crop in Colombia. Of the 22 cases for which LRC coefficients were calculated, 18 were competitive, many of them highly competitive. Santander, Bolívar, Chocó y, and Magdalena are particularly competitive areas in cacao. The country still struggles to produce more than the domestic market requires, but the potential for considerable growth is clear, especially in light of the dynamic world market for the crop. There is little doubt that world markets, especially in Europe, are placing increasing emphasis on high-quality cacao and often on the organic product.

Colombian cacao is considered to be of good quality according to taste and aroma criteria. All of its production has been certified as cacao "fino y de aroma" by the International Cocoa Organization. Only 7% of the world's cacao production is of that quality, and of that production 70% comes from Colombia, Ecuador, Venezuela, and Peru. However, chocolate manufacturers have their own criteria for quality, and quality in cacao is an elusive concept. It depends not only on the variety planted but also on soils, crop management, and postharvest fermentation and drying (Lima, Almeida, Nout, & Zwietering,

2011). The role of each of these factors has not been fully sorted out. And of course the flavor of the resulting chocolate depends on manufacturing techniques, amount of sugar added, and yet other considerations.

However, the principal problem for cacao in the country is the low yields obtained by the vast majority of farmers. There exist a number of cases with significantly higher yields that can serve as demonstrations for the rest of the sector. Postharvest problems are also present, especially the lack of proper fermentation and drying, as are crop diseases, but the diseases can be managed with field techniques similar to those that improve yields. The most important need is consistent training in these practices in all cacao-growing areas.

World markets for cacao continue to grow and, as of this writing, it appears that demand will outpace supply for a number of years to come. In Colombia, the domestic market has been the dominant outlet for production, with emphasis more on quantity than on quality. Ghana, Ivory Coast, Indonesia, and Nigeria have been world leaders in cacao production and export, and in the Americas, Brazil and Ecuador are the principal exporters. Colombia lags far behind those countries in cacao exports. However, in recent years, this has been changing, and nongovernmental organizations and other entities are working with farmers to improve quality, and sales to European and North American specialty chocolate companies are beginning to take off. The organic market, which offers a considerable price premium to producers, is of increasing importance. In Latin America, the Dominican Republic, Costa Rica, Honduras, Panama, and Nicaragua are among the leaders in exporting organic cacao. For small producers, there are three routes of access to export markets: through direct on-farm sales to exporters, through sales to small-scale intermediaries who in turn sell to exporters, and through sales by farmer associations to exporters.

Colombia now has about 35,000 growers of cacao [statistics from the National Cacao Federation (FEDECACAO)]. Other participants in the cacao value chain include intermediaries, processors, exporters, producer associations, FEDECACAO, and government institutions. Research is carried out both by the government and the large national chocolate companies. For the domestic market, intermediaries sell to agents commissioned by the national chocolate companies, who receive commissions based on the volumes they manage. They determine prices on the basis of size of beans, humidity, presence of impurities, number of beans per 100 g of fruit, and other factors, but not based on characteristics that could be directly related to flavor of the final product. According to estimates by CORPOICA, at the time of the original study 75% of the country's cacao production went to the two large chocolate manufacturers via these commissioned agents, and the rest to small processors or in marginal quantities to the export market. On the domestic market, the main cacao products are chocolate drinks, followed by chocolate truffles, candies, pastries and ice creams, and powdered chocolate. Chocolate butter is used to some extent in cosmetics and pharmaceuticals.

Exports of cacao processed into chocolates and other products (customs codes 18010010, 18040000, 18061000, 18062000, 18063100, 18063200, and 18069000) increased by 8.7% annually between 1991 and 2006, mostly to Andean countries. This was faster growth than for any other processed food product. These exports represent 95% of all the exports of the cacao–chocolate value chain, with only 5% corresponding to cacao beans exported for other countries to convert into chocolate. Cacao imports, in beans and semiprocessed forms and mainly from Ecuador, have been increasing rapidly to feed Colombia's industry of chocolate products. For these reasons, increasing cacao production for domestic processing will be an important market for smallholder producers, but the emerging specialty cacao market that offers higher returns cannot be ignored. On average, domestic prices have been below the corresponding international prices (source: Agrocadenas), another indicator that Colombia's cacao is competitive in world markets.

As of the date of the original study, Colombia had 20,000 hectares certified as devoted to organic and other environment-friendly modes of production, but coffee was the main crop grown in these areas. There is some organic cacao planted also, but the amount is not known. Other crops in these areas include hearts of palm, oil palm, sugarcane, bananas, mango, medicinal plants, and some fruits.

The total area planted in cacao in Colombia was 104,000 has. in 2011,[7] up from 83,525 has. in 2005 (FEDECACAO). This growth confirms the crop's profitability for farmers, but a major problem is inefficient crop management. Yields average about 450 kg per hectare. It is almost entirely a smallholder crop. Average farm sizes are a little over 3 hectares, and less than 2% of the cacao plantings cover more than 100 has. According to a 2004 census of cacao producers, crop management techniques were at a low level for about 78% of the producers, at a medium level for about 22%, and at a high level for less than 1% (Departamento Administrativo Nacional de Estadística, Colombia, 2004b). In general, planting densities are less than optimal and farmers only carry out harvesting and the most basic tasks of disease control and pruning. Cacao can be propagated through seeds, tissue culture, or grafting. The last of these is the most widely recommended, and producers are urged to graft between the more productive trees and the others on their own plots. Cacao harvests are carried out a number of times during the year, with peak activity in April–June and November–January.

Postharvest activities include sorting the crop to remove the pods that do not have an appropriate degree of ripeness, cutting them open and removing the cacao beans, and fermenting, drying, cleaning, and packing the beans. This stage is very important because it is when the characteristic aroma and flavor of cacao is set in the beans. Fermentation typically takes 5–6 days. The drying lowers not only the water content but also the acidity and astringency of the beans. Drying can be done under the sun or artificially, with the caution that the temperature never should rise above 60°C. For classifying the product, FEDECACAO recommends taking samples of the beans and cutting them open longitudinally to observe their quality characteristics.

Fermentation is a natural catabolic process of incomplete oxidation ⟩that occurs in many food products through the actions of microbial agents, to arrive at the desired final organic form of the product. The objectives are to improve quality, preserve the product, change the color of the bean, facilitate removal of the cuticle or pulp, and allow the emergence of the chocolate flavor.

Among the principal factors that reduce yields are diseases of the cacao plant, especially *moniliasis*, which is caused by the fungus *Monilia*. It can cause losses of 20−60% of the production. Another important disease is witch's broom, which occurs in hotter, more humid zones, and the plant is vulnerable to yet other infections and insect infestations.

Problems: Genetic Material and Production Practices

More than half of the cacao grown in the country is not hybrid (not an improved variety), and many farms use planting material that is not certified. For this reason, cacao's yields are often low, as is its resistance to diseases, the production obtained is not uniform and the quality is inferior to that of competing countries such as Ecuador, which has hybrids of excellent quality.

In the face of these problems the approach of CORPOICA and FEDE-CACAO (as of the date of the original case study) has been to clone varieties and specimens that are highly productive and exercise strict control over the dissemination of genetic material. However, field experts maintain that the controls are so strict, and the resources to make planting material available are so limited, that in reality farmers have little access to the better material.

According to CORPOICA,[8] one of the main problems for the cacao sub-sector is the advanced age of the trees, which increases their susceptibility to diseases and infestations. The Cacao Census of 1998 showed that 30% of the cacao trees were more than 20 years old and 56% were between 7 and 20 years old. Although renovation of the trees has taken place, the planting density remains low, at less than 600 trees per hectare according to FEDECACAO, when the ideal density is at least 1000 trees per hectare. For its part, Agro-cadenas estimates that only 0.6% of the labor put into cacao plots is devoted to replanting.

As part of the same syndrome of smallholders, they tend to concentrate their management of the trees in the first years of growth and then cease the labors of crop tending. They return to them only in moments when prices jump significantly. The problem of the sparse planting densities is exacerbated by the fact that around 60% of the trees turn out to not bear fruit, or very little. Naturally this is a reflection of poor choice of genetic material which, among other things, leads to trees that have low tolerance of infestations and diseases. Partly because of these issues, recent years have witnessed an upswing in *moniliasis* and witch's broom. There is a danger that Colombia could suffer an episode like that experienced by Brazil a few years ago when its cacao yields fell by more than a third because of witch's broom. Also, in the face of these pathogens smallholders generally are neither aware of appropriate dosages of pesticides nor aware of crop management techniques that also can help control the infections.

Problems: Postharvest and Marketing

The main postharvest issue for cacao is inadequate fermentation and drying of the crop. Each producer seems to have his or her own approach, with the result of quite variable qualities of cacao, and because of poor drying techniques often the cacao becomes moldy. As in the case of crop management issues, there are producers who use appropriate techniques, but postharvest is another area of deficiencies for the majority of producers. Some of the intermediaries who buy cacao from farmers store it with other kinds of crops or merchandise, which is another reason why the quality of the final product is lower than it could be. This shows the need for a complete value chain approach for cacao, as indicated in the discussion with selected marketing agents reported in Section 10.6.5.

Apart from the storage issue, in general, cacao marketing is free of problems except for the lack of sufficient volumes delivered to the market reliably. For exports, in particular, this is an important limitation.

Avenues of Solution for Cacao

The cacao varieties utilized in Colombia are capable, with proper selection of genetic material and crop management, of producing four times the current levels of yields. One of the priorities is to develop a viable supply chain for planting material based on asexual and vegetative propagation of high-quality genetic material.

The other priority is more effective programs in transfer of technology to producers. A promising start has been made in the form of use of the Farmer Field Schools approach. Experience with this approach in the Municipio of Rionegro shows its potential for improving crop management through empowerment of producers. They learn to establish their own tree nurseries and to multiply the trees on their farms that have the highest yields, eliminating those that yield poorly or not at all. They also learn how to improve their postharvest management, including how to build their own solar dryers at a cost of about US$200 per farm. Farmers become "owners" of the process of technological improvement, and they themselves become agents of technology transfer to other farmers. Unfortunately, to date there has neither been a clear decision to widen the application of this extension methodology nor one to test and apply an alternative approach.

If yields and therefore production can be increased, and postharvest management improved for significant numbers of farmers, Colombia appears well positioned to supply cacao to the US market, which represents 30% of all cacao grinding worldwide. Transportation costs favor Colombia over African and Asian producers, along with the recognition of Colombia's quality of "fino y de aroma." Major chocolate producers in the United States have shown interest in Colombian cacao. The organic market is small but increasing rapidly, and its price premiums are substantial.

In addition to the economic benefits that higher yields can confer to smallholders, cacao's need to be cultivated in agroforestry systems will provide valuable benefits to the environment.

Application of Value Chain Filters

- *Seeds and breeds*: inferior and inconsistent varieties are widely used. A systematic, widespread program of dissemination of superior planting materials, as well as training farmers in improving their own varieties through grafting from on-farm options, is needed.
- *Production*: lack of crop tending in general is a major problem. Above all, careful pruning is needed for productivity and disease resistance, and farmers need training in the techniques. Also, agrochemicals are often not applied appropriately when they are needed to control infections of the plant.
- *Harvesting*: no issues.
- *Postharvest*: fermentation and drying techniques are often deficient, and correct procedures in these areas are critical to cacao quality.
- *Storage*: intermediaries sometimes store cacao with other products, and this practice needs to be changed.
- *Transportation*: no issues.
- *Processing*: no issues.
- *Market development*: Markets exist, including for organic cacao, if volumes can be increased and quality levels raised.
- *Value chain governance*: training of intermediaries as well as farmers is required.
- *External and policy*: no issues.

The Value Chain Quality Assessment Matrix for Colombian Cacao

The overall quality matrix score for cacao, at 18, indicates important issues to be resolved in the value chain, especially since the score per market is 9 (Table 10.13). Most of the issues concern genetics, production, and postharvest crop management. Some producers are well ahead of others in resolving them, and for the leading producers who have good links to the marketing chain, the score in the table would be considerably lower. This circumstance illustrates the strong potential of this crop, reinforced by its very competitive LRC scores.

TABLE 10.13 Colombia: Value Chain Quality Assessment Matrix for Cacao

Value Chain Stage	Domestic Markets		Export Markets	
	Fresh	Processed	Fresh	Processed
1. Seeds and breeds		2GE		2GE
2. Production		2CM, CH		2CM, CH
3. Harvesting				
4. Postharvest		2TA		2TA
5. Storage		TA		TA
6. Transportation				
7. Processing				
8. Market development				
9. Value chain governance		LC		LC
10. External and policy				
Sum of scores		9		9

Total score on quality issues: 18. GE refers to deficiencies in the genetic material. In production, CH refers to chemical residues on the berries, and CM to issues in crop management. TA refers to diminution of taste quality owing to inadequate fermentation, drying, and storage; and LC to the lack of coordination between rural buyers and producers. The value "2" in some cells indicates an issue that can block sales to better markets if not resolved.

10.7.2 Blackberries

The Field Findings

The Crop and Its Markets

The leading market for Colombia's blackberry producers is the domestic one, which is experiencing rapid growth of consumption of fresh and processed berries. Colombia has minimal presence on the world market for this crop, exporting no more than 7 tons a year almost entirely to the United States. The predominant species of blackberries (*Rubus glaucus*, of the *Rosaceae family*) cultivated in Colombia is *castilla*. This concentration restricts somewhat the possibilities for widening the domestic market expansion by satisfying a wider range of tastes, but more than 375 species and species aggregates are known so the concentration has been a matter of producer choice.

An advantage of blackberries from the viewpoint of farmers is that they generate income on a very frequent basis, yielding two or three harvests per week. They are competitive on a cost basis, but their returns could be improved substantially through better crop management and postharvest management. It will be important to make improvements in these areas to become more competitive on export markets.

According to the first census of 10 promising fruits for the processing industry (Departamento Nacional de Estadística, Colombia, 2004a), blackberries are the most widely cultivated of those crops in geographical coverage, area planted and number of farms. They are a long-cycle crop. Of the 9195 hectares in production in Colombia in 2005, the majority had been in the ground for more than 3 years and around 10% for more than 10 years. The average area planted in blackberries per farm is 0.38 ha., and the average yield is 8.6 t/ha. These yields are not far below those of Mexico (11.2 t/ha.) and the United States (10 t/ha.), which are the world leaders in productivity for this crop. In limited planting areas and with high crop densities and good crop management, yields of up to 30 t/ha. have been attained in the cases analyzed for this study, which shows the high competitive potential of the crop. Blackberries typically share the farm with crops like tree tomatoes, passion fruit, peas, and corn, but the berries account for a disproportionate amount of family income. Nationally, their production has increased more than threefold in 13 years.

About 87% of the area planted in blackberries is found in 7 of 14 departments with the crop, in localities between 1200 and 3200 m of altitude. They are the Andean departments of Cundinamarca (around Bogotá), Santander, Huila, Valle, Antioquia, North Santander, and Tolima. (Best yields are obtained in localities between 1800 and 2400 m of altitude).

Smallholders have selected seven blackberry species or varieties, choosing the ones that are best adapted to their production zones. The *castilla* species (*R. glaucus Benth*) is the most important commercially, representing 93% of the planted area.[9] It probably was selected originally from among the wild species in Colombia.

Blackberries are vulnerable to insects and diseases that undercut their yields. The most common are the fungi *Oidium lycopersici* and *Peronospora sparsa*, which can be controlled by proper pruning, and fruit flies and worms. The main phytosanitary problems are fruit rot (*Botrytis cinerea*) and anthracnose (*Colletotrichum gloeosporioides*), which can be controlled with good crop management and fertilization.

Upon leaving the farm, the blackberry value chain continues with farmer sales primarily in wholesale markets in county seats (39%), to marketing intermediaries who include blackberries in their regular transactions (27%), and to other intermediaries (29%). Supermarkets purchase 2.4% of the crop directly from farmers, and the processing industry, 3.1%. Only 0.2% is purchased by exporters. For transportation to the first sales site, 65% is via motorized vehicle and 35% takes place on foot and by animal power.

Blackberry competes with other fruits in the fresh market and in processing, especially for juices and pulp, and principally with passion fruit. An important component of the demand for blackberries comes from food processing industries, which use it not only for processed fruit but also as an input into dairy-based products, pastries, cereals, animal feed, and other products. There is also a small amount of demand from the homeopathic product sector and from industries that produce fruit and fruit essences and extracts. Forecasts are that the demand for juices will be growing by around 10% per year.

The blackberry processing industry is oligopolistic with two firms representing over 70% of the market, and one of the two recently purchased operations of the other one. From the viewpoint of farmers, a possible advantage of this market structure is that the largest company is positioning itself to enter more aggressively into export markets, taking advantage of Colombia's counterseasonal blackberry production periods vis-à-vis the United States and Europe. (Details of these and other companies' fruit operations are found in the original case study.) The fruit already has admissible status in the United States and does not face import tariffs.

The Colombian processing industry requires fruit with brix content between 6.8 and 8° but does not pay a premium for other dimensions of quality. Producers and intermediaries are required to deliver the fresh fruit in plastic boxes with a capacity of 25 kg, and frozen fruit is not accepted for processing.

Problems: Production Practices

The lack of proper crop management is evident in many cases. For example, weeding is not done on 17% of the planted area and pruning not done on 19%, while 51% of the area is planted without support poles and 52% without the support of dirt mounds around the plant's base. Without support poles (leaving the crop closer to the ground) the problems are lower planting density, greater vulnerability to fungus infections, and greater difficulty in pruning, among other problems. The cost of wooden and plastic poles is the reason many smallholders do not use them. Another issue is the habit of leaving the pruned material on the ground, which increases the incidence of diseases for the plants. Without pruning, parts of the plants do not receive adequate air, phytosanitary problems are worse, yields and fruit quality are lower, and harvesting becomes more difficult. This circumstance means average yields remain far below the plant's ceiling of approximately 25 tons per hectare. In zones where rains are more irregular, the availability of irrigation needs to be increased.

Many farmers also apply agrochemicals without reference to dosage guidelines and often overdose, responding to evidence of diseases rather than making the applications in a preventive mode. This practice raises production costs and reduces access to export markets because of the presence of chemical residues. During the rainy season fumigation is needed every 8–10 days. However, often it is done too close to harvest time. Although agrochemicals represent 25–30% of direct costs, producers often use them excessively for fear of obtaining low yields. However, the use of organic inputs appears to be increasing. Now on 16% of the blackberry area, fertilization is done organically, and on 36%, it is done with combinations of organic and inorganic material. Similarly, organic material is increasingly used for phytosanitary control, although still on a small scale, and cultivation practices are deficient even on organic farms.

Sufficiency of labor can be an issue in blackberry cultivation since it is harvested all year around. This reinforces its dependence on family labor, although at some moments labor is hired from off the farms. Access to financing for inputs also is a perennial issue. Some farms reviewed in this study planted beans with blackberries to provide an additional source of cash flow. Some plantings have reached an age, up to 25–30 years, at which yields are significantly lower, and capital is needed for replanting.

Problems: Postharvest and Processing

Often the berries are sold without removing stems, even though this means losing a price premium of 12%. Farmers are wary of removing them because of the danger of damaging the fruit, leading to its rejection on the part of buyers. However, with adequate technical assistance this problem can be overcome.

There is little uniformity in the containers used for the harvested berries. Plastic containers of 12.5 kg capacity are recommended to avoid having fruit crushed from the weight of excess layers, and also so that the same container can be used for transportation thereby avoiding unnecessary transfers of the fruit. It is further recommended that the fruit be placed in smaller containers within these boxes. However, this is often not done, especially when the fruit is transported on foot or by animal, and receptacles vary in size. About 11% of the farmers use wooden boxes.

Harvested fruit is not classified by 96% of the producers, which results in lots with a low proportion of first-quality fruit. Only 3% of the production surveyed qualified as first quality; the difference between the price for this quality and second quality is about 30%. The transport and storage stages of the blackberry value chain frequently do not have cold facilities, which affects quality significantly. Blackberries do not store easily, but storage is often required. The recommended temperature range is −0.5–0°C with a relative humidity of 90–95%. In these conditions, blackberries can be stored satisfactorily for 3–4 days. Expanding the cold chain for this crop is an urgent priority, all the more so to be able to penetrate export markets more fully.

Problems: Marketing

From the viewpoint of farmers, the main issue for marketing is the lack of stability in the marketing chains. The intermediaries change frequently and with them the conditions of sale also change: prices, quantities purchased, and periods before payments are made. Few intermediaries are specialized in blackberries since it is a highly perishable crop. Producer organizations for crops like blackberries generally do not exist, so farmers' negotiating power is weak, as are channels for providing technical assistance for them.

Avenues of Solution for Blackberries

Blackberries are a crop on the upswing, and the majority of the cases analyzed in Track 1 showed competitive levels of LRCs. The crop is a significant generator of value added and its processing industries are efficient. Production experiences to date in Colombian blackberries show that the productivity ceiling is high, and so with adequate technical assistance the volumes of production could be raised substantially. However, this would require greater penetration of international markets. Without that, significant production increases would be likely to reduce the prices received by farmers. Organic production is also increasing, but to date it lacks certification, which blocks farmers from receiving the full benefit of selling organic production.

On the production side, more appropriate use of agrochemicals is a priority for this crop, all the more so if exports are to be increased. Another priority is better crop management including pruning and the use of stakes for support. The latter will help increase yields, help the crop take root faster, and decrease the crop's vulnerability to diseases. On the postharvest side, training of producers to remove the stems would bring them important increases in incomes. More broadly, it also would be worthwhile to explore other varieties beyond *castilla*, to broaden the market by introducing different flavor profiles, for example, the variety known in Colombia as French blackberry.

Application of Value Chain Filters

- *Seeds and breeds*: no obstacles, but testing of other varieties would be worthwhile.
- *Production*: inappropriate application of agrochemicals, cultivation without supports for plants, insufficient pruning.
- *Harvesting*: no issues.
- *Postharvest*: Farmer reluctance to remove stems.
- *Storage*: more cold storage facilities are needed.
- *Transportation*: refrigerated transportation needed; better containers for transportation needed.
- *Processing*: no issues.
- *Market development*: if chemical residue issues can be contained and yields increased, firms are poised to export more.
- *Value chain governance*: instability of marketing channels in rural areas.
- *External and policy*: no issues.

The Value Chain Quality Assessment Matrix for Colombian Blackberries

The overall score for blackberries is midrange at 17 (Table 10.14). Most of the issues concern production, storage, and transportation. The markets exist, but the problems are on the supply side. Per market, the average score in the table is 8.5, showing that important issues remain to be resolved. However, as the text indicates, some producers are well ahead of others in resolving them, and for the leading producers who have good links to the marketing chain the score in the table would be considerably lower. This circumstance illustrates the strong potential of this crop.

10.7.3 Lulo (*Naranjilla*)

The Field Findings

The Crop and Its Markets

Lulo (*Solanum quitoense*), or *naranjilla* for its similarity in appearance to orange (*naranja*), originated in the Andean valleys close to the Equator. It is

TABLE 10.14 Colombia: Value Chain Quality Assessment Matrix for Blackberries

Value Chain Stage	Domestic Markets		Export Markets	
	Fresh	Processed	Fresh	Processed
1. Seeds and breeds				
2. Production	CH, CM	CH		
3. Harvesting				
4. Postharvest	TA	TA		
5. Storage	HH, SP	HH, SP		
6. Transportation	HH, SP, PK	HH, SP, PK		
7. Processing				
8. Market development				
9. Value chain governance	LC	LC		
10. External and policy				
Sum of scores	9	8		

Total score on quality issues: 17. In production, CH refers to chemical residues on the berries, and CM to issues in crop management. TA refers to perceived physical quality issues (stems left in the fruit), HH to damage from high heat and humidity in the value chain, and SP to spoilage; PK signifies lack of appropriate packaging from the farm onward, and LC to the lack of coordination between rural buyers and producers.

produced at altitudes from 800 to 1500 m above sea level in temperate, humid climates with temperatures ranging from 16 to 24°C and with rainfall requirements between 1000 and 3000 mm. The fruit is balloon shaped from 4 to 8 cm in diameter, weighing 40–60 g.

Because it is an Andean product, world production of lulo is limited to Colombia (44,400 tons) and Ecuador (15,500 tons). International trade in fresh lulo consists of Colombian imports from Ecuador to fill the deficit in the domestic market and to provide raw material for Colombia's export of frozen lulo pulp and other processed forms of lulo to the United States and Europe. However, Colombia's (small) export quantities are not known because this product does not have its own customs code at the 10-digit level.[10] The team members for this study have had contact with exporters who ship processed lulo to the United States, Spain, Belgium, England, Aruba, Haiti, and Jamaica, and they say the international demand is far in excess of what they can supply. Fresh lulo has not yet gained admissibility into the US market. However, Colombia is working toward that, and the Center for Phytosanitary Excellence has completed risk analysis for the crop.

The salient characteristics of lulo are its flavor, aroma, and color of its pulp, together with its nutritional content, which is high in vitamins A and C and in iron. Owing to these properties and its multiple uses in processing industries, it has become a fruit with significant possibilities in both domestic and international markets. Another common use of it in importing countries is as a decorative element in table centerpieces and sometimes in baskets with other tropical fruits. In Colombia households commonly consume lulo in various forms including ice creams, yogurt, juices, salads, and desserts.

Marketing channels for lulo are diverse, with about 47% of farmers selling to wholesalers in the county seat, 30% to other wholesalers, 18% to collection points, 3% directly to processors, 2% to supermarkets, and 1% to exporters. Prices fluctuate in response to supply conditions.

Lulo is a highly competitive and profitable crop. Of the 20 cases evaluated quantitatively, 18 proved to be competitive, a number of them highly competitive. Production has been growing in response to increasing demand for processed lulo, mainly as a juice. Its cultivation is spread widely through 19 departments in Colombia. Crop management practices for lulo and its competitiveness vary markedly across regions. The most competitive departments are Risaralda, Cesar, and Cundinamarca, although Huila is the largest producer, followed by Valle del Cauca, Boyacá, and Nariño. Several varieties are grown. A new variety preferred by processing firms (La Selva) has been developed by researchers, but so far it has not gained much acceptance among producers because it is better suited for the processing industry rather than for fresh consumption.

Lulo is a smallholder crop. The average area planted in it per farm is 0.65 hectares. In Huila, Risaralda, and Santander the areas planted are somewhat larger, averaging close to a hectare, whereas in Nariño and North Santander the average areas are 0.37 ha. and 0.27 ha., respectively. It is mostly a monoculture crop, although it is sometimes interspersed with coffee or other crops. Seeds are from retained stocks for two-thirds of the farmers, and specialized nurseries supply only 7% of the seeds. Weekly harvests begin 8−10 months after planting and continue for 2 years, although the plant can live up to 5 years. Chemical fertilizer and pesticides are the main weapons for plant growth and control of pests and diseases. The use of toxic chemicals is a frequent issue. Only 4% of the farmers use organic techniques for pest and disease management. Postharvest, 58% of the producers classify the crop and 67% use vehicles for transport to the point of sale.

Problems: Genetic Material and Production Practices

Lulo yields have remained at around 8 t/ha., and often lower, although the technological ceiling is at least double that. Reasons include the practice of producers of storing their own seed instead of purchasing quality seed from agrodealers or nurseries. And when farmers select seeds to be set aside for the next year, it is important to choose the healthiest, most productive plants, and then after the fruit is fermented and washed, select the best seeds from deep within the fruit and dry them in open air. Inadequate provision of shade for lulo also is a factor that limits yields, and this goes to the question of appropriate cropping systems for it.

A more complete research agenda is needed for this crop, including development of varieties with attractive qualities for consumers, identification of improved approaches for controlling pests and diseases, development of better systems for seed multiplication, development of integrated pest management (IPM) systems for this crop, and research on cropping systems with shade.

Problems: Postharvest and Processing

Up to 44% of the producers place the harvested fruit in bags instead of boxes. They do not wax the fruit. As a result of these practices, about 30% of the crop is lost in the postharvest stage. And few producers are linked into more commercial value chains in which supermarkets and other buyers can provide assistance regarding quality controls.

Problems: Marketing

Fresh lulo does not have admissibility for the US market. For processed lulo, and eventually for fresh lulo when admissibility is obtained, the main handicap is lack of sufficient supply. This is turn reflects the lack of integration of the value chains, i.e., the lack of contracting between processors or exporters on the one hand and producers on the other hand. Producer organization is also weak for this crop. To supply the market with 200 tons per month, equivalent to a container a week, would require the participation of 300 producers, and lulo farmers are not yet organized in groups of that size.

Avenues of Solution for Lulo

Lulo's potential is strong. However, as an amazonic fruit lulo requires shade. It does not thrive in as a sole crop or in systems with annual crops. This is a limitation in adapting it to the needs of modern agriculture and represents a priority for research on the crop. Another urgent priority is the development of varieties with greater resistance to pests and diseases and of effective methods of IPM for the crop. Its intensity in use of labor gives an advantage to smallholders, but in the longer run it may constitute a barrier to increased production, and accordingly research also should focus on varieties and cultivation methods that are less demanding of labor.

Application of Value Chain Filters

- *Seeds and breeds*: inadequate access to clean, productive seed, and inadequate seed selection on the part of farmers.
- *Production*: inappropriate application of agrochemicals, lack of sufficient shade in some cases.
- *Harvesting*: no issues.
- *Postharvest*: Use of inappropriate packaging materials and insufficient attention to waxing the fruit.
- *Storage*: no issues.
- *Transportation*: Better packaging needed in all the transportation of the fruit.
- *Processing*: no issues.
- *Market development*: No issues because supply is the constraint. However, in the longer run, as supply increases, control of agrochemical residues can become a limitation to fresh lulo export.
- *Value chain governance*: marketing channels are informal with little transmission to farmers of information on quality criteria and required quantities.
- *External and policy*: no issues.

The Value Chain Quality Assessment Matrix for Colombian Lulo

The overall score for lulo is fairly strong at 13 (Table 10.15). Most of the issues that need to be addressed to realize its strong competitive potential concern seeds, production, postharvest, and transportation. Markets exist and the constraints are on the supply side. A more complete research agenda is needed for this crop, and if varietal and crop management issues can be resolved, its future is bright given the strong demand. Better integration of value chains is also needed. Export market constraints are not indicated in the

TABLE 10.15 Colombia: Value Chain Quality Assessment Matrix for Lulo

Value Chain Stage	Domestic Markets		Export Markets	
	Fresh	Processed	Fresh	Processed
1. Seeds and breeds	GE	GE		
2. Production	CH, CM	CM		
3. Harvesting				
4. Postharvest	PK	PK		
5. Storage				
6. Transportation	SP, PK	SP, PK		
7. Processing				
8. Market development				
9. Value chain governance	LC	LC		
10. External and policy				
Sum of scores	7	6		

Total score on quality issues: 13.

matrix because the sector still struggles to satisfy domestic demand, but if supply increases sufficiently so that more exports can be contemplated, then the issues indicated here will apply to those markets also.

10.7.4 Mango

The Field Findings

The Crop and Its Markets

Mango originated in South Asia and spread from there to the rest of the world. It appears to have been introduced to Central and South America by the Portuguese. Its scientific name is *Mangifera indica*, and it is the most important fruit in the family *Anacardiaceae*, which has 64 genera including cashew and plums. Varieties present in India are not cultivated commercially because of their high fiber content and low quality, but they have been used as genetic material for the commercial varieties developed in Florida. Varieties from Indochina and the Philippines are grown commercially in some parts of the Western Hemisphere, for example, the Manila variety in Mexico and the Cecil in Cuba. Those developed in South America tend to be fibrous, and they include the Madame Francis cultivated in Haiti, the Julie cultivated in Trinidad, and the Itamarca cultivated in Brazil.

The most widely grown varieties throughout the world are those developed in Florida. They include Haden, Irwin, Keitt, Kent, Palmer, and Tommy Atkins. The most common variety in Colombia is the Hilacha, which has a fibrous, stringy fruit. Additional varieties in the country are Mariquita or Chancleta, Sufrida, Tommy Atkins, Haden, Kent, Keitt, Van Dyke, Ruby, Zill, and quite a few others. About 65% of Colombia's mango production is the common variety Hilacha and other local varieties. Tommy Atkins mango receives the highest prices on the domestic market, but those prices are the most volatile. Harvests are susceptible to damage from heavy rains, which creates significant fluctuations in supply and hence in prices.

Mangoes are the mostly widely consumed fruit in the world. Recent studies suggest significant health benefits (see, for example, Schattenberg, 2014, citing research of Dr. Susanne Talcott at Texas A&M University; and, for the food scientist, Barnes, Krenek, Meibohm, Mertens-Talcott, & Talcott, 2016). Most of Colombia's mango production, however, goes to the domestic market. Producers sell to rural wholesalers, collection points, producer associations (which are few), and exporters. Although in a few cases farmers sell directly to processing industries, in general intermediaries play a strong role in mango marketing. Processors absorb 30% of the production; they are concentrated in the departments of Cundinamarca, Valle del Cauca, Antioquia, and Santander.

Mango is versatile. The processed products include marmalades, pulp, baby food preparations, and juices. Many other forms and derivatives of mango can be produced, including mango slices in syrup, canned mango, sherbets and ice creams, flavored gelatins, frozen mango cubes, vinegars and sauces, and dehydrated green mango slices. The nonedible parts of the fruit can be processed into ingredients for soaps and starch and for chicken feed and pectin for concentrates to be fed to cattle. Mango juice is the second most widely consumed juice in Colombia, after that of blackberry and ahead of that of lulo.

To date, Colombia's mango exports are all in processed form since this fruit does not yet have admissibility to the large US market in fresh form. Export values have been in the range of $1 million to $3 million per year (as of the original case study). World market trends toward natural and healthy products with fragrances and exotic tastes favor increasing exports of mango, especially when processed into juices and other products.

Colombia's mango production has been growing relatively rapidly. The main producing departments are Cundinamarca and Tolima. In eight of the nine mango cases studied, the crop turned out to be competitive on cost-price grounds, which augurs well for continued expansion of production. Another positive factor is that Colombia's average mango yields, at over 12 tons/ha., are above the world average. Yields of up to 30 tons/ha. are not uncommon in Colombia.

Mango trees require good light, exposure to the air, and humidity in the range of 40–60%. Irrigation is not particularly important for this crop but is most needed in early stages of growth. Trees propagated from seeds require little care and have a long life, but the quality of the fruit usually is inferior. Grafted trees produce better quality fruit but need more attention to yield on a commercial scale. Harvests can begin in the trees' third or fourth year, and larger harvests are obtained from the fifth year onward. They can continue to yield for 30 years or more, depending on how they are managed. To provide income flows to producers and while the trees are developing, other fruits such as pineapple, passion fruit, and papaya can be grown between the rows of mangoes, and annual crops such as corn, beans, cassava, and vegetables can also be planted for that purpose.

Three types of farmers cultivate mango in Colombia: (1) smallholders, who are the vast majority, with trees of the common varieties and who do not manage the crop in any way but rather simply collect mangoes when they ripen; (2) farmers of various sizes of holdings who work with their trees (pruning, protecting the roots, controlling plant diseases, etc.); and (3) the few commercial, large-scale mango producers who have better varieties of mango and who employ advanced crop management techniques—these are entrepreneurs or groupings of farmers.

Problems: Genetic Material and Production Practices[11]

The most fundamental problem for mango production in the country is that smallholders view the crop with an extractive mentality, with an approach of gatherers rather than cultivators. They seek immediate sources of income without a prior investment of labor or cash. They are generally not aware of the needs of the crop in terms of soils, water, and environmental conditions, or of the sources, doses, and recommended timing of inputs required to achieve higher levels of productivity and quality, even though they may be aware of these concerns for other crops. Also, their varieties are those that grew spontaneously or were available in the area, and often not the higher value varieties. They do not use the recommended methods of propagation and allow their trees to grow very high, which complicates the management of crop diseases and makes difficult the pruning for greater productivity. The majority of farmers with mangoes devote most of their time to other activities, such as cultivating other crops, livestock raising, and working off the farm.

Mango is vulnerable to crop diseases and infestations. It is estimated that 30% of the mango plots are infested with fruit fly and 25% with diseases such as *anthracnose* (*C. gloeosporioides*), stem rot, and others.

Harvesting typically is done manually, usually by shaking the tree until the fruit falls to earth or by using a long hook to make the fruit fall. As a result, the fruit is often bruised, and its vulnerability to microbial deterioration increases. The wounds opened in the fruit intensify the production of ethylene, which in turn affects the healthy fruits.

At times mango is harvested before the fruit has matured, with the consequences that it does not reach consumers with an appropriate degree of ripeness, its sensitivity to low temperatures increases, and it dehydrates rapidly, which leads to wrinkled fruit. At other times the harvest takes place too late, which reduces the shelf life of the fruit and makes it more vulnerable to attacks by microorganisms and to physical damage, all of which reduce its commercial value. In general, the harvests are not carried out in a timely manner.

Harvest losses also are incurred through improper use of tools, cuts made in the fruit during harvesting, and detaching of the stem (pedicle), all of which give rise to dehydration, which in turn causes bacterial contamination and white spots that affect the fruit's appearance. According to the Ministry of Agriculture (Ministerio de Agricultura y Desarrollo Rural, 2006), 38% of the problems in mango are attributable to crop management, 32% to issues with planting or propagating the trees, and 16% to inadequate plant health measures. Fumigation of mango trees typically is done only when other crops need it. This makes it clear that the principal issues in mango are at the production level.

Taken together, deficiencies in crop management and problems in the postharvest phase generate losses of about 30% of this fruit among smallholder producers. And for the fruit that is not lost, its poor quality obliges these farmers to sell in markets that pay low prices.

Problems: Postharvest and Processing

After harvest, mangoes are rarely protected from sun and rain while awaiting collection or in transport. Also, since mangos are more resistant to bruising than some other kinds of fruit, producers pay little attention to the type of container used for harvesting and transportation. They also tend to pile up the fruit too much, causing damage to the fruit from compression from the weight of pile, and buffers are not put under or around the piles to cushion the fruit from vibrations during transport. Its shelf life is reduced by the long distances frequently traveled between farms and collection centers.

Often the fruit is not cleaned adequately to remove resin residues. They are spread on the fruit more or less extensively depending on how it is harvested and the degree of the fruit's ripeness at the time of harvest. The usual practice is to wipe off the residue with a rag, whereas the recommended method is to immerse them in an alum solution with a fungicide added, prior to cooling. Evidently the majority of smallholders are largely unaware of the best practices for this crop.

By the same token, cooling of the harvested crop generally is not carried out at appropriate temperatures, which are in the range of 20—25°C for best maturation of the mangoes. The consequences include slow maturation, loss of color of the flesh and the skin, drying out of the flesh, and changes in flavor.

Also, infections like anthracnose sometimes are not apparent until the fruit is harvested and in a mature state. At times producers and shippers are not aware of its presence until it arrives at an export destination, at which time sales prices are discounted sharply.

Problems: Marketing

The fundamental marketing problems are (1) low quality of the fruit, which regularly limits producers' possibilities of participating in more remunerative segments of the market, and (2) low and variable levels of supply, which mean that producers cannot develop long-term relationships with key markets (Plan Frutícola Nacional, 2006). Lack of knowledge of plant management and treatments for hygiene issues also represents a limitation for export markets, particularly with respect to the fruit fly and chemical residues on the fruit.

According to farmers themselves, there are few producer organizations that can promote the adoption of efficient and competitive models of production and marketing via training in best practices. Also, the lack of sustained access to better markets limits the opportunities for access to production credit.

Avenues of Solution for Mango

The multiple options for processing mango into different kinds of products and its apparent health benefits are reasons for optimism about the future of the crop, beyond its sales as a fresh fruit. The main challenge is to convert smallholders from gatherers of this fruit into cultivators of it through training in best practices for the production, harvest, and postharvest stages of the value chain. Mango produced in the existing, mostly artisanal manner will continue to have a steady but low-value domestic market. The need is to raise quality and ensure reliable supplies in engage more fully in higher value markets, both domestically and in export. Among other things, this will require better organization of producer groups that are involved in mango production.

The following recommendations are applicable to the postharvest, transportation, and marketing phases of the mango value chain:

1. Postharvest management of the crop can be improved with simple covers that shade it from high temperatures and protect it from rain. Equally, tarps can be placed over the crop when it is transported in open trucks.
2. Timing of harvests can be improved by using tables of indexes of ripeness of the fruit and that take into account the time required for transport and storage until the product reaches consumers.
3. Farmers need to be made more aware of the physical characteristics of the fruit that are preferred by consumers, including shape and color and firmness of the flesh, and its preferred level of acidity—along with ability to recognize when to harvest to optimize these characteristics.
4. Techniques for harvesting the fruit without damaging it can be improved with simple options such as long poles with blade tips and bags attached to

them to catch the fruit. It is recommended to harvest fruit with stems at least 15 cm long to minimize dehydration and chances of infections.

5. Use of lined, plastic boxes is recommended for transporting the fruit, and it is also recommended that not more than three layers of the fruit be placed in the boxes. Transfer and other handling of the fruit should be minimized.

6. Some commercial buyers for fresh and processed markets recommend quality parameters along the lines of Section 2.4.7 (especially the third paragraph in that subsection). These include color (yellow-orange), taste, fragrance, absence of any hints of fermentation, freshness, firmness, and uniformity; lack of presence of fungi or indications of decomposition; and lack of impurities. Among other things, satisfying these standards requires careful selection of the fruit by farmers.

On the production side, training producers in control of anthracnose, the principal disease threat, is very important, especially in ways that can reduce costs incurred in agrochemicals. (The Tommy Atkins and Kent varieties are relatively resistant or tolerant of this disease.) Methods of control involve cutting down diseased fruit and burning it, better spacing of trees, pruning and cutting back foliage, and thinning the fruit on the tree. When recourse to pesticides is necessary, it is important that producers are informed of the best timing and methods of application, from viewpoints of productivity, environmental protection, and public health. Many producers tend to apply agrochemicals indiscriminately.

Application of Value Chain Filters

These filters refer to the production of mango for high-value markets, both domestic and export.

- *Seeds and breeds*: inadequate seed selection on the part of farmers and too much dependence on lower quality varieties.
- *Production*: lack of pruning and other crop tending; inappropriate application of agrochemicals.
- *Harvesting*: improvised harvesting techniques that damage the fruit are widespread; timing of harvests often is suboptimal regarding degree of ripeness of the fruit.
- *Postharvest*: Inappropriate methods of cleaning resins off the fruit, inappropriate detaching of stems, lack of covers for protection against sun and rain.
- *Storage*: storage often occurs at inappropriate temperatures.
- *Transportation*: packaging for transport in ways that damage the fruit.
- *Processing*: no issues.
- *Market development*: No issues because supply of good-quality fruit is the constraint. In the longer run, lack of fruit fly management and inadequate control of agrochemical residues can become a limitation to fresh mango export.

- *Value chain governance*: Lack of producer organization for learning best practices. Most marketing channels are informal with little information provided to farmers on crop management for quality.
- *External and policy*: no issues.

The Value Chain Quality Assessment Matrix for Colombian Mango for High-Value Markets

The overall score for mango in Table 10.16, at 40 (average of 10 per market), indicates the presence of serious problems to be overcome for the majority of mango producers. Again, the underlying and cross-cutting issue is the need to convince smallholders that it is worthwhile for them to cultivate mangoes and handle them according to best practices in the postharvest phase, instead of simply gathering them and throwing them in boxes with little attention to the damage caused to quality of the fruit. Mango is a product with significant competitive potential, and world markets for it are good. Colombia is positioned to become a major player in those markets, but a well-organized, systematic effort at training smallholders will be required.

TABLE 10.16 Colombia: Value Chain Quality Assessment Matrix for Mango for High-Value Markets

	Domestic Markets		Export Markets	
Value Chain Stage	Fresh	Processed	Fresh	Processed
1. Seeds and breeds	GE	GE	GE	GE
2. Production	CH, CM, PD	CM, PD	2CM, PD	CM, PD
3. Harvesting	TA		TA	
4. Postharvest	TA, SP	TA, SP	TA, SP	TA, SP
5. Storage	HH	HH	HH	HH
6. Transportation	SP, PK	SP, PK	SP, PK	SP, PK
7. Processing				
8. Market development				
9. Value chain governance	LC	LC	LC	LC
10. External and policy				
Sum of scores	11	9	11	9

Total score on quality issues: 40.

10.7.5 Rubber

The Field Findings

The Crop and Its Markets

Colombia is both a producer and a net importer of rubber, which indicates possibilities for import substitution provided that the competitiveness of the country's production can be maintained in larger quantities and the supply can be stable. The postharvest stage is critical for obtaining rubber of international quality through appropriate coagulation of the latex in the field. The world market for natural rubber has seen oversupply for many years, but this situation appears to be coming to an end.

Colombia has agroclimatic conditions that are propitious for natural rubber in extensive areas. It grows in low tropical zones (sea level to 1300 m altitude), requires temperatures in the range of 20–30°C, evenly distributed rainfall year around totally about 150–250 cm, high humidity (60–80%), and approximately 1500 h of bright sunshine a year. Wind speeds should be less than 50 km/h, and soils can range from very acid (pH of 0.4) to slightly alkaline (pH of 7.5). A priority need in Colombia is for research on developing varieties that are resistant to diseases of this plant, especially South American leaf blight, which is caused by the fungus *Microcyclus ulei*. Additionally, it will be important to develop barrier zones around each rubber growing area, which helps block the spread of the leaf blight.

Since rubber trees have a strong and branching root system, for their best growth they require deep soils that are fertile and permeable. These conditions can be natural or developed artificially with drainage, irrigation, erosion control, cover crops, and cropping systems.

The competitiveness of rubber production is based on multiple factors, including the genetic material, knowledge of the agroecological conditions where it is planted, the planting system used, management of the crop during its productive lifetime, the type of tree tapping employed, and the type of postharvest treatment. The importance of providing technical assistance to producers in all these phases and for these factors should not be underestimated.

All the cases studied in Colombia turned out to be competitive, although there were differences in the growing practices. For example, in the major rubber-producing department of Caquetá, located in the Amazonic region of the country along the southern border (Fig. 7.1), rubber is planted in an agroforestry system with cacao and timber trees, whereas in other departments different systems are used.

Natural rubber can be obtained from several tree species such as *Hevea brasiliensis*, *Hevea benthamiana*, *Hevea camargoana*, *Hevea camporum*, *Hevea guianensis*, and *Hevea microphylla*. However, *H. brasiliensis* is the only species cultivated throughout the world for the purpose of obtaining its latex.

Raw rubber drawn from other plants is contaminated by a mixture of resins that have to be extracted before the latex can be utilized.

Rubber trees originated in the Amazon River watershed and were taken to Asia where they tended to perform better because this continent is free of the fungus that causes South American leaf blight. Today, Asia is the main producer of rubber by a large margin. The tree grows to heights of 10−20 m. It typically produces for 25 years after a 7-year gestation period. When planted, branching of the tree is confined to heights above 2.5 m so that the trunk is smooth in the parts tapped for latex. Tapping is done through vertical cuts in the bark that ascend in a corkscrew pattern to avoid damaging the tree.

Rubber trees are often planted in agroforestry systems that generate returns to farmers during the long gestation period of rubber. Coffee producers in Colombia say that rubber in their plots gives very good returns. Transitional or semipermanent crops also help tide farmers over the initial years after planting rubber; in particular, they use papaya, pineapple, cacao, corn, and passion fruit. Another product of rubber plantings is the trees themselves when they cease to yield rubber, as the wood is sold for use in construction.

Colombia does not have an exportable surplus of rubber in spite of favorable natural conditions for it. Opinions about the quality of Colombian natural rubber are not unanimous. Some believe that the quality does not comply with accepted international standards that also are requirements for its use in domestic industries such as tire manufacture. Others think that the obstacle to its greater use in domestic manufacturing are its limited volume and fluctuating supply. The types of natural rubber produced in Colombia are: ribbed smoked rubber (RSS) (50% of the production), natural rubber crepe (NRCX) (25%), and technically specified rubber (TSR) (25%). Colombia's production of the TSR, when it is free of impurities, is of the midlevel technical grade TSR20, which is the main grade traded on the Singapore Commodity Exchange (Accenture Strategy, 2014, p. 19) and for which international prices are commonly quoted.

The principal buyers of RSS and rubber residues (in the field buckets) are the manufacturers of glues. NRCX, both domestic and imported, is used for hoses, rubber bands, hair buns, ties for flower bunches, soles of shoes, and other products. NRCX has displaced RSS for some uses. Imported liquid rubber latex is used for gloves, balloons, condoms, and other products. Imported synthetic rubber takes up 60% of the domestic market for manufactures based on rubber.

Three levels of rubber buyers exist in Colombia. The first level consists of large importers, basically tire producers such as Goodyear and Michelin. Together they import 3000−4000 metric tons of TSR20 every 3−5 months, a portion of which they sell to enterprises on the second level. At the second level are enterprises such as Mack Lubricants that import and also sell to third-level companies. Some other companies at the second level include Cauchosol,

Eterna y Escobar, and Martínez. At the third level are smaller companies that use rubber in their manufacturing but do not import it regularly. In 2006, five companies (Goodyear, Michelin, Bridgestone, Firestone, and Eterna) represented about 82% of the assets and sales in the domestic rubber market. These and other companies using rubber as a raw material are located principally in Bogotá, Antioquia, and Valle del Cauca.

Worldwide, 70% of natural rubber goes into the production of tires. Consequently, the rapid growth of demand for automobiles, above all in China and other Asian countries, has helped natural rubber recover some of the market it lost to synthetic rubber. It also has led to improvements in techniques of producing natural rubber.

The types of natural rubber that have gained market share are the TSR grades 5, 10, 15, 20, and 50.[12] On an average, 57% of Colombia's natural rubber imports have been of TSR and 30% of natural rubber latex. The main exporters to Colombia are Indonesia, Guatemala, and Singapore. Colombia's exports of processed rubber products, which are significant in the country's balance of payments, go mainly to Ecuador, Venezuela, México, Brazil, and Chile. Tires are the leading export among processed rubber products.

Colombia's production of natural rubber supplies less than 10% of the needs of the domestic market. In 2006 the area planted in natural rubber appears to have been about 20,783 hectares, although estimates vary considerably regarding rubber areas and production. According to Agrocadenas, the country has about 263,000 hectares that are optimal for planting rubber, on grounds of both climate and soils. What is striking about rubber production in Colombia is that most of the producers are smallholders who use family labor, in sharp contrast with the very large rubber plantations in Southeast Asia. The production of these smallholders is heterogeneous and of varying quality, and it supplies the small and medium rubber processing industries in Colombia. In some cases rubber-producing smallholders have banded together in associations such as ASOHECA in Caquetá. And there are a few large-scale plantations in Meta and Orinoquia. The average annual rubber yield in the country is 1.28 tons per hectare. Caldas is the leading department in this regard with average yields of 1.50 t/ha. After Caquetá, the leading departments in the production of natural rubber are Santander, Antioquia, Meta, and Putumayo, in that order.

Caquetá has the country's only plant for producing TSR20 rubber, operated by ASOHECA, but it operates at well below capacity owing to the limited supply of rubber coagulate in that region. In addition, the rubber blocks it produces do not yet meet international standards and hence do no satisfy the requirements of large domestic tire producers such as Michelin and Goodyear. ASOHECA is working with farmers on good agricultural practices for natural rubber and to increase volumes, and to improve its own processing.

The production of NRCX, which is used in the domestic shoe industry, is concentrated in the Department of Meta, mostly in a single large plantation.

RSS is produced in Caldas, Santander, Tolima, Guaviare, Antioquia, Meta and Caquetá.

Propagation via grafting produces better quality rubber trees than use of seeds. Key element of the process of grafting is establishment of a small "clonal garden" (typically 1 m square with a plant density of 8000), which produces the genetic material to be grafted onto rootstock, and a nursery for producing the rootstock from seeds.

Once the trees are yielding, latex (a white- or cream-colored liquid that has particles of rubber) is collected by "bleeding" (tapping) the trees with cuts in the bark made as described earlier, with a special knife. The trees produce latex most rapidly at the end of the dry season. It flows through tin channels hung from the trees into plastic or metal buckets. Making the cuts correctly is a delicate matter and influences the yields of the trees. This is one of the aspects of production that most can benefit from technical assistance. The first year in production yields about 255 kg of latex and yields increase by about 200 kg per year until reaching a peak of 1.2 tons. However, multiple factors influence yields.

Once collected, the latex is treated on the farm according to the form of the product to be sold. It can be concentrated latex (60% liquid) or solid rubber. For the former, since latex tends to coagulate quickly, it is necessary to add an anticoagulant with an ammonium base, which can serve to maintain the latex in liquid form for the desired length of time. The solid latex can take various forms such as sheets, crepes, and/or granules, which are the most important.

To produce a rubber sheet, once the liquid latex is collected from the farms it is *filtered* with the aim of eliminating pieces of bark, insects, leaves, bits of latex in early coagulation, flowers, and other extraneous elements. Afterward it is *diluted* by adding water to the latex to bring it to the desired concentration. Then the hydrocarbon particulates of rubber dispersed in the latex are brought together in what is termed *acidification*, and finally the process proceeds to coagulation. In this stage latex, water and acid are mixed in a recipient and left for 12−24 h, resulting in a spongy coagulated product. Subsequently, a *sheet* is produced using a machine that passes over the rubber coagulate 6 to 8 times until a sheet 1.5−2 cm thick is attained. For *drying* the rubber, sheets are hung from bamboo rods over two thick wires, in the shade and with good air circulation. Complete drying takes 8−15 days. Finally, the rubber is molded into pellets, bunches or sheets. When the aim is to produce rubber coagulate, the efforts after the tapping focus on acidification in buckets, and 3 or 4 days later the coagulate is put into packing containers and sold. This can produce raw material of top quality. In Colombia, coagulate is produced in Caquetá and is used to produce TSR20.

The principal diseases of rubber plants include the earlier-mentioned South American leaf blight, which is the most serious disease for rubber in the Americas, pink disease (*Corticium salmonicolor*), mancha areolada (*Thanatephorus cucumeris*), black stripe (*Phytophthora palmivora*), cancer of the trunk (*Phytophthora* spp.), and *anthracnose*. Among the infestations that attack the plant are carpenter ants, the worm known in Colombia as *gusano cachón* that has voracious larvae, termites, and the hairy worm.

South American leaf blight can leave rubber plants completely without leaves, killing the ends of the branches, and killing the entire tree gradually, reducing latex production. Among the factors that can slow the advance of this disease are drier climates and soft breezes that dry the leaves rapidly. Chemical control of the diseases is recommended but only in the nurseries. The most effective approach is the development of resistant varieties and their propagation.

Problems: Genetic Material and Production Practices

The consensus of field experts and the study carried out by Colombia's Servicio Nacional de Aprendizaje (SENA, 2006) is that the country's rubber production is far from uniform and does not comply with technical specifications in the markets. The majority of the producers are smallholders spread through several regions, with very old trees, inadequate management of them, and lack of technical knowledge about cultivating rubber. The results are low yields and a level of quality that does not appeal to industrial users of natural rubber.

In areas with the old trees, soils were not treated prior to planting, little fertilizer was applied, and little effort was made to control infections and infestations, with the result that a high percentage of the trees have been lost. On average, only 290 trees have survived per hectare, of 400–500 planted. Also, these conditions have typically added another year to the initial period before yields can be obtained.

On occasions rubber has been planted on deforested soil or in areas recently used for ranching, which is not recommended because the soil has been compacted by the weight of livestock roaming over it and soil fertility and organic material have been lost through runoff of rains. A current practice is for smallholders to plant rubber trees between rows of basic food crops but without leaving enough space for the rubber and reducing the luminosity it needs in the early years of its growth. A consequence is a longer period before the trees reach sufficient maturity to be tapped. Another problem is the tendency of many farmers to combine rubber cultivation with livestock, which leads to damage to the trees' roots from the movement of the livestock.

Rubber yields have been the dominant criterion for selection of varieties to be planted, leaving aside secondary criteria that can also be important. These criteria include quality of the latex, resistance of the trees to damage from wind, their adaptability to soils that are poor in nutrients, the quickness of their recovery from the cuts in the bark and tapping, their resistance to diseases and infestations, and their ability to survive the dry season in good health. Fertilization is usually done without technical guidance regarding the appropriate fertilizer nutrient combinations for each zone and soil type.

Many producers prefer to plant the trees with roots exposed rather than in bags, which increases the chances of losing trees significantly. Of those planted with roots exposed, around 30% are lost. The manner of planting, densely in simple furrows, often means that when the trees grow sufficiently it is difficult for light to penetrate through the network of foliage and humidity is retained around the trees, with the result of lower or delayed yields and more vulnerability to diseases. As can be seen, a large number of the issues with rubber cultivation occur during planting and the early years after planting.

There are several problems regarding the practices for tapping the latex, principally the following:

- Farmers often do not stick to a strict program of frequency of making the cuts and tapping.[13] If it is not possible to make the cuts on a given day, owing to rains or other reasons, they usually do not replace the day or hours lost, with the consequence of diminution of future production.
- Little attention is paid to the time of day in which the tapping is carried out; the greatest production of latex can be obtained in the morning owing to the stronger flow of resin within the tree in those hours.
- Cutting is done in a way that damages the bark, from which regeneration should be occurring. The result is scars or knots in the bark, which make future cuttings difficult after that section of the tree has recovered. Regeneration can take 5 years or more, during which time the cuts should be made on the other side of the tree or ascending in a corkscrew pattern, as mentioned earlier.
- Sometimes farmers try to stimulate the flow of latex, but this results in drying up of the tree's productive capacity.

In addition to these issues, control of diseases and infestations is quite deficient, especially for the devastating South American leaf blight. Farmers know too little about it, and resistant varieties are not developed.

Another common phenomenon is the existence of several different varieties in the same rubber plot, obliging the farmer to initiate tapping at different moments. In Caquetá, 90% of the older rubber holdings have multiple varieties, with trees that have lost their leaves next to trees in good condition, complicating the cultivation of the plot.

Problems: PostHarvest and Processing

During tapping some farmers allow residues of bark, fungicide, and stimulants to fall into the channels and collectors for the latex, lowering its quality, especially for production of coagulate. Often sufficient care is not taken to avoid humidity where the latex is processed, and hence fungus gets into it along with other extraneous material. When the latex is collected on the farm, it is recommended to place it in containers appropriate for rinsing it, to remove impurities, and drying it, but this is often not done.

In the vast majority of the cases, the postharvest processing on farm produces rubber sheets instead of coagulate, but the latter is needed for the production of the TSR rubber, which is in most demand by the large rubber processing companies in the country. However, the costs of equipment for producing TSR rubber are high and the supply of coagulate for it fluctuates and frequently is not of sufficient quality. In the countries that are world leaders in rubber, the production of TSR is done in facilities located near the rural collection centers, but in Colombia, the raw material for TSR has to be transported long distances before being processed.

Problems: Marketing

The fundamental problems for marketing natural rubber in Colombia are lack of sufficient volumes, unreliable supply, and poor quality. The lack of sufficient farmer training and the types of postharvest treatment equipment used are major obstacles to improving quality. However, as pointed out later, greater efforts can be made to develop industries that would utilize Colombia's existing types and quality of natural rubber.

Avenues of Solution for Rubber

Colombia has very favorable climate conditions for the production of rubber. It has extensive swaths of territory that are appropriate for this crop, especially in the Amazonía, Orinoquía, the lower and marginal coffee zones, and the Magdalena. To take advantage of this potential, quality issues need to be addressed above all, and at both the farm level and in processing. Better equipment and greater research and extension efforts are needed for the entire value chain from farmer to industrial buyer.

Testing of rubber varieties in different conditions and development of new ones is also needed, and above all varieties that are more resistant to plant diseases and with the aim of producing better quality and uniform raw material for rubber industries. ASOHECA has worked with plots for observing and comparing the performance of different varieties, opening possibilities of eventually developing Colombian varieties. It is especially important to test rubber varieties in zones with a dry season of at least 4 months, with humidity of around 65% in that season and heavy rainfall in the other seasons. These conditions help reduce the threats of diseases and infestations. CORPOICA considers that such zones are found in Orinoquía, some parts of Magdalena medio, and some areas of Antioquia and Córdoba. The incidence of the South American leaf blight is reduced in these areas because each year 95% of the foliage is regenerated during the dry season in conditions that are not conducive to that disease because of the low humidity.

On the side of extension programs, clearly there is a pervasive lack of knowledge of proper management of this crop. Producers need technical assistance during the entire lifetime of a project until the trees are in full production. Along with this, investments are needed in more advanced processing facilities so that efforts can be made to obtain the international certification for TSR rubber, which is also needed for sales of natural rubber to the large tire companies in the country.

It has been suggested that there are other rubber-based industries that merit consideration because they could use the prevailing quality of natural rubber. They include foam rubber pillows and mattresses and other foam products and chewing gums. There also may be potential markets for Colombia's raw material in the areas of medical and pharmaceutical products, parts for autos and engineering industries, sporting goods, conveyor belts, footwear, bags for ice and water, carts, gloves, band-aids, raincoats, diving suits, erasers, runners and baizes, and glues, among others. Many of these industries use sheet rubber and rubber residues, which still represent a significant share of Colombia's natural rubber production, even though efforts are being made to produce quality coagulate. These rubbers also deliver better prices to farmers than quality coagulate does. Finally, rubber seeds are used in animal feed for cattle, chickens, and swine, after being harvested and having outer coverings removed, and being grilled and ground up. The oil extracted from the seeds is used in manufacturing paints, varnishes, and some kinds of soaps. Hence more creativity in promoting rubber-based manufacturing and marketing of Colombia's natural rubber might be able to strengthen this sector for small farmers, but it is uncertain how the rubber industry will respond.

In other words, rubber might present a special case in which, instead of adapting Colombia's production to the specifications of international markets, national markets could be developed that use the types and qualities of natural rubber currently produced in the country. However, basic issues such as the frequent presence of impurities would have to be addressed, and the interest of these industries in expanding or establishing operations would have to be researched.

Application of Value Chain Filters

- *Seeds and breeds*: inadequate varietal testing and development efforts, inadequate varietal selection by farmers, insufficient use of grafting of better varieties on known types of rootstock.
- *Production*: planting on soils that are too compacted or depleted of nutrients, lack of fertilization in early stages of tree growth, cropping systems that do not give the trees sufficient access to light and that increase humidity around the trees, bringing livestock too close to the rubber trees, which damages the roots.
- *Harvesting*: inefficient cutting and tapping procedures that often reduce the trees' productivity, inappropriate management of the flow of latex for avoidance of contamination with multiple kinds of impurities.

- *Postharvest*: lack of sufficient protection of the harvested latex once in containers; frequently inappropriate types of containers are used.
- *Storage*: no issues.
- *Transportation*: distances to processing centers are too great.
- *Processing*: upgrading of processing facilities is needed.
- *Market development*: domestic markets can be developed for industries that use Colombia's sheet rubber and rubber residue granules, in addition to improving quality to meet requirements of TSR markets.
- *Value chain governance*: More organizations of rubber producers are needed to deliver more effective training on management and harvesting of this crop.
- *External and policy*: no issues.

The Value Chain Quality Assessment Matrix for Colombian Rubber

All the rubber production cases analyzed in Track 1 are highly competitive. However, this is a case in which the Track 2 analysis contradicts the Track 1 analysis and casts strong doubts on the crop's competitiveness under present conditions. On the cost-price side, the markets for rubber are complex and the markets for TSR offer lower prices than the markets for other types of rubber. Colombia's competitiveness is stronger in the latter than in the former, but still there are major issues in crop management and postharvest procedures for the vast majority of rubber producers.

The potential export market is very distant in time, so the scores in Table 10.17 refer only to the domestic market, part of which demands natural rubber of export quality. The overall score for rubber in Table 10.17, at 22 (average of 11 per market), indicates the presence of serious problems to be addressed for the majority of rubber producers. The score could be higher (less competitive) if the extent of crop management and postharvest issues were reflected. The lack of training, and the lack of producer organizations for that purpose, are at the root of the problems, which range from inadequate choice of varieties to the need for better grafting practices and crop management, and much improved harvesting practices. There may be a valuable role here for contract farming, as a way of delivering technical information to farmers and providing them the incentive of stable markets, as per Section 6.2. As noted, there could be room on the agroindustrial side for development of industries that can use the types of natural rubber produced by most smallholders in Colombia.

10.7.6 Plantains

The Field Findings

The Crop and Its Markets

Plantain (*Musa paradisíaca*) is a member of the banana family that is primarily oriented to the domestic market, but it is also the second largest fruit

TABLE 10.17 Colombia: Value Chain Quality Assessment Matrix for Rubber

Value Chain Stage	Domestic Markets		Export Markets	
	Sheet, Residues	Coagulates	Sheet, Residues	Coagulates
1. Seeds and breeds	2GE	2GE		
2. Production	CM, PD, TA	CM, PD, 2TA		
3. Harvesting	IM	2IM		
4. Postharvest	PK	PK		
5. Storage				
6. Transportation				
7. Processing	PR	PR		
8. Market development	NM	CE		
9. Value chain governance	LC	LC		
10. External and policy				
Sum of scores	10	12		

Total score on quality issues: 20. The 2GE score refers to both lack of varietal development (especially for disease resistance) and the lack of sound seed selection and grafting practices by farmers. For production, TA refers to lack of uniformity of the natural rubber. LC refers to the need for more numerous and stronger producer organizations for rubber.

export, after bananas. Its exports are mainly to Hispanic consumers in the United States, Belgium, and Holland (to be distributed from the latter throughout Europe). Exports are shipped by sea, principally in the form of green plantains, in containers with temperatures controlled in the range of 7.8−8.3°C. The fruit is packed in 50- and 25-pound boxes.

Unlike bananas, plantains are cooked before being eaten, when either green or very ripe, to convert their starch into sugars. They are fried, boiled, steamed, or roasted. Plantain is also processed into snacks, flours, and animal feed concentrates and is used as an input into other processed foods. Plantain processing activities are increasing in the country.

Plantains begin to yield harvests in about 18 months after planting. Typically they continue producing for 6−15 years, although with proper care of the plant the productive period can extend to 20 years and more. They require temperatures between 18 and 32°C, precipitation between 1800 and 2500 mm a year, and altitudes between sea level and 1500 m. The crop grows best in soils with a pH between 6.0 and 7.5. According to the Colombian Banana

Growers Association (AUGURA), the principal variety grown in the country is Hartón, and others are Dominico, Dominico Hartón, Popocho, Cachaco, and Pelipita. Each variety tends to be associated with a particular region of the country: Hartón principally in the Eastern plains (*Llanos orientales*, or Orinoquia) and in the Atlantic Coast; it is the preferred variety in both domestic and international markets. The variety Dominico Hartón is grown in the central coffee zone, because it adapts well to altitudes above 1000 m. The varieties Cachaco and Pelipita are also cultivated in the *Llanos orientales*, and they are consumed mostly at the local level.

In Colombia, 87% of the plantain is planted in cropping systems with coffee, cacao, cassava, and fruits. An advantage of such systems is that plantain provides needed shade for coffee. For smallholders it is an important food crop as well as a source of additional income. It is harvested weekly, biweekly, or monthly, depending on the crop management system. From the viewpoint of family food security, the fact that it can be harvested year around makes it an important addition to smallholder systems that include crops like coffee, whose harvests occur only a few months of the year.

The plantain value chain has four basic links: farmers, marketing agents, processors, and exporters. Marketing channels are segmented by region with little national integration. To some degree the value chain links also provide advisory services to farmers as well as production inputs. The main problem in the plantain value chain is protecting the crop against diseases, since the cost of treatments is high from a farmer's viewpoint. Postharvest management and storage are also deficient, and improving them would require an even tighter integration between producers and retail markets or export channels.

Both yields and LRC values differ spatially across Colombia, mainly as a function of soils and climates. Among the 16 production vectors examined, nine fell in the uncertain range of competitiveness, with LRC values between 0.93 and 1.08. The remaining seven were competitive with values between 0.60 and 0.89. Plantains are grown throughout Colombia in temperate and warm zones. The areas of greatest concentration of plantain are in Antioquia, the Urabá (bordering Panama), and the coffee-growing zones. The most competitive areas are the foot of the mountains as they drop into the *Llanos orientales* and the inter-Andean zones including areas with coffee. However, plantains in the former area are more affected by two principal diseases of this plant, the *Sigatoka Negra* and the *Moko*. The production cases that used more agrochemicals tended to have the lowest degree of competitiveness. This underscores the need for increased efforts to apply natural methods for controlling diseases and pests, above all if farmers wish to increase production for export.

For smallholders plantain represents a profitable option, even without much crop management or investment: 80% of farms with plantain have been 1–5 ha of land, and 15% have 5–15 hectares. From the data analyzed, on average a hectare of plantain generated over US$1500 per year. A farm can manage 4 ha of plantain with family labor, and it will generate incomes at the level of two minimum salaries per year. More productive management systems for plantain can be even more profitable. In most parts of the country, plantain yields are around 7−9 tons per hectare, and about 15 tons in the case of Meta (where the Eastern Plains begin). The country's average yield in 2005 was 7.7 tons/ha. and according to FAO that was the world's fourth highest average yield for plantain. Only Venezuela, Peru, and Ghana had higher average yields.

The recommended planting density in monoculture systems is 1666 plants/ha. It should not be greater than 3300 plants/ha. to avoid excessive competition among the plants for light. In coffee-growing areas the plantain yields per hectare are lower, but that is because its planting densities are lower, since it is part of a cropping system. Per plant, the plantain yields in coffee areas are higher than in other areas.

Plantain processing activities are concentrated in large cities like Bogotá, Medellín, Cali, and Barranquilla. The processing firms include Frito Lay, McCain, Congelagro, Productos Yupi S.A., and Kopla. Frito Lay processed around 400 tons a week at the time of the original study. Quindío is the main area that produces plantain for processing, owing to the high Brix level of the ripe plantains grown there. It was thought this was due to the nature of the soils in that area, but further studies showed it is attributable to the fertilizer doses, above all boron. Yields and input use are low in Quindío, but the high quality of the product compensates for that. All growing areas are equally suitable for supplying the plantains that are processed green. Cauca is another department that produces high-quality plantain, with more sophisticated crop management.

Colombia has three research centers devoted to the genetic improvement of plantain, born of private initiative and supported by CIAT. In these centers, propagation of the plant is carried out through thermotherapy.

Problems: Genetic Material and Production Practices

Residues of pesticides and other agrochemicals are considered one of the most important issues with plantain, for both human health and the environment. The FAO has pointed out that developing methods of IPM for plantain will be key for its sustainable cultivation. The design of IPM methods depends on various factors, including location within the country, cropping system, and the variety planted. In Caldas, the Colombian Agricultural Institute (ICA) and CORPOICA have developed a successful IPM approach, but it has not yet been disseminated or adapted to other areas. One of the principal ways in which diseases are spread is through the plantlets used as genetic material via transplanting. For that reason, the propagation of genetic material in vitro and certified by ICA (or other organization) is recommended. However, this material can be expensive for smallholders who grow plantains for home consumption or local markets, and multiplication of the material by this process is slow. Greater planting density can reduce the crop's vulnerability to diseases, by lowering the dampness of the leaves, but at the same time it tends to lessen the weight of the fruit bunches thus lowering the harvest's suitability for the more demanding markets.

Disinfecting agricultural tools and boots after each harvest is another way to reduce the spread of plant diseases, but this practice is not widely followed. For improving IPM it will be important to gather more information about the weeds and other nearby crops that are hosts for the diseases, along with development of methods of early detection of the diseases. Control of diseases absorbs a large part of the labor devoted to this crop, and that makes it more costly to produce than would be the case under more effective methods of disease control.

Problems: Postharvest and Processing

Postharvest practices include collecting the bunches on the farm, selecting the acceptable ones and cutting off extraneous material, classifying and washing them, disinfecting with a fungicide on the crown, rinsing, and packing in boxes appropriate for the market destination. The losses in this crop management and transportation are estimated at 12% of the harvested volume, much of it due to careless handling on farm and in transport. Postharvest management guidelines are not available to farmers. However, given the importance of the exports markets for plantain, and the processing industry uses of it for snacks, flour, and precooked foods, farmers are forced to learn better postharvest practices.

Problems: Marketing

The geographical fragmentation of production and markets for plantain has given rise to a situation common to perishable products, in which marketing intermediaries structure and manage the relationship between farmer and markets. However, the growth of supermarkets and formal wholesale markets is providing another option to producers, which takes the form of de facto contracts with fixed prices but also with strict requirements regarding deliveries and quality of the product. Equally, the increasing use of plantain in processing industries is providing a defined and stable market with its own quality requirements.

Plantain exports traditionally were centered in the Urabá because of the infrastructure developed there for banana exports, but recently, production for export is spreading throughout the country. Of the wholesale plantain price in the United States, approximately 55.5% is the producer's price, 13.2% represents shipping costs, and 31.2% goes to marketing margins. This price structure provides incentives to farmers. As more producers adopt better phytosanitary control practices, export markets will open up for them also.

Avenues of Solution for Plantains

Improved technologies of production are, without doubt, high on the agenda for plantains. The Ministry of Agriculture has emphasized this point and has laid out general elements of this agenda, as follows:

- Development of genetic material that is more resistant to disease and pests.
- Development of crop management techniques that reduce the incidence of disease.
- Inspection, vigilance, and control: strengthening controls on the use of pesticides.
- Dissemination and certification of IPM methods for managing disease and pests.

Another evident problem is the need for development of a more appropriate institutional framework for this crop. The fruit and vegetable fund does not function as well as the corresponding funds for coffee, sugar, oil palm, and rice, for example, owing to the fact that it has to support a large array of fruit and vegetable crops widely scattered through the country.

In spite of the issues affecting plantain, its potential in Colombia is strong for the following reasons:

- There is considerable export experience with this crop and export demand is practically unlimited.
- Domestic markets are seeking greater supplies of plantain, and demand for it by processing industries is increasing
- Plantain's cropping systems are suitable for smallholders.
- It is produced competitively in Colombia's climate and ecosystems.

Application of Value Chain Filters

- *Seeds and breeds*: development of more resistant varieties is needed; farmers need to be more selective in use of planting material to reduce the spread of diseases.
- *Production*: disease and pest management techniques are largely deficient, with the result that the product's quality for markets is reduced and diseases and pests are not combatted effectively—and at high cost in labor.
- *Harvesting*: no issues.
- *Post-harvest*: careless handling on farm of harvested bunches is commonly observed; guidelines for good post-harvest practices on farm are not available to farmers.
- *Storage*: no issues.
- *Transportation*: careless packing and handling in transport results in loss of some of fruit.
- *Processing*: no issues.
- *Market development*: no issues; markets exist provided that quality standards can be met and deliveries can be made reliable.

- *Value chain governance*: an effective research and extension institution is needed for this crop along the lines of comparable institutions for some other crops in the country.
- *External and policy*: no issues.

The Value Chain Quality Assessment Matrix for Plantains

The overall VQA score for plantains is 18, but per market it is only 6, which indicates a high degree of competitiveness by Track 2 criteria. The issues to be resolved are clearly defined and programs can be mounted to deal with them. Even without resolving them, plantains are an important crop both for smallholder food security and for commercial markets, both domestic and external (Table 10.18).

TABLE 10.18 Colombia: Value Chain Quality Assessment Matrix for Plantains

	Domestic Markets		Export Markets	
Value Chain Stage	Fresh	Processed	Fresh	Processed
1. Seeds and breeds	GE	GE	GE	
2. Production	CH, PD	CH, PD	CH, PD	
3. Harvesting				
4. Postharvest	SP	SP	SP	
5. Storage				
6. Transportation	PK	PK	PK	
7. Processing				
8. Market development				
9. Value chain governance	LC	LC	LC	
10. External and policy				
Sum of scores	6	6	6	

Total score on quality issues: 18. The LC indication in value chain governance refers to the need for a product-specific research and extension organizations, as exists for some other crops in Colombia.

ENDNOTES

1. The bulk of the empirical work on this study was carried out by Henry Samacá Prieto and Héctor J. Martínez Covaleda, with the assistance of Sandra Acero Walteros and Elkin Pardo. Charles Richter provided helpful comments on drafts.

2. Lulo, *Solanum quitoense*, is called naranjilla in some countries. It is a fruit with a citruslike flavor that is used mainly for juice that is popular in Colombia. It has a short shelf life that inhibits its use as a fresh fruit.

3. Coffee's international price is notoriously volatile, and its competitiveness depends strongly on the price level at the time of the assessment.

4. These are the corridors defined by USAID's MIDAS project.

5. In this context, main marketing agents are exporters or buyers for supermarkets or (in a few cases) for processing plants, or marketing agents that sell directly to them. This value chain differs from the traditional one. For example, the main marketing agents do not make purchases in the traditional urban wholesale markets but rather from contracted lower level agents or directly from farmers.

6. There are varied theories about the origin of cacao, depending on the type of cacao under discussion.

7. The Consejo Nacional Cacaotero consistently reports larger areas planted.

8. http://www.CORPOICA.org.co/Libreria/libropreg.asp?id_libro=3&id_capitulo=11.

9. The other species or species aggregates include *Rubus bogotensis, Rubus notingensis, Rubus poephyromallus, Rubus floribundus, Rubus giganteus*, and *Rubus nubigenus*. Each has its preferred zones, altitudes, and characteristics.

10. In Colombia's custom classifications, fresh lulo is lumped together with other fruits in the "other category" of 08.10.90.90.00 and processed lulo is placed in the "other category" of 20.08.99.90.90.

11. Much of the information in this subsection and the next one is based on CORPOICA (2007).

12. The technically specified rubbers represent a new form of producing and packaging natural rubber, through the application of quality control systems that are clearly established in the international context. This kind of rubber was developed in Malaysia in 1965 in response to industry's need for a product with invariant levels of viscosity and processing characteristics, and above all with more rigorous technical specifications than those found in ribbed smoked rubber or natural rubber crepe. The production of TSR rubbers is governed by ISO standard 2000, ISO/TC 115 and TCR's own norms. TSR rubbers are classified by origin (SMR is Standard Malaysian Rubber, SIR is Standard Indonesian Rubber, SSR Standard Singapore Rubber), types (according to standards set by buyers: SP, Superior Processing—80% latex and 20% vulcanized; AC, Anticrystalizing, composed of natural and synthetic rubber; and others) and degrees of quality that depend on color, the percentage of impurities, plasticity, content of nitrogen and volatile material, content of ash, and viscosity. The degree of quality are 5, 10, 15, 20 y 50: TSR 5 is the best since it has less than 0.05% of impurities and is color is light. TSR 10 has a light brown color, TSR 15 is brown, TSR 20 is dark brown, and TSR 50 very deep brown.

13. In Colombia there are three alternative programs of cutting and tapping: d/2, every 2 days the same tree is cut; d/3, common in medium-size plantations, for tapping every third day; and d/4, common in large plantations or areas where field labor is scarce, for tapping every fourth day.

Part IV

Concluding Remarks

Chapter 11

Assessing Agricultural Competitiveness and Its Determinants

As can be seen in this book's discussions, competitiveness is a multifaceted, dynamic concept. In tropical agriculture, what decision makers, farmers, and investors would like to know is whether a product has a reasonable chance of being competitive, i.e., of earning a profit, in the international context—as an import substitute or as an export—and what improvements may be needed to ensure its competitiveness.

To answer these questions it is useful to carry out analysis of both the cost-price dimension and the quality dimension of competitiveness, either along the two-track lines of this book or through variants of these approaches. It should be emphasized that in many cases these kinds of analyses are only starting points. Before promoting or investing in a product it may be necessary to dig deeper into issues that need to be resolved, especially those that affect farm yields and product quality, and have particularly thorough conversations with field experts and market representatives. It may also be necessary to review the economic policy framework and ask what is likely to change in the future, and also what policies may need to be modified to remove obstacles to long-run competitiveness (LRC).

From a cost-price viewpoint, both ex ante and ex post indicators are useful. The latter are particularly helpful in identifying trends that indicate competitiveness, through structured reviews of export and import time series (Chapter 3). However, those trends may be influenced by policy measures that are not sustainable in the long run. Also, it may be difficult to bring the ex post calculations down to the level of specific products because of the nature of the product groupings in customs classifications of trade data. The ex ante indicators of LRCs abstract from transitory policy influences and market distortions, but they are limited to static snapshots. Their usefulness can be enhanced through calculations of cost-price competitiveness for different production locations in a country, for different technologies of production (including those proposed for new projects), and under different scenarios on factor prices, outputs, and product prices. This kind of sensitivity analysis, illustrated in Chapters 9 and 10, provides insights into the circumstances that

The Competitiveness of Tropical Agriculture. http://dx.doi.org/10.1016/B978-0-12-805312-6.00011-8

309

may help a product become competitive, or lose competitiveness. Incorporation of the spatial dimension into the analysis is also valuable not only to identify the most competitive locations but also because competitiveness can vary from farm to farm, and having more data points provides a larger sample for assessing it.

Cost-price competitiveness is not sufficient for a product to be produced profitably. The material in the book, especially in Chapter 6, illustrates the importance of overcoming quality issues for fulfilling competitive potential. These issues are very diverse and are specific to products and country contexts, and therefore the book proposes a qualitative assessment framework that may be adapted to the conditions that characterize a given locality and set of products.

Policy dialogs can be launched on the basis of the ex ante calculations of competitiveness, as in Section 7.9 of the book. Is policy support for agriculture aligned with the country's inherent competitiveness? How can it be made more supportive of competitiveness, in areas such as agricultural research, credit policy, and land tenure? How does the exchange rate affect competitiveness (Chapter 9)? What would be the effects on total employment and incomes of shifting that support in the direction of the more competitive crops and livestock products?

Clearly competitiveness at the product level is influenced by national factors such as distance from major world markets, infrastructure development (particularly internal transportation and cold chains), and macroeconomic policy. Regional markets also can be important as arenas in which a country's products are competitive when they may not be in wider world markets, as in the cases of El Salvador and Rwanda. Climate also plays an important role. Arabica coffees can be grown only at higher altitudes such as those in East Africa, Colombia, Mexico, Sumatra, and Central America, but which Vietnam and Brazil do not possess for the most part. The range of altitudes in Kenya, Rwanda, Colombia, and Guatemala, to mention only a few countries, favor the cultivation of other high-value crops that cannot be grown in the lowland tropics. Grenada, a center of nutmeg production, is in danger of losing its main international buyers because hurricanes too often destroy the plantations. This situation is a reminder than climate change is posing risks and new challenges for competitiveness. For example, it is driving Arabica coffees to even higher altitudes than before, and it is spreading more widely the reach and virulence of crop and animal diseases and infestations.

Beyond these basic considerations a number of broad conclusions emerge from the studies reported here. First, often tropical countries have their strongest competitive advantage in high-value or niche crops that individually may be small in the overall picture but that taken together can be important for the country and for farm families. These crops typically generate 5—10 times more income per hectare than basic food crops do. Colombia is a case in which they are much more competitive than most basic grains (Chapter 7). This was also a finding of an earlier study by the author and colleagues for Mexico (Bassoco & Norton, 1983). However, policy may lag this finding and concentrate on support for noncompetitive crops that will have a slow or nil growth rate in the future. Therefore giving policy support to them will drag down the sector's overall rate of income and employment growth (Chapter 7).

Second, these high-value crops are labor intensive, which is one reason they are competitive in countries with large amounts of underemployed and relatively inexpensive labor. However, they can also be capital intensive, and above all they are intensive in knowledge (Chapter 9). Fulfilling the competitive potential of tropical agriculture means dealing successfully with a large variety of agronomic and postharvest challenges (Chapters 8 and 10). This requires continuous access to sources of information through the multiple channels that make up what has been called the Agricultural Knowledge and Information System. More effective agricultural extension approaches are part of such a system, particularly when they utilize information and communication technologies and respond to the concerns of farmers. Integration of farmers and farmer groups into value chains is another way to ensure that relevant information reaches them. Development of social capital in the form of farmer cooperatives and associations also plays a valuable role in facilitating farmer access to information.

Hence value chains are important conveyors of technical information and finance to farmers as well as access to markets, although they do not always function as hoped. Without them smallholders have great difficulty in translating their cost competitiveness into profitable, durable activities. In the final analysis, all the elements of knowledge and information systems are important tools for meeting the quality challenges of competitiveness. Seen in this light, the interviews with agricultural marketing agents in Colombia, which argued for incorporating all value chain links into agricultural development projects, may offer lessons for project design. Their opinion is that projects should generate enabling activities for all the actors in a value chain, not only for primary producers. Marketing intermediaries are agents of change and can mobilize other forces for change for producers. They also provided special insights into the export marketing business, which in the end is an important force for improvement of farmer productivity and quality through provision of knowledge, inputs, finance, and market incentives.

Chapter 12

Competitiveness in a Development Perspective

An inverse relationship between farm size and productivity has been documented throughout the world over the years (Carter, 1984), more recently in Nepal (Thapa, 2007) and Rwanda (Ali & Deininger, 2014). Therefore the search for competitiveness can be a quest for products and production methods that favor smallholders. Most of the high-value crops in this volume's case studies are primarily smallholder crops.

However, for these farmers a move away from staple foods and entirely into high-value crops can be risky. Higher incomes from coffee or cacao or macadamia or other valuable crops can improve family nutrition because farmers will be able to buy more food than they could by planting only staple foods. However, if a high-value crop fails because of bad weather, disease, or pests, or if its volatile prices take a turn for the worse, then their only alternatives may be hunger or increased indebtedness for the farm family. In addition, crops like coffee bring in income only for a few months each year. Producers at or near subsistence may not be able to set aside enough income to feed their families for the rest of the year. A buyer from a major coffee company once told this author that he wanted farmers to specialize in coffee alone, to become more skilled at producing that crop. His vision was that higher coffee yields and quality would earn enough income to last the families through the year, but the recent epidemic of coffee leaf rust put holes in that approach.

For these reasons, a venture into more competitive lines of production by smallholders usually means devoting only a portion of their land to higher value crops, or adopting mixed cropping systems that include food crops. Another variable of relevance for family nutrition is the portion of crop income that goes to women in the households, who are known for devoting a higher share of proceeds to family welfare. In the coffee—black pepper cropping system in southern India, for example, the income earned from black pepper belongs to the woman in the household. When this kind of arrangement holds, it can reduce the risks associated with adoption of high-value crops. It also empowers women in families and communities.

In broad terms, a smallholder family needs the *capacity to respond* to opportunities and the *opportunities* themselves. These paths to development

The Competitiveness of Tropical Agriculture. http://dx.doi.org/10.1016/B978-0-12-805312-6.00012-X
313

may be termed the *inner route* and the *outer route*, respectively. The capacity to respond includes physical assets (land, access to water, livestock, etc.) and above all human capital: knowledge and the capability of acquiring more knowledge. In addition, strengthening farmer resilience, especially in the face of climate change, is an increasingly important part of the inner route. Opportunities include access to markets for their harvests, to credit for inputs, to high-quality genetic material, and the like. Berdegué and Fuentealba (2011) present a framework similar to this in their treatise on smallholder agriculture. Participation in value chains, whether through farmer groupings or via contract agriculture or other arrangements for linking up with value chains can strengthen the inner route by providing training to enhance human capital endowments. Value chains can also facilitate the outer route by providing opportunities for access to markets, finance and genetic material, and other inputs. Supermarkets have greatly expanded their roles in developing countries in recent years, and they often are leaders of value chains, sometimes buying directly from smallholders, farmer associations, and cooperatives, and providing them training for quality control in the process.

Competitiveness analysis, both quantitative and qualitative, can assist in the assessment of options for the inner and outer route for smallholders by contributing additional criteria for judging crops or livestock products and value chains that development projects and programs are considering introducing to given rural communities. It adds discipline for making those choices with an eye to their economic sustainability in the face of market requirements.

Value chains are creations of private initiatives, involving actors who find common interest in collaborating for bringing products to final markets or to the processing stage. How can value chains be strengthened by supportive policies? The challenges are more difficult for niche products than for commodities. Public sector efforts to meet these challenges have tended to concentrate on fostering export value chains rather than domestic ones. The difficulties of expanding these value chains have been described well by Martincus and Carballo (2010):

> Developing countries lag behind in terms of export diversification. In particular, their exports tend to be concentrated in homogeneous commodities. This is particularly true for Latin American economies.... Diversifying into...more complex goods represents an important challenge. Differentiated goods are heterogeneous both in terms of their characteristics and their quality. This interferes with the signaling function of prices thus making it difficult to trade them in organized exchanges. In short, information problems involved in trading differentiated products are more severe than those faced with when trading more homogeneous goods.

For the case of Costa Rica's export promotion agency PROCOMER, Martincus and Carballo found that it succeeded in broadening the export markets for products already exported, but it did not have an effect on the number of products exported. At the same time, they point out that for the case of Chile, Álvarez and Crespi (2000) found that the export promotion agency PROCHILE succeeded in increasing the number of destination markets reached by Chilean exporters and also in diversifying the mix of products exported.

Peru's export promotion agency Sierra Exportadora has had extraordinary success in developing some value chains for smallholders in the country's Andean highlands. It has promoted the development, quality improvement, and export of many products. For the case of cheese, Sierra Exportadora reviewed the capacity of hundreds of cheese makers in the highlands and selected the most promising examples. The initial focus was on training for good manufacturing processes in cheese. Advisory assistance from Europe and Uruguay was obtained for this stage of the work. Then international markets were explored, initially in Brazil, and assistance was provided to implement the standards for a new label, Terrandina Peruvian Dairy, and successful efforts were made to gain market acceptance for the label. As a result, many Andean dairy farmers now can sell milk to a profitable and expanding cheese operation that exports.

Sierra Exportadora has supported the development of many other products, including coffee, cocoa, avocadoes, quinoa and other grains, berries, and other fruits. It has provided assistance for new processing technologies, opening markets, training producers, and generally catalyzing significant growth in Andean production for high-value markets. This experience shows that purposive efforts can indeed strengthen value chains, although success is never guaranteed.

In another notable case, a program in Panama funded by the Inter-American Development Bank, called PROCOMPETITIVDAD, set up an independently administered fund to provide venture capital grants for up to 13 different activities for small agriculture-related businesses. These activities included visits to explore markets in importing countries and learn about their standards, contracting technical assistance for production, bringing buyers into Panama, investing in modernization or expansion of facilities, attending short courses on product hygiene and other quality issues, and so forth. Proposals were reviewed by an independent panel of experts, mostly from outside the country, and funds were disbursed within 30 days to the winning proposals.[1] Evidently there is room for much creativity in efforts to improve value chains.

Important as value chains are, often there is no substitute for strengthened farmer organizations. Indeed, many export buyers will not deal with smallholders except through cooperatives and other organizations in the field. The famed financing organization for small farmers in developing countries, Root Capital, provides technical assistance for strengthening cooperatives as a precondition for providing finance. Grameen Foundation has developed software to improve cooperatives' delivery of extension services to their members by making use of cell phones and iPads.

What is sometimes missing from these efforts is work to restructure cooperatives along entrepreneurial lines, as a rural variant of the corporate model, so that there is a secure link between effort and reward. Small farmer cooperatives are still vulnerable to takeover by small groups within the membership, and corrupt management and dissolution. Special legislation usually is required to provide them a more durable structure, so that there is a defined framework under which such "corporate cooperatives" can be registered. Often it takes the form of amendments to a country's code of commerce rather than its land tenure or rural legislation. The basic thrust is to make members shareholders in the entity, and to give them the right to sell their sales, so that they will have value. However, sometimes the organization may wish to stipulate, for example, the first right of refusal on the part of remaining members for purchase of those shares.

The variants of such legal structures can be many, but this kind of reform goes hand-in-hand with efforts to provide training in entrepreneurship for smallholders and their families. De facto, smallholders are businesspersons, with the special circumstance that failure to cover costs can mean hunger for their families. Entrepreneurial abilities are especially important for awareness of costs, constructing business plans, penetrating markets, and producing outputs that are acceptable in markets. It can be especially important to train rural youth in entrepreneurship, including in the many rural services activities, from machinery repair and machinery services provision to advisory assistance to rural finance to paraveterinary occupations to input suppliers and many other options. Here the focus on strengthening farm organizations and developing entrepreneurship comes full circle back to agricultural knowledge and information systems. It is a common complaint that rural youth find agriculture an unattractive occupation, and farming populations are aging. In Africa, the average age of farmers is 60 (Food and Agriculture Organization of the United Nations, 2014, pp. 27−29, p. 2). Sustainability of competitiveness and increases in rural incomes will require incorporating more youth into farming, especially into the more technical aspects. The interest of young persons in agriculture can be sparked by introducing them to the quality challenges of production, the market challenges, the techniques of water management and irrigation, ways to clone genetic material and to select high-quality seeds, the management of dairy goats, ways to produce organic pest controls and produce organically, postharvest processes, and the many other technical facets of high-value agriculture. In this regard, extension and other information systems are called to focus more on youth, as well as on women, and technical agricultural institutions in rural areas need to be strengthened and their curricula updated for the quality challenges of today's agriculture.[2] Public−private partnerships, shown to be effective in various spheres, also can be applied to agricultural education. Agricultural enterprises and nongovernmental organizations working in the field can put their experience to use in vocational schools and other training entities. In the end, it is human capital that matters most for competitiveness.

ENDNOTES

1. Venture capital funds have a considerable history already in Uganda and are spreading in Africa. For a review of issues concerning those funds, see Miller et al. (2010).
2. See the discussion of rebranding agriculture in schools in IFAD, CTA and FAO, *Youth and Agriculture* (2014).

References

Accenture Strategy. (2014). *Extracting value from natural rubber trading markets: Optimizing marketing, procurement and hedging for producers and consumers.*

Alexandratos, N., & Bruinsma, J. (June 2012). *World Agriculture towards 2030/2050, the 2012 Revision. ESA Working Paper No. 12-03.* U. N. Rome: Food and Agriculture Organization.

Ali, D. A., & Deininger, K. (February 2014). *Is there a farm-size productivity relationship in African agriculture? Evidence from Rwanda. Policy Research Working Paper 6770.* Washington, DC: The World Bank, Development Research Group.

Álvarez, R., & Crespi, G. (2000). Exporter performance and promotion instruments: Chilean empirical evidence. *Estudios de Economía, 27*(2) [Universidad de Chile. Santiago, Chile].

Amann, E., Baer, W., Trebat, T., & Villa, J. M. (July 2014). *Infrastructure and its role in Brazil's development process. Working Paper No. 10, International Research Initiative on Brazil and Africa (IRIBA).* UK: School of Environment, Education and Development, The University of Manchester.

Antholt, C. H. (1998). Agricultural extension in the twenty-first century. In C. K. Eicher, & J. M. Staatz (Eds.), *International agricultural development* (3rd ed.). Baltimore: The Johns Hopkins University Press.

Asenso-Okyere, K. (2013). An African view of EU and US agricultural policy reforms. In *2012 global food policy report.* Washington, DC: International Food Policy Research Institute.

Azzouzi, E., Laytimi, A., & Abidar, A. (2007). Effect of incentive policy on performance and international competitiveness of greenhouse tomatoes, Clementine mandarins, Maroc-late oranges, and olive oil in Morocco. *Acta Agriculturae Scandinavica, Section C – Food Economics, 4,* 172–180.

Baffes, J., Tschirley, D., & Gergely, N. (2009). Pricing systems and prices paid to growers. In D. Tschirley, C. Poulton, & P. Labaste (Eds.), *Organization and performance of cotton sectors in Africa.* Washington, DC: The World Bank.

Balassa, B. (1965). Trade liberalization and revealed comparative advantage. *Manchester School of Economic and Social Studies, 33,* 99–123.

Balassa, B. (July 1974). *Estimating the shadow price of foreign exchange in project appraisal* (Vol. 26 (2)). Oxford Economic Papers.

Ball, V. E., Butault, J.-P., San Juan, C., & Mora, R. (2010). Productivity and international competitiveness of agriculture in the European Union and the United States. *Agricultural Economics, 41,* 611–627. http://dx.doi.org/10.1111/j.1574-0862.2010.00476.x.

Barirega, A. (March 18, 2014). Potential for value chain improvement and commercialization of Cape gooseberry (*Physalis peruviana* L.) for livelihood improvement in Uganda. *Ethnobotany Research and Applications, 12,* 131–140. http://dx.doi.org/10.17348/era.12.0.131-140.

Barnes, R. C., Krenek, K. A., Meibohm, B., Mertens-Talcott, S. U., & Talcott, S. T. (2016). Urinary metabolites from mango (*Mangifera indica* L. cv. Keitt) galloyl derivatives and in vitro hydrolysis of gallotannins in physiological conditions. *Molecular Nutrition & Food Research, 60,* 542–550. http://dx.doi.org/10.1002/mnfr.201500706.

Bassoco, L. M., & Norton, R. D. (1983). A quantitative framework for agricultural policies. In R. D. Norton, & M. Leopoldo Solís (Eds.), *The book of CHAC: Programming studies for Mexican agriculture.* Baltimore, MD: Johns Hopkins University Presss.

Bebek, U. G. (August 30, 2011). Consistency of the proposed additive measures of revealed comparative advantage. *Economics Bulletin, 31*(3), 2491−2499.

Bennet, G. S. (2009). *Food identity preservation and traceability: Safer grains.* CRC Press.

Berdegué, J. A., & Fuentealba, R. (January 2011). Latin America: the state of smallholders in agriculture. In *Conference on new directions for smallholder agriculture* (pp. 24−25). Rome: International Fund for Agricultural Development (IFAD).

Bervejillo, J. E., Alston, J. M., & Tumber, K. P. (December 2011). In *The economic returns to public agricultural research in Uruguay. RMI-CWE Working Paper Number 1103. Robert Mondavi Institute Center for Wine Economics, CA.*

Block, S., & Peter Timmer, C. (1994). *Agriculture and economic growth: Conceptual issues and the Kenyan experience.* Cambridge, MA: Harvard Institute for Economic Development.

Bordey, F. H., Launio, C. C., Beltran, J. C., Litonjua, A. C., Manalili, R. G., Mataia, A. B., et al. (November 2015). Game changer: is PH rice ready to compete at least regionally? *Rice Science for Decision Makers.* ISSN: 2094-8409, 6(1).

Bourne, M. C. (2014). Food security: postharvest losses. In N. K. van Alften (Ed.), *Encyclopedia of food and agricultural systems.* Academic Press, ISBN 9780444525123.

Bruno, M. (January−February 1972). Domestic resource costs and effective protection: clarification and synthesis. *Journal of Political Economy, 80*(1), 16−33.

Burfisher, M. E. (Ed.). (January 2001). *The road ahead: Agricultural policy reform in the WTO − summary report. Agricultural Economics Report No. 797. Market and trade economics division, economic research service.* Washington, DC: U.S. Department of Agriculture.

Carnahan, M. (2015). Taxation challenges in developing countries. *Asia & the Pacific Policy Studies, 2*(1), 169−182.

Carter, M. R. (1984). Identification of the inverse relationship between farm size and productivity: an empirical analysis of peasant agricultural production. *Oxford Economic Papers, 36,* 131−145.

Chen, C., Yang, J., Yang, J., & Findlay, C. (April 2008). Measuring the effect of food safety standards on China's agricultural exports. *Review of World Economics, 144*(1), 83−106.

Comisión de Ajuste de la Institucionalidad Cafetera. (2002). *El café, capital social y estratégico. Revista Nacional de Agricultura, SAC. No. 934.*

COMPETE Project. (November 2009). *Staple food value chain analysis, country report.* Malawi: USAID.

Consorcio Latinoamericano y del Caribe de Apoyo a la Investigación y el Desarrollo de la Yuca −CLAYUCA (n.d.). La yuca como instrumento de la reactivación agrícola en Colombia. CIAT and Ministerio de Agricultura y Desarrollo Rural. Cali, Colombia.

CONSULTPLAN. (1998). *Diagnóstico de los distritos de riego de mediana y gran escala en operación* (Vol. II). Bogotá: Instituto Nacional de Adecuación de Tierras, Subprograma de Servicios Complementarios.

Contraloría General de la República, Colombia. (2001). *Informe especial: La parafiscalidad agropecuaria, Bogotá.*

Contraloría General de la República. (2002). *Modelo, política e institucionalidad agropecuaria y rural. Bogotá.*

CORPOICA. (2007). *Evaluación, caracterización y cultivares de mango criollo presentes en las regiones de occidente, norte y centro del país; para mejorar la competitividad del cultivo de mango en Colombia. Bogotá.*

Corporación Colombia Internacional, CCI. (December 2000). *Perfil de Producto: Frutas Procesadas. Bogotá.*

CRECER and Grameen Foundation. (October 2015). *Perfil de proceso de beneficio húmedo. Asociación Barillense de Productores —ASOBAGRI. Ficha Técnica. Guatemala, Guatemala.*

Dabbene, F., Gay, P., & Tortia, C. (2013). Traceability issues in food supply chain management: a review. *Biosystems Engineering.*

Dalum, B., Laursen, K., & Villumsen, G. (1998). Structural change in OECD export specialization patterns: de-specialization and 'stickiness. *International Review of Applied Economics, 12,* 447—467.

Damiani, O. (April 2001). *Diversificación de la agricultura y reducción de la pobreza rural: Cómo los pequeños agricultores y asalariados rurales son afectados por la introducción de cultivos no-tradicionales de alto valor en el Noreste de Brasil. Proyecto Conjunto INDES — Programa Japón. Serie Documentos de Trabajo I-31JP. Washington, DC.*

Departamento Administrativo Nacional de Estadística, Colombia. (2004a). *Censo nacional de diez frutas agroindustriales y promisorias, Resultados. Bogotá.*

Departamento Administrativo Nacional de Estadística, Colombia. (2004b). *Documento metodológico de cacao. Bogotá.*

Dethier, J.-J., & Moore, A. (April 2012). *Infrastructure in developing countries: An overview of some economic issues. ZEF-Discussion Papers on Development Policy No. 165.* University of Bonn.

Dorneles, T. M., de Lima Dalazoana, F. M., & Schlindwein, M. M. (2013). Análise do índice de vantagem comparativa revelada para o complexo da soja sul-mato-grossense. *Revista de Economia Agrícola, 60*(1), 5—15. (São Paulo. Jan./jun).

Dunn, E. (2005). *AMAP BDS knowledge and practice task order lexicon.* microNOTE No. 6. Washington, DC: USAID.

Ekboir, J. (2012). Facilitating smallholders' access to modern marketing chains. In *Agricultural innovation systems, an investment sourcebook. Agriculture and rural development.* Washington, DC: The World Bank.

Estur, G., Poulton, C., & Tschirley, D. (2009). Quality control. In D. Tschirley, C. Poulton, & P. Labaste (Eds.), *Organization and performance of cotton sectors in Africa.* Washington, DC: The World Bank.

Ffafchamps, M. (2004). *Market institutions in sub-Saharan Africa: Theory and evidence.* Boston, MA: MIT Press.

Flaherty, K., Stads, G.-J., & Srinivasacharyulu, A. (2013). *Benchmarking agricultural research indicators across Asia-Pacific. ASTI synthesis report.* Washington, DC and Bangkok: International Food Policy Research Institute and Asia-Pacific Association of Agricultural Research Institutions.

Florio, M. (2014). *Applied welfare economics, cost-benefit analysis of projects and policies.* New York: Routledge.

Food and Agriculture Organization of the United Nations, FAO. (May 2014). *Contribution to the 2014 United Nations Economic and Social Council (ECOSOC) Integration Segment.* Rome (pp. 27—29).

GebreMariam, S., Amare, S., Baker, D., Solomon, A., & Davies, R. (2013). *Study of the Ethiopian live cattle and beef value chain.* Addis Ababa: ILRI Discussion Paper No. 23.

Gergely, N. (2009). Cost efficiency of companies, overall sector competitiveness, and macro impact. In D. Tschirley, C. Poulton, & P. Labaste (Eds.), *Organization and performance of cotton sectors in Africa.* Washington, DC: The World Bank.

Gorton, M., Davidova, S., & Ratinger, T. (2000). The competitiveness of agriculture in Bulgaria and the Czech Republic vis-à-vis the European Union (CEEC and EU agricultural competitiveness). *Comparative Economic Studies, 42*(1), 59—86.

Grewal, B., Grunfeld, H., & Sheehan, P. (2012). *The contribution of agricultural growth to poverty reduction. ACIAR Impact Report No. 76.* Canberra: Australian Centre for International Agricultural Research.

Grubel, H., & Lloyd, P. J. (1975). *Intra-industry trade: The theory and measurement of international trade in differentiated products.* New York: John Wiley and Sons.

Gustavsson, J., Cederberg, C., Sonesson, U., van Otterdijk, R., & Meybeck, A. (2011). *Global food losses and food waste: Extent, causes and prevention.* Rome: United Nations Food and Agriculture Organization (FAO).

Hallem, D., Liu, P., Lavers, G., Pilkauskas, P., Rapsomanikis, G., & Claro, J. (2004). *The market for non-traditional agricultural exports. FAO commodities and trade technical paper.* Rome: United Nations Food and Agriculture Organization.

Heckscher, E. F., & Ohlin, B. (1991). *Heckscher-Ohlin trade theory. Translated, edited, and introduced by Harry Flam and M. June Flanders.* Cambridge, MA: MIT Press.

Hemme, T., Mohi Uddin, M., & Ndambi, O. A. (2014). Benchmarking cost of milk production in 46 countries. *Journal of Reviews on Global Economics, 2014*(3), 254−270.

Hoen, A., & Oosterhaven, J. (2006). On the measurement of comparative advantage. *The Annals of Regional Science, 40*(3), 677−691.

Hoff, K., & Stiglitz, J. E. (1995). Introduction: imperfect information and rural credit markets − puzzles and policy perspectives. *The World Bank Economic Review, 4*(3), 1990 [Reprinted in: From the World Bank Journals, Selected Readings. The World Bank, Washington, DC].

Hurley, T. M., Pardey, Philip G., & Rao, X. (November 2014). *Returns to food and agricultural R&D investments worldwide, 1958−2011.* St. Paul, MN: International Science & Technology Practice & Policy (InSTePP) Brief.

Hurley, T. M., Rao, X., & Pardey, P. G. (2014). Re-examining reported rates of return to food and agricultural research and development. *American Journal of Agricultural Economics,* 1−13. http://dx.doi.org/10.1093/ajae/aau047.

IFAD (International Fund for Agricultural Development), CTA (Technical Centre for Agricultural and Rural Cooperation), and Food and Agriculture Organization of the United Nations (FAO). (2014). *Youth and agriculture, key challenges and concrete solutions.* Rome, ISBN 978-92-5-108475-5.

Ilyas, M., Mukhtar, T., & Javed, M. T. (2009). Competitiveness among Asian exporters in the world rice market. *The Pakistan Development Review, 48*(4), 783−794.

de Janvry, A., & Sadoulet, E. (February 2010). Agricultural growth and poverty reduction: additional evidence. *World Bank Research Observer, 25*(1).

Jooste, A., & van Schalkwyk, H. D. (2001). Comparative advantage of the primary oilseeds industry in South Africa. *Agrekon: Agricultural Economics Research, Policy and Practice in Southern Africa, 40*(1), 35−44.

Karim, I. E. E. A., & Ismail, I. S. (October 2007). Potential for agricultural trade in COMESA region: a comparative study of Sudan, Egypt and Kenya. *African Journal of Agricultural Research, 2*(10), 481−487.

Kiet, N. T., & Sumalde, Z. M. (2006). Comparative and competitive advantage of the shrimp industry in Mekong River Delta, Vietnam. *Asian Journal of Agriculture and Development, 5*(1), 57−79.

Kim, R., Larsen, K., & Theus, F. (2009). Introduction and main messages. In K. Larsen, R. Kim, & F. Theus (Eds.), *Agribusiness and innovation systems in Africa.* Washington, DC: Agriculture and Rural Development, World Bank.

Komarek, A. M. (2010). Crop diversification decisions: the case of vanilla in Uganda. *Quarterly Journal of International Agriculture, 49*(3), 227−242.

Kondo, J. (May 2000). Strengthening partnership in agricultural research for development in the context of globalization. In *Global Forum on Agricultural Research (GFAR), Latin America and the Caribbean Regional Forum, Document No. GFAR/00/25. Dresden, Germany* (pp. 21−23).

Kumar, N. R., & Rai, M. (2007). Performance, competitiveness and determinants of tomato export from India. *Agricultural Economics Research Review, 20*, 551−562.

Lagman-Martin, A. (February 2004). *Shadow exchange rates for project economic analysis: Toward improving practice at the Asian Development Bank. ERD Technical Note No. 11.* Manila: Asian Development Bank.

Laibuni, N., Waiyaki, N., Ndirangu, L., & Omiti, J. (January 26, 2012). Kenyan cut-flower and foliage exports: a cross country analysis. *Journal of Development and Agricultural Economics, 4*(2), 37−44. http://dx.doi.org/10.5897/JDAE11.050.

Lancaster, K. J. (April 1966). A new approach to consumer theory. *The Journal of Political Economy, 74*(2), 132−157.

Laursen, K. (1998). In *Revealed comparative advantage and the alternatives as measures of international specialization. DRUID Working Paper No. 98−30. Danish Research Unit for Industrial Dynamics (DRUID)*.

Le, Q.-P. (August 2010). Evaluating Vietnam's changing comparative advantage patterns. *ASEAN Economic Bulletin, 27*(2), 221−230. http://dx.doi.org/10.1355/ae27-2e.

Leavitt Partners and Eurofins. (August 2012). *US food safety modernization Act: Overview and impact for importers and exporters*.

Legg, J., Attiogbevi Somado, E., Barker, I., Beach, L., et al. (2014). A global alliance declaring war on cassava viruses in Africa. *Food Security, 6*, 231−248. http://dx.doi.org/10.1007/s12571-014-0340-x.

Lima, L. J. R., Almeida, M. H., Nout, M. J., & Zwietering, M. H. (2011). *Theobroma cacao L.,* "the food of the gods": quality determinants of commercial cocoa beans, with particular reference to the impact of fermentation. *Critical Reviews in Food Science and Nutrition, 51*(8), 731−761. http://dx.doi.org/10.1080/10408391003799913.

Lynam, J., & Theus, F. (2009). Value chains, innovation and public policies in African agriculture: a synthesis of four country studies. In K. Larsen, R. Kim, & F. Theus (Eds.), *Agribusiness and innovation systems in Africa*. Washington, DC: World Bank Institute and Agriculture and Rural Development, The World Bank.

Mahrizal, Lanier Nalley, L., Dixon, B. L., & Popp, J. S. (May 2014). An optimal phased replanting approach for cocoa trees with application to Ghana. *Agricultural Economics, 45*(3), 291−302. http://dx.doi.org/10.1111/agec.12065.

Mark, J., Brown, F., & Pierson, B. J. (1981). Consumer demand theory, goods and characteristics: breathing empirical content into the Lancastrian approach. *Managerial and Decision Economics, 2*(1), 32−39.

Martincus, C. V., & Carballo, J. (August 2010). Export promotion activities in developing countries: what kind of trade do they promote? In *IDB Working Paper Series No. IDB-WP-202.* Washington, DC: Inter-American Development Bank, Integration and Trade Sector.

Masimbe, S., Gahizi, F., Musana, B., Sangano, J., & (G&N Consultants, Ltd.). (October 2008). *A survey report on the status of horticulture in Rwanda. Prepared for the Rwanda Horticulture Development Authority (RHODA)*.

MDF Training and Consultancy. (August 2014). *End-of-project evaluation of the implementation and results of the CFC/FLIPA project, Ecuador/Colombia. Report CFC/FIGOOF/30. Common Fund for Commodities. La Paz, Bolivia*.

Meade, B., Baldwin, K., & Calvin, L. (December 2010). *Peru: An emerging exporter of fruits and vegetables. Report FTS-345−01, Economic Research Service. U.S.* Washington, DC: Department of Agriculture.

Mellor, J. (May 2000). *Faster more equitable growth: The relation between growth in agriculture and poverty reduction. CAER II Discussion Paper No. 70.* Cambridge, MA: Harvard Institute for International Development.

Melo, C. J., & Hollander, G. M. (2013). Unsustainable development: alternative food networks and the Ecuadorian Federation of Cocoa Producers, 1995–2010. *Journal of Rural Studies, 32,* 251–263.

Milian, S. L. (June 29, 2014). *Cardamom —the 3Gs- green gold of Guatemala. Gain Report No. GT-1404. USDA Foreign agricultural Service.* Washington, DC: Global Agricultural Information Network.

Miller, C., Richter, S., McNellis, P., & Mhlanga, N. (2010). *Agricultural Investment Funds for Developing Countries.* Rome: United Nations Food and Agriculture Organization.

Ministère des Affaires Etrangères, France. (2002). *Memento de l'agronome. CIRED – GRET.*

Ministerio de Agricultura y Desarrollo Rural, Colombia. (2002). La cadena de cacao en Colombia. *Observatorio de Competitividad – Agrocadenas. Bogotá.*

Ministerio de Agricultura y Desarrollo Rural, Colombia. (September 2006). *Agenda de Investigación, Innovación y Desarrollo tecnológico del Sector Agropecuario Colombiano – Avances 2003–2005. Bogotá.*

Minten, B., Randrianarison, L., & Swinnen, J. F. (2009). Global retail chains and poor farmers: evidence from Madagascar. *World Development, 37*(11), 1728–1741. http://dx.doi.org/10.1016/j.worlddev.2008.08.024.

Misión Paz. (2001). *Desarrollo agropecuario y rural: La estrategia. Germán Holguín Zamorano, Director.* Cali, Colombia: Universidad Icesi.

Mpagalile, J., Ishengoma, R., & Gillah, P. (2009). Tanzania: sunflower, cassava and dairy. In K. Larsen, R. Kim, & F. Theus (Eds.), *Agribusiness and innovation systems in Africa.* Washington, DC: Agriculture and Rural Development, World Bank.

Murekezi, C., & Van Asten, P. J. A. (October 2008). Farm banana constraints in Rwanda: a farmer's perspective. Banana and plantain in Africa: harnessing international partnerships to increase research impact. *Conference Book of Abstracts,* 5–9.

Mwasha, N. A., & Kweka, Z. (2014). Tanzania in the face of international trade: the analysis of revealed comparative advantage from 2009 to 2012. *International Journal of Business and Economics Research, 3*(1), 15–28. http://dx.doi.org/10.11648/j.ijber.20140301.13.

Norton, R. D. (2004). *Agricultural development policy: Concepts and experiences.* London: John Wiley & Sons.

Norton, R. D. (2014). Policy frameworks for international agricultural and rural development. In N. K. van Alfen (Ed.), *Encyclopedia of food and agricultural systems.* Academic Press, ISBN 9780444525123.

Norton, R. D., & Álvaro Balcázar, V. (November 2003). A study of Colombia's agricultural and rural competitiveness. In *Report prepared for the United Nations Food and Agriculture Organization.* Bogotá: The World Bank, and the United States Agency for International Development.

Norton, R. D., & Angel, A. (September 2004). *Ventajas Comparativas en la Agricultura Salvadoreña.* Antiguo Cuscatlán, El Salvador: Salvadoran Foundation for Economic and Social Development (FUSADES).

Norton, R. D., & Argüello, R. (2007). La Competitividad Agrícola y Forestal de Colombia. *Tomo I. Bogotá: Preparado para el proyecto MIDAS de la Agencia de los Estados Unidos para el Desarrollo Internacional a través de J. E. Austin Associates.*

Norton, R. D., Argüello, R., Samacá, H., & Martínez, H. (2008). *Una perspectiva de la competitividad agrícola en Colombia.* Bogotá: USAID in Cooperation with the Government of Colombia.

Oberthür, T., Läderach, P., Jürgen Pohlan, H. A., & Cock, J. (Eds.). (2012). *Specialty coffee: Managing quality.* International Plant Nutrition Institute.

Odame, H., Musyoka, P., & Kere, J. (2009). Kenya: corn, tomato and dairy. In K. Larsen, R. Kim, & F. Theus (Eds.), *Agribusiness and innovation systems in Africa.* Washington, DC: World Bank Institute and Agriculture and Rural Development, the World Bank.

On the Frontier Group (OTF Group). (June 2006). *National strategy for Rwandan Horticulture, Strategy development overview. Kigali, Rwanda.*

Pal, S., & Byerlee, D. (2006). India: the funding and organization of agricultural R&D — evolution and emerging policy issues. In P. G. Pardey, Julian M. Alston, & R. R. Piggott (Eds.), *Agricultural R&D in the developing world: Too little, too late?* Washington, DC: International Food Policy Research Institute.

Plan Frutícola Nacional. (2006). *Desarrollo de la Fruticultura en Cundinamarca Bogotá. Octubre.*

Ploetz, R. C., Zentmyer, G. A., Nishijima, W. T., Rohrback, K. G., & Ohr, H. D. (Eds.). (1994). *Compendium of tropical fruit diseases, 1994.* The American Phytopathological Society.

Poulton, C., Labaste, P., & Broughton, D. (2009). Yields and returns to farmers. In D. Tschirley, C. Poulton, & P. Labaste (Eds.), *Organization and performance of cotton sectors in Africa.* Washington, DC: The World Bank.

Poulton, C., & Tschirley, D. (2009). A typology of African cotton sectors. In D. Tschirley, C. Poulton, & P. Labaste (Eds.), *Organization and performance of cotton sectors in Africa.* Washington, DC: The World Bank.

Pray, C. E., & Nagarajan, L. (2013). Private investments on the rise in Africa. In N. Beintema, & G.-J. Stads (Eds.), *Is Africa investing enough? Chapter 5 in: 2013 global food policy report.* Washington, DC: International Food Policy Research Institute.

Prowse, M. (February 2012). Contract farming in developing countries — a review. *A Savoir 12.* Agence Française de Développment (AFD).

Ramaswami, B., Singh Birthal, P., & Joshi, P. K. (February 2005). *Efficiency and distribution in contract farming: The case of Indian poultry growers. Discussion Papers in Economics No. 05—01.* Delhi: Indian Statistical Institute, Planning Unit.

Ravallion, M., & Datt, G. (January 1996). How important to India's poor is the sectoral composition of economic growth? *The World Bank Economic Review, 10*(1).

Reardon, T., Timmer, P., & Berdegue, J. (2004). The rapid rise of supermarkets in developing countries: induced organizational, institutional, and technological change in agrifood systems. *The Electronic Journal of Agricultural and Development Economics, 1,* 168—183.

Reardon, T., Barrett, C. B., Berdegué, J. A., & Swinnen, J. F. M. (2009). Agrifood industry transformation and small farmers in developing countries. *World Development, 37*(11), 1717—1727. http://dx.doi.org/10.1016/j.worlddev.2008.08.023.

Republic of Rwanda, Ministry of Agriculture and Animal Resources and Ministry of Commerce and Industry. (July 2008a). *Rwanda national coffee strategy 2009—2012. Kigali.*

Republic of Rwanda, Ministry of Agriculture and Livestock Resources. (December 2008b). *Strategic plan for the transformation of agriculture in Rwanda — Phase II. Kigali, Rwanda.*

Ricardo, D. (1817). *On the principles of political economy and taxation.* Albemarle-Street, London: John Murray.

Rooney, L. W. (December 2010). Virtues of sorghum: utilization and supply chain management. Presented at the sorghum food enterprise and technology development in Southern Africa workshop. In *International sorghum and millet collaborative research support project (INTSORMIL).* Lusaka, Zambia: USAID.

Rutherford, D. D., Burke, H. M., Cheung, K. K., & Field, S. H. (2016). Impact of an agricultural value chain project on smallholder farmers, households, and children in Liberia. *World Development, 83*, 70−83. http://dx.doi.org/10.1016/j.worlddev.2016.03.004.

Salike, N., & Lu, B. (2015). An examination of Nepal's export choice based on revealed comparative advantage. *NRB Economic Review*, 75−89.

Sanderson, T., & Ahmadi-Esfahani, F. Z. (2011). Climate change and Australia's comparative advantage in broadacre agriculture. *Agricultural Economics, 42*, 657−667.

Sarker, R., & Ratnasena, S. (December 2014). Revealed Comparative Advantage and Half-a-Century Competitiveness of Canadian Agriculture: A Case Study of Wheat, Beef, and Pork Sectors. *Canadian Journal of Agricultural Economics, 62*(4), 519−544. http://dx.doi.org/10.1111/cjag.12057.

Schattenberg, P. (2014). *Research studies show mango may help prevent breast cancer. AgriLife Today. May 21, 2014*. College Station, Texas: Texas A&M University.

Serin, V., & Civan, A. (2008). Revealed comparative advantage and competitiveness: a case study for Turkey towards the EU. *Journal of Economic and Social Research, 10*(2), 25−41.

Servicio Nacional de Aprendizaje (SENA), Colombia. (2006). Estudio de caracterización ocupacional del sector del caucho natural en Colombia. *SENA − Mesa Sectorial del Caucho. Bogotá. Julio.*

Shriver, J. (October 26, 2015). Responding to the climate crisis through crop diversification. *Daily Coffee News by Roast Magazine.*

Sicard, T. L. (2003). *Factores tecnológicos y ambientales en la competitividad de la agricultura colombiana. Report prepared for Norton and Balcázar (2003). Bogotá. marzo.*

da Silva, C. A., & de Souza Filho, H. M. (2007). *Guidelines for rapid appraisals of agrifood chain performance in developing countries. Agricultural Management Marketing and Finance Occasional Paper No. 20.* Rome: FAO.

Solow, R. M. (1970). *Growth theory, an exposition.* Oxford University Press.

Squire, L. (1989). Project evaluation in theory and practice. In H. Chenery, & T. N. Srinivasan (Eds.), *Handbook of development economics* (Vol. II). Elsevier Science Publishers.

Stads, G.-J., Beintema, N., Perez, S., Flaherty, K., & Falconi, C. (2016). *Agricultural research in Latin America and the Caribbean: A cross-country analysis of institutions, investment and capacities.* Washington, DC: ASTI Led by IFPRI and the Inter-American Development Bank.

Tanzi, V., & Zee, H. (March 2001). *Tax policy for developing countries. Economic Issues 27.* Washington, DC: International Monetary Fund.

Thapa, S. (October 2007). The relationship between farm size and productivity: empirical evidence from the Nepalese mid-hills. In *Contributed paper prepared for presentation at the 106th seminar of the European Association of Agricultural Economics − Pro-poor development in low income countries* (pp. 25−27). Montpellier, France: Food, agriculture, trade, and environment.

The World Bank. (1993). *The East Asian miracle, economic growth and public policy.* New York: Oxford University Press.

Timmer, C. P. (2009). *A world without agriculture, the Structural transformation in historical perspective.* Washington, DC: American Enterprise Institute for Public Policy Research.

Timmer, C. P. (December 1997). *How well do the poor connect to the growth process? CAER II Discussion Paper No. 17.* Cambridge, MA: Harvard Institute for International Development.

Trienekens, J. H. (2011). Agricultural value chains in developing countries a framework for analysis. *IFAMA, International Food and Agribusiness Management Review, 14*, 2.

Trienekens, J. H., van der Vorst, J. G. A. J., & Verdouw, C. N. (2014). Global food supply chains. In Neal K. Van Alfen (Ed.), *Encyclopedia of Food and Agricultural Systems.* Academic Press, ISBN 9780444525123.

Trienekens, J., & Willems, S. (2007). Innovation and governance in international food supply chains: the cases of Ghanaian pineapples and South African grapes. *IFAMA, International Food and Agribusiness Management Review, 10*, 4.

Tsakok, I. (1990). *Agricultural price policy, a practitioner's guide to partial equilibrium analysis.* Ithaca, NY: Cornell University Press.

TTC Insights. (November 25, 2015). *What are specialty coffee roasters talking about?*

Turner, A., & Norton, R. D. (May 2009). *Evaluation of crops for watershed development in Rwanda. Report prepared for the World Bank through J. E. Austin Associates, Inc.*

United Nations Food and Agriculture Organization. (October 12−13, 2009). *Global agriculture towards 2050. How to feed the world 2050.* Rome: High-Level Expert Forum.

United Nations Food and Agriculture Organization. (2011). *Global food losses and food waste: Extent, causes and prevention. Study conducted for the International Congress SAVE FOOD! at Interpack 2011, Düsseldorf.* Germany. Rome: FAO.

United Nations University. (2014). *World Risk Report 2014.* Institute for the Environment and Human Security.

Uribe, M. C. (2002). Carga impositiva y reforma tributaria: Racionalidad y eficiencia. *Revista Nacional de Agricultura, 935* (SAC. Bogotá. julio-diciembre).

Van Asten, P. J. A., Florent, D., & Apio, S. (2008). Opportunities and constraints for dried dessert bananas export in Uganda. In *Banana and plantain in Africa: Harnessing international partnerships to increase research impact. Conference book of abstracts, 5−9 October.*

Van Asten, P. J. A., Mukasa, D., & Uringi, N. O. (2008). Farmers earn more money when banana and coffee are intercropped. In *Banana and plantain in Africa: Harnessing international partnerships to increase research impact. Conference book of abstracts. 5−9 October.*

Vanek, J. (May 1959). The natural resource content of foreign trade, 1870−1955, and the relative abundance of natural resources in the United States. *Review of Economics and Statistics, 41*, 146−153.

Vanitha, S. M., Kumari, G., & Singh, R. (2014). Export competitiveness of fresh vegetables in India. *International Journal of Vegetable Science, 20*(3), 227−234. http://dx.doi.org/10.1080/19315260.2013.789812.

Vollrath, T. L., & De Huu, V. (December 1988). *Investigating the Nature of World Agricultural Competitiveness.* U.S. Department of Agriculture, Economic Research Service. Technical Bulletin No. 1754.

Vollrath, T. L. (1991). A theoretical evaluation of alternative trade intensity measures of revealed comparative advantage. *Review of World Economics (Weltwirtschaftliches Archiv), 127*(2), 265−280.

Webb, R. (2013). *Conexión y despegue rural.* Lima, Marzo: Instituto del Perú. Universidad de San Martín de Porres.

Webber, C. M., & Labaste, P. (2010). *Building competitiveness in Africa's agriculture, a guide to value chain concepts and applications.* Washington, DC: Agriculture and Rural Development, The World Bank.

Wiggins, S., & Keats, S. (May 2013). *Leaping & learning: Linking smallholders to markets.* London: Agriculture for Impact, Imperial College, and Overseas Development Institute (ODI).

Williams, C. N., Uzo, J. O., & Peregrine, W. T. H. (1991). *Vegetable production in the tropics.* London: Longman Press, ISBN 0-582-60609-8.

World Bank. (February 2009). *Awakening Africa's sleeping giant: Prospects for commercial agriculture in the Guinea Savannah zone and beyond. Agriculture and rural development unit, sustainable development network.* Washington, DC: Africa Regional Office, World Bank.

Yercan, M., & Isikli, E. (2007). International competitiveness of Turkish agriculture: a case for horticulture products. *Acta Agriculturæ Scandinavica Section C, Food Economics, 4*(3), 181−191.

Yu, R., Cai, J., & Leung, P. (2009). The normalized revealed comparative advantage index. *The Annals of Regional Science, 43*(1), 267−282.

Zapata, P., Zapata, J. L., Alegría Saldarriaga, C., Mauricio Londoño, B., & Cipriano Díaz, D. (2002). Manejo del cultivo de la uchuva en Colombia. In *Boletín Técnico, Programa Nacional de Transferencia de Tecnología Agropecuaria*. Rionegro, Antioquia, Colombia: CORPOICA.

Index

Printed in the United States
By Bookmasters